T0192217

STRUCTURAL ANALYSIS OF MULTI-STOREY BUILDINGS

STRUCTURAL ANALYSIS OF MULTI-STOREY BUILDINGS

Karoly Zalka

CRC Press is an imprint of the
Taylor & Francis Group, an **informa** business

First edition published by CRC Press 2012

Second edition published 2020
by CRC Press
Taylor & Francis Group
6000 Broken Sound Parkway NW, Suite 300
Boca Raton, FL 33487-2742

First issued in paperback 2021

ISBN 13: 978-1-03-204301-2 (pbk)
ISBN 13: 978-0-367-35025-3 (hbk)
ISBN 13: 978-0-429-32937-1 (ebk)

DOI: 10.1201/9780429329371

Visit the Taylor & Francis Web site at
http://www.taylorandfrancis.com

and the CRC Press Web site at
http://www.crcpress.com

Visit the eResources: https://www.crcpress.com/9780367350253

Publisher's Note
The publisher has gone to great lengths to ensure the quality of this reprint but points out that some imperfections in the original copies may be apparent.

In memory
of
Lajos Kollár

Contents

Notations

CAPITAL LETTERS

A cross-sectional area; area of plan of building; floor area; corner point
A_a area of lower flange
A_b cross-sectional area of beam
A_c cross-sectional area of column
A_d cross-sectional area of diagonal bar in cross-bracing
A_h cross-sectional area of horizontal bar in cross-bracing
A_f area of upper flange
A_g area of web
A_o area of closed cross-section defined by the middle line of the wall sections
B plan breadth of the building (in direction y)
B_l local bending stiffness for sandwich model
B_0 global bending stiffness for sandwich model
C centre of vertical load/mass; centroid
E modulus of elasticity
E_c modulus of elasticity of columns
E_d modulus of elasticity of diagonal bars in cross-bracing
E_h modulus of elasticity of horizontal bars in cross-bracing
E_w modulus of elasticity of shear walls
F concentrated load (on top floor level); frame
F_{cr} critical concentrated load (on top floor level)
$F_{cr,x}$ critical concentrated load (on top floor level) for sway buckling in direction x
$F_{cr,y}$ critical concentrated load (on top floor level) for sway buckling in direction y
$F_{cr,\varphi}$ critical concentrated load (on top floor level) for pure torsional buckling
F_g full-height global bending critical load (for concentrated top load)
F_l full-height local bending critical load (for concentrated top load)
F_t Saint-Venant torsional critical load (for concentrated top load)
F_ω warping torsional critical load (for concentrated top load)
G modulus of elasticity in shear
(GJ) Saint-Venant stiffness
$(GJ)_e$ effective Saint-Venant stiffness
H height of building/frame/coupled shear walls/core
I second moment of area
I_{ag} auxiliary constant
I_b second moment of area of beam

I_c second moment of area of column
I_f sum of local and global second moments of area
I_{flange} second moment of area of the flange of I-beam
$I_{f\omega}$ total warping constant
I_g global second moment of area of the columns of the frame
$I_{g\omega}$ global warping constant
I_h second moment of area of horizontal bar in cross-bracing
I_x second moments of area with respect to centroidal axis x
I_{xy} product of inertia with respect to axes x and y
I_y second moments of area with respect to centroidal axis y
I_w second moment of area of shear walls
I_ω warping (bending torsional) constant; local warping constant
$I_{\omega x}$ auxiliary constant
J Saint-Venant constant
$\bar{J}$ supplementary Saint-Venant constant
K shear stiffness of frame; shear critical load
K^* shear stiffness/shear critical load of coupled shear walls
K_b full-height global shear stiffness; global shear critical load
K_b^* full-height global shear stiffness/shear critical load of coupled shear walls
K_c local shear stiffness related to the columns; local shear critical load
K_d shear stiffness representing the effect of the diagonal bars in cross-bracing
K_e effective shear stiffness
K_h shear stiffness representing the effect of the horizontal bars in cross-bracing
L width of structure; plan length of building (in direction x)
M bending moment
M_i concentrated mass at the ith floor level; bending moment share on ith unit
M_t torsional moment
N total applied uniformly distributed vertical load; normal force; axial load
N_{cr} critical load (for uniformly distributed floor load)
$N_{cr,x}$ critical load for sway buckling in direction x
$N_{cr,y}$ critical load for sway buckling in direction y
$N_{cr,\varphi}$ critical load for pure torsional buckling
N_f local bending critical load of frame
N_h homogeneous solution
N_g full-height global bending critical load
N_l full-height local bending critical load
N_p particular solution
N_t Saint-Venant critical load
N_w bending critical load of shear walls
$N_{y\varphi}$ coupled sway-torsional critical load
N_ω warping critical load
$N(z)$ vertical load at z
O shear centre
Q uniformly distributed floor load [kN/m^2]
S "overall" lateral stiffness; shear stiffness for sandwich model
S_ω global torsional stiffness
T shear force
W shear wall

SMALL LETTERS

a length of wall section; stiffness ratio
a_i stiffness ratio
a_0, a_1, a_2 coefficients for cubic equation
b length of wall section; stiffness ratio
b_i stiffness ratio
b_w width of diagonal strip for infill
b_0, b_1, b_2 coefficients for cubic equation
c length of wall section
c_i stiffness ratio
c_1 stability coefficient/critical load factor
d length of wall section; length of diagonal; depth of beam; deflection
d_{ASCE} maximum deflection recommended by ASCE
e location of centroid (with bracing cores)
e^* location of centroid (with bracing cores)
e_o location of shear centre (with bracing cores)
f frequency; auxiliary constant; number of frames and coupled shear walls
f_b lateral frequency associated with local bending stiffness
f_f lateral frequency of frame
f_g lateral frequency associated with global bending stiffness
f_s lateral frequency associated with the effective shear stiffness
$f_{s'}$ lateral frequency associated with the "original" shear stiffness
f_t torsional frequency associated with the Saint-Venant torsional stiffness
f_w lateral frequency of shear walls/cores
f_x lateral frequency in direction x
f_y lateral frequency in direction y
$f_{y\varphi}$ coupled lateral-torsional frequency
f_φ frequency of pure torsional vibration
f_ω torsional frequency associated with the warping torsional stiffness
g gravity acceleration
h height of storey; length of wall section
h^* different storey height between ground floor and first floor
i summation index for bracing units
i_p radius of gyration
j summation index for columns
k non-dimensional parameter; torsional parameter; summation index
k_s non-dimensional parameter for stability analysis
k_φ non-dimensional torsion parameter for frequency analysis
l width of bay; span of connecting beams (with cores)
l^* distance between shear wall sections for coupled shear walls
m number of shear walls/cores/wall sections; mass; length of beam section
m_z intensity of torsional moment
n number of columns/walls; number of storeys
p intensity of uniformly distributed vertical load on beams
q intensity of shear flow; intensity of axial load; eigenvalue
q_i apportioner
q_ω torsional apportioner
r reduction factor for beam stiffness

r^* reduction factor for beam stiffness with coupled shear walls
r_f mass distribution factor for the frequency analysis
r_s load distribution factor for the stability analysis
s non-dimensional stiffness ratio for bracing unit; effectiveness factor; vertical distance of connecting beams with partially closed U-core
s_i width of shear wall section; non-dimensional stiffness ratio
s_f effectiveness factor for frequency analysis
s_φ torsional effectiveness factor
t wall thickness; distance of shear centre and centroid; time; perpendicular distance of bracing unit from shear centre; distance of column axis from the centroid of cross-sections with frames
t_b thickness of connecting beam with partially closed U-core
t_b^* reduced thickness
t_f wall thickness of flange
t_w wall thickness of web
u horizontal deflection in direction x
u_{max} maximum horizontal deflection in direction x
v horizontal deflection in direction y
v_o horizontal deflection of the shear centre in direction y
v_{max} maximum horizontal deflection in direction y
v_φ horizontal deflection caused by torsional moment around the shear centre
v_1 horizontal motion in direction y
w wind load [kN/m]
w^* wind load [kN/m^2]
x horizontal coordinate axis; horizontal coordinate
$\bar{x}$ horizontal coordinate axis; coordinate in coordinate system $\bar{x}-\bar{y}$
x_c coordinate of the centroid in the x-y coordinate system of the shear centre
x_i coordinate of the shear centre of the *i*th bracing unit
x_{max} location of maximum translation
$\bar{x}_i, \bar{y}_i$ coordinates of the shear centre of the *i*th bracing unit in the coordinate system $\bar{x}-\bar{y}$
$\bar{x}_o$ coordinate of the shear centre in coordinate system $\bar{x}-\bar{y}$
y horizontal coordinate axis; horizontal coordinate
$\bar{y}$ horizontal coordinate axis; coordinate in coordinate system $\bar{x}-\bar{y}$
y_b deflection due to bending deformation
y_c coordinate of the centroid in the x-y coordinate system of the shear centre
y_i coordinate of the shear centre of the *i*th bracing unit; deflection due to interaction
y_o coordinate of shear centre
$\bar{y}_o$ coordinate of the shear centre in coordinate system $\bar{x}-\bar{y}$
y_s deflection due to shear deformation
y^* deflection of "new" frame
z vertical coordinate axis; vertical coordinate

GREEK LETTERS

α eigenvalue; critical load parameter; auxiliary parameter
α_s eigenvalue; critical load parameter for the sandwich model with thin faces
α_φ eigenvalue; torsional critical load parameter

β part critical load ratio; auxiliary parameter
β_s part critical load ratio for the sandwich model with thin faces
β_φ part torsional critical load ratio
Δ displacement; difference
η frequency parameter for lateral vibration
η_φ frequency parameter for pure torsional vibration
γ weight per unit volume
κ auxiliary parameter
λ global critical load ratio
ν Poisson ratio
ω_1, ω_2 auxiliary constants
ω circular frequency
φ rotation
φ_{max} maximum rotation
Ω_1, Ω_2 auxiliary constants
Ψ auxiliary constant
ρ mass density per unit volume; cross-sectional constant
τ_x, τ_y eccentricity parameters for the three-dimensional analysis

1

Introduction

When multi-storey buildings are investigated, two completely different approaches are available for the structural engineer. Choosing an analytical model is the preferred choice of those who rely on "conventional" methods of analysis. This approach often leads to simple, closed-form, albeit normally approximate solutions. Relying on discrete models offers the possibility of carrying out a more detailed but at the same time often fairly time-consuming analysis. Because of the complexity of discrete models, using a computer for the analysis is a must.

A couple of decades ago approximate methods played a very important and normally dominant role in the structural design of large and complex structures as often, because of the lack of computer power, it was not feasible, or practical, or sometimes possible, to carry out an "exact" analysis. Then more and more powerful computers with more and more sophisticated programs started to become available to wider and wider structural engineering communities. Such programs made it possible to deal with very big and complex structures. These programs make life fairly easy for the structural engineer. However, using complex computer programs may also have disadvantages. The more and more sophisticated and user-friendly programs may create an atmosphere when the structural engineer relies on them too much and finds less incentive to acquire an in-depth knowledge of the behaviour of the structure. And the lack of in-depth knowledge might easily lead to uneconomic or unsafe structures.

It may become tempting to pass the responsibility of the structural analysis on to the computer and then to accept the results without doubt. This may especially be the case when the structure is large and its three-dimensional behaviour is complex. Even the question "Do we need hand calculations at all?" can emerge. The answer, however, for several reasons, is "Yes, we need relatively simple hand calculations". Simple hand calculations offer a useful way of making checks on the results given by the computer. Such checks are very important because when the structural engineer handles a great number of input and output data and evaluates the results, it is easy to overlook something or to make mistakes. Once a mistake is made, it may be difficult to find it. The term "Computer Aided Disaster", or CAD for short, may be used as an eye-catching phrase or a sensationalist session title at conferences but the warning is justifiably on the wall: one avoidable catastrophe would be one more than can be accepted. The significance of the independent verification of the computer-based results cannot be overemphasised (Brohn, 1996; Smart, 1997; MacLeod, 2005).

Experience shows that the old verdict "This result must be correct as it was given by the computer" can still be heard. Even when it is *obvious* to the knowledgeable that the result in question is incorrect. A quick check using a back-of-the-envelope calculation could often remedy the situation in minutes.

But there are other advantages of developing and applying simple hand calculations. When such methods are developed, structural elements of secondary importance (e.g. partitions and other non-load-bearing structural elements) are normally ignored and the investigation centres on dominant aspects and neglects phenomena of secondary importance. As a consequence, a simple method with fewer aspects to concentrate on can give a clearer picture of the behaviour emphasising the most important key characteristics of the structure. This is also helpful in developing structural engineering common sense. Understanding the contributions of key structural characteristics is especially important with large and complex structures.

Perhaps the best way to tackle the task of the structural analysis of multi-storey buildings is to employ both approaches: at the preliminary design stage simple hand methods can quickly help to establish the main structural dimensions and to point to efficient bracing system arrangements. More detailed computer-based analysis can follow. Before the final decision is made, it is essential to check the results of the computer analysis and confirm the adequacy of the key elements of the bracing system. Here, again, suitable simple methods can be very useful.

This book is concerned with the structural analysis of multi-storey buildings whose bracing system consists of frames, coupled shear walls, shear walls and cores. Such structures are generally large, contain a great number of structural elements and behave in a three-dimensional manner. Using the analytical approach, relatively simple models can be created for the analysis.

The continuum method will be used which is based on an equivalent medium that replaces the whole building. The discrete load and stiffnesses of the building will be modelled by continuous load and stiffnesses. This approach makes it possible to use analytical tools to produce relatively simple, closed-form solutions to the resulting differential equations and eigenvalue problems.

The fact that the methods in the book are all based on continuous models has another advantage. When the results of a finite element analysis (based on discrete models) are checked, it is advantageous to use a technique that is based on a different approach, i.e., on continuous medium.

Structural analysis is normally carried out at two levels. The structural engineer has to ensure that (a) the individual elements (beams, columns, floor slabs, etc.) are of adequate size and material to carry their load and (b) the structure as a whole has adequate stiffness and the bracing system fulfils its main role to provide sufficient stability to the building.

The book does not deal with individual structural elements. Its aim is to present simple analytical methods for the complex global analysis of whole structural systems in the three main structural engineering areas. Assuming three-dimensional behaviour, closed-form solutions will be given for the maximum rotation and deflection, the fundamental frequency and the critical load of the building.

Whenever methods of analysis are developed, certain assumptions have to be made. These assumptions reflect a compromise: they help to create relatively simple methods but at the same time they ensure that the results are of adequate accuracy. Accordingly, it will generally be assumed that the structures are

- at least four storeys high with identical storey heights
- regular in the sense that their characteristics do not vary over the height
- sway structures with built-in lower end at ground floor level and free upper end

and that

- the floor slabs have great in-plane and small out-of-plane stiffness
- the deformations are small and the material of the structures is linearly elastic
- P-delta effects are negligible.

Structural engineering research and practice often see researchers/structural designers who have specialized in one area with limited knowledge elsewhere. Designers are often reluctant to deal with theoretical matters; researchers often have little practical knowledge (or attitude); those dealing with stress analyses are sometimes ignorant of stability matters; people engaged in earthquake engineering may not be very good at the optimisation of bracing systems, etc.

This book offers a unified treatment for the different structures (frames, coupled shear walls, shear walls and cores, and their assemblies) and also for the different types of investigation (deflection, rotation, frequency, stability). The same terminology will be used throughout, and it will be shown that these seemingly independent areas (deformations, frequencies, critical loads—or stress, dynamic and stability analyses) are in fact very closely related. In addition, the global critical load ratio links them to the performance of the bracing system in a rather spectacular manner.

Although real multi-storey buildings seldom develop planar deformation only, Chapter 2 (dealing with the planar analysis of individual bracing units) is probably the key chapter of the book in the sense that it introduces most of the characteristic stiffnesses that will be used for the three-dimensional investigations of whole systems later on. It is also shown here how the complex behaviour can be traced back to the local bending, global bending and shear deformations (and their torsional equivalents) of the bracing system. All the characteristic types of bracing units are covered here: sway- and infilled frames, frames with cross-bracing, coupled shear walls, shear walls and cores.

Three-dimensional behaviour is the subject of Chapters 3, 4, 5 and 6. The investigations in Chapter 3 centre on buildings subjected to lateral load and the main aim is to present simple, closed-form solutions for the maximum deflection and rotation of the building. It is spectacularly shown how the key contributors to the resistance of a multi-storey building—the bending and shear stiffnesses, and their interaction—influence the performance of the bracing system. Chapter 4 deals with the frequency analysis of buildings. Closed-form formulae and tables make it possible to calculate the lateral and torsional frequencies of the building. The coupling of the lateral and torsional modes can be taken into account by a simple summation formula or, if a more accurate result is needed, by calculating the smallest root of a cubic equation. The often neglected but very important area of stability is covered in Chapter 5. In using critical load factors, simple (Euler-like) formulae are presented for the lateral and torsional critical loads. The combined sway-torsional critical load is obtained using a summation formula or calculating the smallest root of a cubic equation.

Chapter 6 introduces the global critical load ratio which is a useful tool for monitoring the "health" of the bracing system. It can be used to show in minutes whether a bracing system is adequate or not, or a more rigorous (second-order) analysis is needed. The global critical load ratio can also be used to assess different bracing system arrangements in order to choose the most economic one. The results of three comprehensive worked examples demonstrate the practical use of the global critical load ratio.

To illustrate the practical use of the methods and formulae presented in the book, nineteen examples worked out to the smallest details are included. The examples range from the deflection or frequency or stability analysis of individual bracing units to the complex deflection and frequency and stability analyses of bracing systems, considering both planar and spatial behaviour. The examples are to be found at the end of the relevant chapter/section.

Numerous approximate methods have been published for the structural analysis of multi-storey structures. Most of them deal with individual bracing units. Some of them can even handle three-dimensional behaviour. However, it is surprising how few, if any, have been backed up with convincing accuracy analysis. Chapter 7 is devoted to the very important but often neglected question of accuracy and reliability. Using 32 individual bracing units at different storey heights, the accuracy of the relevant formulae is demonstrated by comparing the results of the closed-form solutions presented in the book with the results of the "exact" (computer-based) analyses. Altogether 1631 checks are made in two groups. The first group contains 983 individual bracing units whose maximum deflection, fundamental frequency and critical load are determined. The second group contains the three-dimensional bracing systems of 648 multi-storey buildings. Here, too, the maximum deflection, the fundamental frequency and the critical load of these systems are determined. The results demonstrate the applicability and accuracy of the methods presented in the preceding chapters. Information regarding the accuracy of the procedures used in the nineteen worked examples concludes Chapter 7.

Although most of the formulae in the book are of the back-of-the-envelope type, due to the complexity of global three-dimensional analyses, some of the calculations may still seem to be rather cumbersome to carry out by hand. It is very rare, however, that a structural engineer today would wish to do actual hand-calculations, however simple they may be. Convenient spreadsheets and calculation worksheets make it possible to carry out the structural analysis and document its result at the same time in minutes. All the methods presented in the book are suitable for this type of application; in fact the worksheet version of all the nineteen worked examples has been prepared and made available for download. Each worksheet is prepared using both MathCad and Excel. These one- to eight-page long worksheets cover a very wide range of practical application and can also be used as templates for other similar structural engineering situations. Short summaries of the nineteen worksheets are given in the Appendix.

2

Individual bracing units: frames, (coupled) shear walls and cores

The bracing system of multi-storey buildings is normally made up of different units: frames, shear walls, coupled shear walls and cores. They all contribute to the overall resistance of the system but their contributions can be very different, both in weight and also in nature. It is therefore essential for the structural designer to know their behaviour in order that safe and economic bracing systems can be created.

Frames can be considered unique as they have each of the three basic stiffness characteristics that play an important role in the structural analysis of buildings. Their importance is underlined by the fact that the analysis of whole buildings (consisting of frames, coupled shear walls, shear walls and cores) can often be traced back to the investigation of a single frame which, in turn, can be replaced by its equivalent column. It is therefore advantageous to start the investigation with the analysis of frames.

2.1 DEFLECTION ANALYSIS OF RIGID SWAY-FRAMES UNDER HORIZONTAL LOAD

Frames are defined as rigid when all the (bay-size) beams and (storey-high) columns join at nodal points in a rigid manner (i.e. the 90° angle between beam and column at nodal points remains 90° after deformation) and all the supports are fixed at ground-floor level.

The behaviour of frames under lateral load is complex, mainly because they develop a combination of bending and shear deformations (Figure 2.1). Due to the complexity of the problem, designers and researchers have made considerable efforts to develop approximate techniques and methods. Perhaps the best and most widespread method is the continuum method which is based on an equivalent medium that replaces the frame. It is difficult to pinpoint exactly who developed the first continuum model but the method probably surfaced in the early 1940s. In her excellent paper, Chitty (1947) investigated parallel beams interconnected by cross bars, subjected to uniform lateral load, and established the governing differential equation of the problem. In a following paper she applied the method to tall buildings under horizontal load; however, she neglected the axial deformations of the columns (Chitty and Wan, 1948). Scientists from all over the world followed—many of them apparently unaware of the previous efforts—who created and sometimes reinvented and later further developed continuum models (Csonka, 1950; Beck, 1956; Rosman, 1960; Despeyroux, 1972; Stafford Smith *et al.*, 1981; Hoenderkamp and Stafford Smith, 1984; Coull, 1990). Perhaps the most comprehensive treatment of building structures under horizontal load is given by Stafford Smith and Coull (1991). The continuum model has also been applied successfully to the stability and dynamic analyses of buildings (Danay *et al.*, 1975; Rosman, 1981; Rutenberg, 1975; Kollár, 1986; Hegedűs and Kollár, 1999; Zalka, 2000; Potzta and Kollár, 2003).

The procedure presented in the following will result in a very simple and expressive formula for the deflection, identifying three distinctive parts: bending mode, shear mode, and their interaction.

In addition to the general assumptions listed in the Introduction, it will also be assumed that the structures are subjected to uniformly distributed lateral load.

2.1.1 Characteristic deformations

In line with, and using the terminology established in the theory of sandwich structures (Plantema, 1961; Allen, 1969; Hegedűs and Kollár, 1999; Zalka, 2000), the behaviour of a frame may be characterised by three types of deformation and the corresponding stiffnesses. The deformation of a frame is complex but it can be "divided" into three virtual sub-deformations. These sub-deformations are obtained by emphasising one possible type of deformation and neglecting the other two. The *shear* deformation of the frame is characterized by the double-curvature bending of the columns and the beams between two nodal points (Figure 2.1/a). For *global bending*, the structure as a whole unit is considered and the bending of the unit occurs through the axial deformation—lengthening and shortening—of the full-height columns (Figure 2.1/b). *Local bending* is associated with the full-height bending of the individual columns of the frame when the beams purely act as connecting media with pinned ends between the columns (Figure 2.1/c). The corresponding stiffnesses are the shear stiffness, the global bending stiffness and the local bending stiffness.

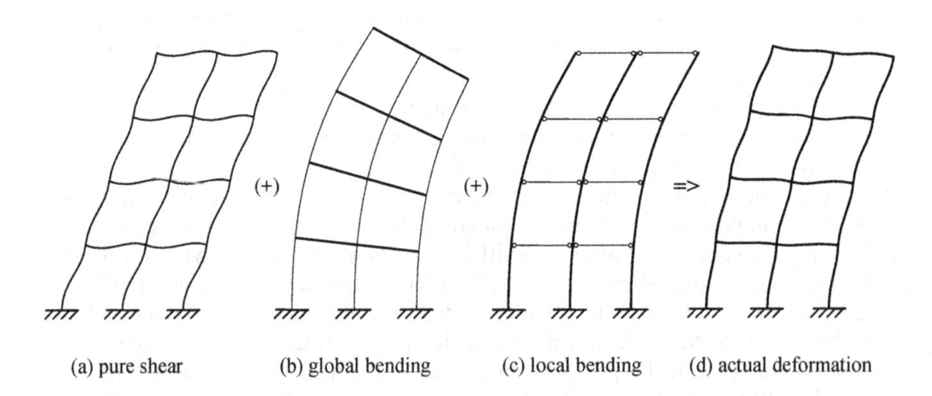

Figure 2.1 Characteristic virtual sub-deformations and the actual deflection shape.

The actual deformation of the frame is a combination of the shear, global bending and local bending deformations (Figure 2.1/d).

From now on, these characteristic deformations will be used, together with the stiffnesses that go with them, not only for the lateral deflection analysis below but also for the rotation analysis later on as well as for the frequency and stability analyses in later chapters.

2.1.2 One-bay, multi-storey frames

For the deflection analysis, consider first the one-bay, multi-storey frame, shown in Figure 2.2/a. The continuum method will be used for the analysis and the frame comprising (storey-high) columns and (bay-size) beams will be replaced with continuous medium. A full-height equivalent column will be created with the three characteristic stiffnesses introduced in Section 2.1.1. As for the load of the frame, we should first consider the way a multi-storey building takes its external load. The original lateral load acts on and is distributed over the façade of the building. This load is transmitted from the façade to the floor slabs in the form of concentrated forces. The floor slabs then transmit these forces to the bracing units. Each bracing unit takes its share from the external load in proportion to its contribution to resisting the external load. One such bracing unit is our one-bay, multi-storey frame which is replaced with its equivalent column subjected to these concentrated forces (Figure 2.2/b).

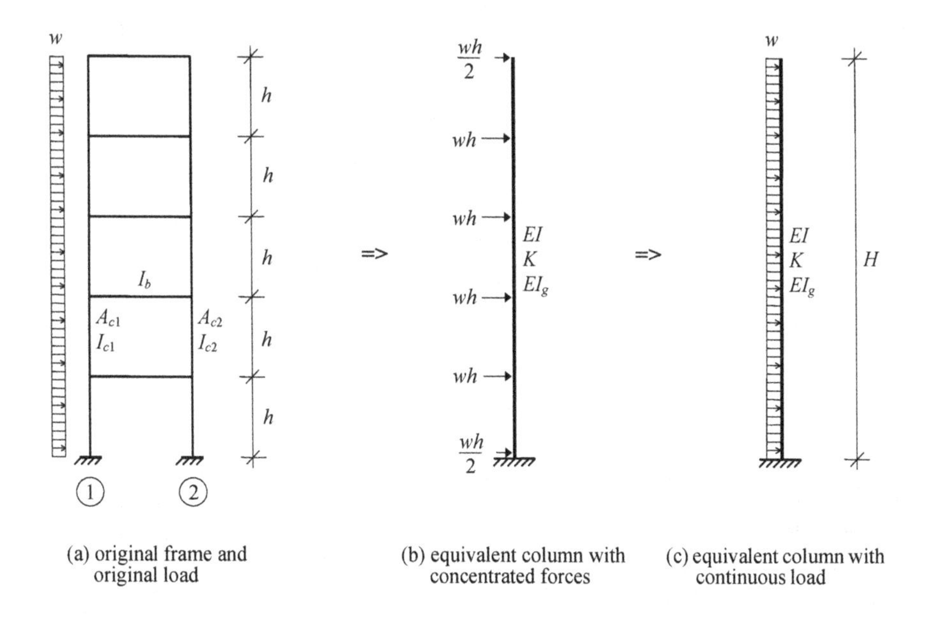

Figure 2.2 One-bay, multi-storey frame with its external load.

To make the load on the equivalent column continuous, this system of concentrated forces is finally distributed over the height of the structure resulting in the uniformly distributed load of intensity w (Figure 2.2/c). It should be noted that no approximation is made here: the original load (Figure 2.2/a) and the load on the equivalent column (Figure 2.2/c) are identical in every sense. (The situation with the frequency and stability analyses will be different.)

The establishment of the stiffnesses of the equivalent column is a more complex task. These stiffnesses will now be introduced, one by one, embedded into the procedure of setting up the governing differential equation of the deflection problem.

The derivation of the governing differential equations below follows Chitty's original derivation (Chitty, 1947) but with structural engineering notation. The solution procedure, however, will be somewhat different. A two-step procedure will be presented that makes it possible to produce simple, expressive and user-friendly formulae for the deflection (Zalka, 2009).

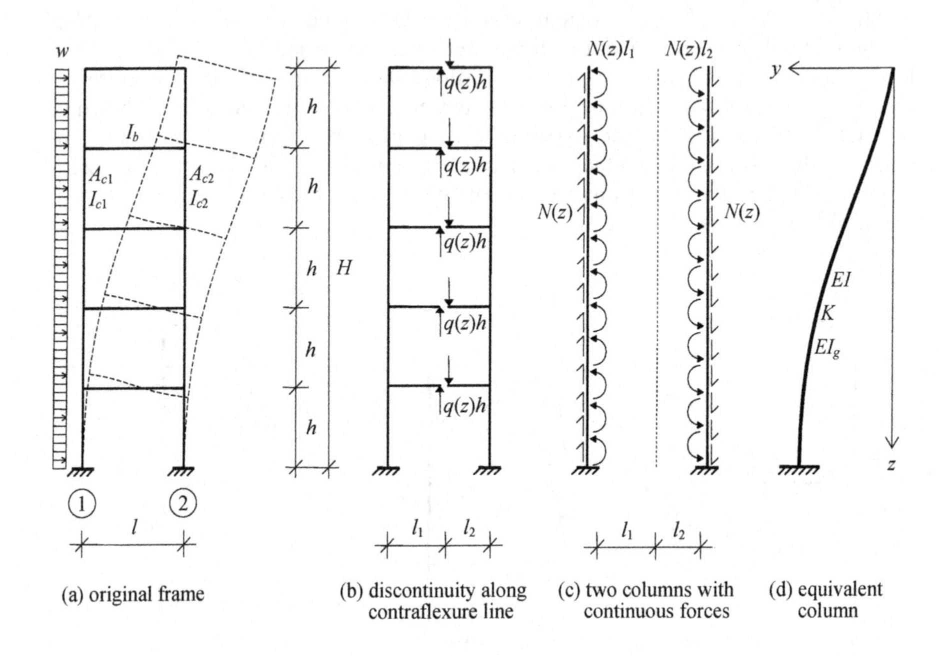

Figure 2.3 Origination of the continuum model.

In the usual manner with the continuum procedure, first the beams are cut at the vertical line of contraflexure. The resulting lack of continuity is compensated for by a shear flow of intensity $q(z)$. This shear flow materializes at floor levels in the form of forces $q(z)h$ (Figure 2.3/b). [Shear flow $q(z)$ itself is shown later in Figure 2.6/c]. It is assumed that there are "enough" numbers of beams so that they can be considered a continuous connecting medium between the columns. As a rule, the technique can safely be applied to structures of at least four-storey height. The shear flow is then transferred to the columns (Figure 2.3/c) in the form of normal force $N(z)$ and bending moments $N(z)l_1$ and $N(z)l_2$. [Normal force $N(z)$ is shown later in Figure 2.6/d.]

Finally, after setting up a differential equation responsible for the lack of continuity in the following sections [c.f. Equation (2.9)], an equivalent column will be created as the continuum model for the problem (Figure 2.3/d). The origin of coordinate system *y–z* is placed at and fixed to the top of the column.

If the beams are cut, relative vertical displacements develop along the line of contraflexure. Three different actions will cause displacement and they will now be considered, one by one, as if they occurred independently of each other.

The relative displacement due to the bending of the columns (Figure 2.4/a) is

$$\Delta_1(z) = y'(z)(l_1 + l_2) = y'(z)l$$

where $y'(z)$ denotes the slope of the columns. Due to the rigid column-beam connections, $y'(z)$ also denotes the slope of the beams.

The displacement is positive when the end of the beam-section belonging to the left column moves downward and the other upward.

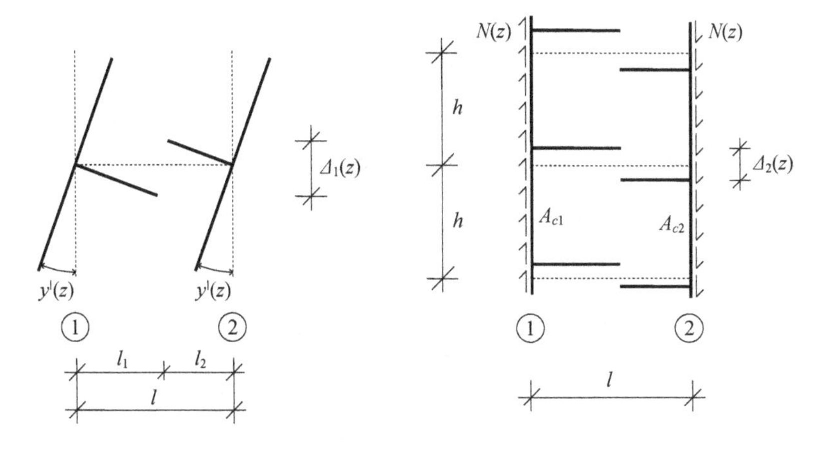

(a) due to the bending of the columns (b) due to the axial deformation of the columns

Figure 2.4 Vertical displacement at contraflexure point.

The axial deformation of the columns (Figure 2.4/b), due to the normal forces originating from the shear forces in the connecting beams, also contributes to the overall relative displacement

$$\Delta_2(z) = -\frac{1}{E}\left(\frac{1}{A_{c1}} + \frac{1}{A_{c2}}\right)\int_z^H N(z)dz$$

where

$$N(z) = \int_0^z q(z)dz$$

is the normal force causing axial deformation in the columns, $q(z)$ is the intensity of the shear flow, A_{c1} and A_{c2} are the cross-sectional areas of the columns, H is the height of the structure and E is the modulus of elasticity. [Normal force $N(z)$ is shown later in Figure 2.6/d.]

Due to the bending of the beams (Figure 2.5), the shear force at contraflexure also develops relative displacement. Assuming that the point of contraflexure is at mid-bay, this relative displacement is

$$\Delta_3^*(z) = -2\frac{q(z)h\left(\frac{l}{2}\right)^3}{3EI_b} = -\frac{q(z)l^3h}{12EI_b} = -\frac{q(z)l^2}{\frac{12EI_b}{lh}} = -\frac{q(z)l^2}{K_b} \tag{2.1}$$

where I_b is the second moment of area of the beams, h is the storey height and l is the size of the bay. The term

$$K_b = \frac{12EI_b}{lh} \tag{2.2}$$

is defined as the stiffness of the beams (distributed over the height).

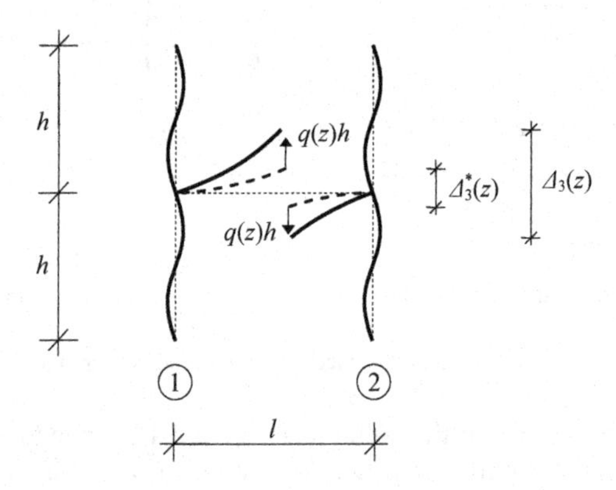

Figure 2.5 Vertical displacement at contraflexure point due to the bending of the beam.

Equation (2.1) only holds when the beams have horizontal tangent to the columns at the nodal points, i.e. when the columns are considered infinitely stiff (dashed line in Figure 2.5). This may be the case with coupled shear walls where the wall sections are often much stiffer than the connecting beams and can prevent the rotation of the beams at nodal points. However, this is not the case with frames where the relatively flexible columns develop double-curvature bending between the beams (solid line in Figure 2.5). It follows that, due to the flexibility of the columns, in the case of frames, Equation (2.1) should be amended and the vertical displacement $\Delta_3^*(z)$ increases to $\Delta_3(z)$ as

$$\Delta_3(z) = -\left(\frac{q(z)l^2}{\frac{12EI_b}{lh}} + \frac{q(z)l^2}{\frac{12EI_c}{h^2}}\right) = -q(z)l^2\left(\frac{1}{K_b} + \frac{1}{K_c}\right)$$

In the above equation the stiffness of the columns (distributed over the height) is defined as

$$K_c = \frac{12EI_c}{h^2} \tag{2.3}$$

and I_c is the second moment of area of the columns.

The *shear stiffness* of the frame (distributed over the height) can now be defined as

$$K = \left(\frac{1}{K_b} + \frac{1}{K_c}\right)^{-1} = K_b \frac{K_c}{K_b + K_c} = K_b r \tag{2.4}$$

In Equation (2.4) the term

$$r = \frac{K_c}{K_b + K_c} \tag{2.5}$$

is also introduced. As $r < 1$ always holds, in relation to K_b, r can be considered as a reduction factor. This reduction factor will play an important role later on.

Replacing K_b in Equation (2.1) with K, the actual relative displacement of the frame, when the bending of both the beams and the columns is taken into account, emerges as

$$\Delta_3(z) = -\frac{q(z)l^2}{K}$$

The above formula is "exact" if the point of contraflexure is at mid-bay, i.e. if the structure is symmetric. However, its accuracy is considered adequate in most practical cases when the cross-sections of the columns are different. [When the stiffnesses of the columns are very different, e.g., the frame connects to a shear wall, a more accurate approach is needed. Formulae for such cases are given elsewhere, e.g. in (Stafford Smith and Coull, 1991).]

The above three relative displacements would develop if the beams were cut. However, the beams of the actual structure are not cut and therefore the sum of relative displacements Δ_1, Δ_2 and Δ_3 must equal zero for the real structure. This condition is expressed by the equation

$$ly'(z) - \frac{q(z)l^2}{K} - \frac{1}{E}\left(\frac{1}{A_{c1}} + \frac{1}{A_{c2}}\right)\int_z^H N(z)dz = 0 \tag{2.6}$$

With

$$N'(z) = q(z) \tag{2.7}$$

and introducing

$$I_g = A_{c1}l_1^2 + A_{c2}l_2^2 = \frac{A_{c1}A_{c2}}{A_{c1} + A_{c2}}l^2 \tag{2.8}$$

as the global second moment of area of the frame, and after differentiating and some rearrangement, Equation (2.6) can be rewritten and the condition for continuity along the line of contraflexure assumes the form

$$y''(z) - \frac{l}{K}N''(z) + \frac{l}{EI_g}N(z) = 0 \tag{2.9}$$

In the above equation EI_g is the *global bending stiffness* of the frame.

The bending of the two full-height columns is considered next, based on the classical relationship

$$y''(z)EI = -M(z)$$

for bending. Because of the connecting beams, the two columns, with their combined second moments of area, are forced to assume the same deflection shape. The external moments (from the horizontal load) are now supplemented by the moments caused by the shear forces along the line of contraflexure (Figure 2.3/b,c) as

$$y''(z)E(I_{c1} + I_{c2}) = -M(z) - (l_1 + l_2)\int_z^0 q(z)dz$$

Introducing the sum of the second moments of area of the columns

$$I_c = I_{c1} + I_{c2} \tag{2.10}$$

as the local second moment of area of the frame, and making use of

$$\int_z^0 q(z)dz = -N(z)$$

and with $N(z)l_1 + N(z)l_2 = N(z)(l_1 + l_2) = N(z)l$, the equation can be rewritten as

$$y''(z)EI_c = -M(z) + lN(z) \tag{2.11}$$

The governing differential equation of deflection is obtained by combining Equations (2.9) and (2.11). Normal force $N(z)$ is obtained from Equation (2.11) as

$$N(z) = \frac{1}{l}\left[y''(z)EI_c + M(z)\right]$$

Substituting this and its second derivative

$$N''(z) = \frac{1}{l}\left[y''''(z)EI_c + M''(z)\right]$$

for N and N'' in Equation (2.9), some rearrangement leads to

$$y''''(z) - \left(\frac{K}{EI_c} + \frac{K}{EI_g}\right)y''(z) = \frac{1}{EI_c}\left(\frac{K}{EI_g}M(z) - M''(z)\right) \tag{2.12}$$

Before the solution of the problem is produced, a small modification has to be made. Detailed theoretical investigations (Hegedűs and Kollár, 1999) show and accuracy analyses (Zalka and Armer, 1992) demonstrate that in the above continuum model the bending stiffness of the columns is somewhat over-represented. (For low-rise structures this over-representation may lead to unconservative results of up to 16%.) This over-representation can easily be corrected by introducing reduction factor r defined by Equation (2.5) in such a way that the second moment of area of the columns of the frame is adjusted by factor r as

$$I = I_c r$$

Accordingly, from this point on, this modified second moment of area will be used and

$$EI = EI_c r \tag{2.13}$$

will be defined as the *local bending stiffness* of the structure.

In order to shorten the formulae, the following—mostly temporary—notation will be used:

$$a = \frac{K}{EI_g}, \qquad b = \frac{K}{EI}, \qquad \kappa = \sqrt{a+b} = \sqrt{\frac{K(I+I_g)}{EII_g}} = \sqrt{\frac{Ks}{EI}}$$

$$s = 1 + \frac{a}{b} = \frac{a+b}{b} = 1 + \frac{I}{I_g} = \frac{I + I_g}{I_g}, \qquad \frac{a}{a+b} = \frac{I}{I+I_g} \tag{2.14}$$

Using the above notation and with $M(z) = wz^2/2$, the short version of Equation (2.12) is

$$y''''(z) - \kappa^2 y''(z) = \frac{w}{EI}\left(\frac{az^2}{2} - 1\right) \tag{2.15}$$

This is the governing differential equation of the frame that has now been replaced by a single cantilever with the corresponding local bending stiffness EI, global bending stiffness EI_g and shear stiffness K (Figures 2.3/d and 2.6/b).

The deflection of the frame can be obtained in two ways. One possibility is to solve Equation (2.15) directly, in one step. Alternatively, the solution can be produced in two steps: first, the solution for the normal force is obtained then, using the formula for the normal force, the deflection is determined. Another aspect of the solution is the choice and placement of the coordinate system. Although the actual result of the procedure obviously does not depend on the solution process and the choice and placement of the coordinate system, the structure and complexity of the result do. The aim is to produce a result that is relatively simple and shows how the different stiffness components influence the magnitude of the deflection of the structure.

After solving the differential equation in the two different ways indicated above and using different coordinate systems, it turned out that the simplest result can be produced when the two-step approach is applied and the origin of the coordinate system is fixed to the top of the equivalent column (Figures 2.3/d and 2.6/b). The main steps of this procedure will now be presented.

In combining Equations (2.9) and (2.11), and with $M(z) = wz^2/2$ and notation (2.14), the governing second-order differential equation for the normal force emerges as

$$N''(z) - \kappa^2 N(z) = -\frac{bwz^2}{2l} \tag{2.16}$$

Two boundary conditions accompany this differential equation. The first condition expresses the fact that the normal force must assume zero at the top of the equivalent column (Figure 2.6/d):

$$N(0) = 0$$

The magnitude of the shear flow at $z = H$ is zero (Figure 2.6/c) as, because of the vertical tangent of the columns at the built-up end, a beam there would not bend. The second boundary condition therefore is

$$N'(H) = q(H) = 0$$

The solution to differential equation (2.16) is sought in the form of

$$N(z) = N_h(z) + N_p(z)$$

where

$$N_h(z) = A\sinh(\kappa z) + B\cosh(\kappa z)$$

is the homogeneous solution and

$$N_p(z) = Cz^2 + Dz + E$$

is a particular solution of the inhomogeneous problem.

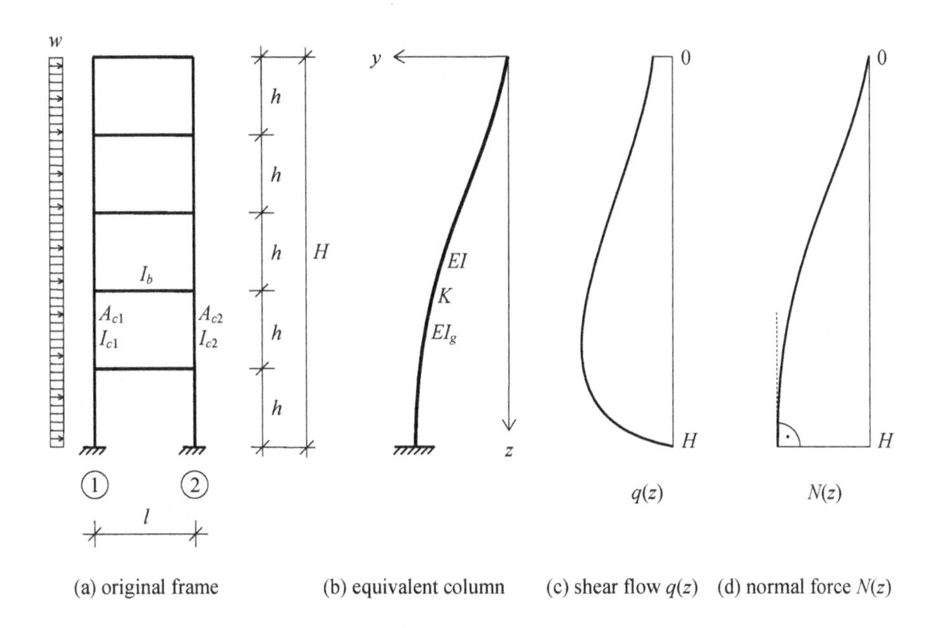

Figure 2.6 One-bay, multi-storey frame F1 (Figure 7.1/a) with shear flow $q(z)$ and normal force $N(z)$.

In substituting $N_p(z)$ and its second derivative for $N(z)$ and $N''(z)$ in Equation (2.16)

$$2C - \kappa^2 Cz^2 - \kappa^2 Dz - \kappa^2 E = -\frac{bwz^2}{2l}$$

constants C, D and E are determined by setting the coefficients of the powers of function z equal of the two sides:

$$C = \frac{bw}{2l\kappa^2}, \qquad D = 0, \qquad E = \frac{bw}{l\kappa^4}$$

With the values of *C*, *D* and *E* now available, the solution assumes the form

$$N(z) = A\sinh(\kappa z) + B\cosh(\kappa z) + \frac{bw}{2l\kappa^2}z^2 + \frac{bw}{l\kappa^4}$$

where the two constants are obtained using the two boundary conditions:

$$N(0) = 0 \qquad \rightarrow \qquad B = -\frac{bw}{l\kappa^4}$$

and

$$N'(H) = A\kappa\cosh(\kappa H) - \frac{bw}{l\kappa^3}\sinh(\kappa H) + \frac{bwH}{l\kappa^2} = 0 \rightarrow A = \frac{bwH}{l\kappa^3\cosh(\kappa H)}\left(\frac{\sinh(\kappa H)}{\kappa H} - 1\right)$$

Knowing the two constants, the equation for the normal force (Figure 2.6/d) is obtained as

$$N(z) = \frac{wb}{l\kappa^2}\left(\frac{\sinh(\kappa H)\sinh(\kappa z)}{\kappa^2\cosh(\kappa H)} - \frac{H\sinh(\kappa z)}{\kappa\cosh(\kappa H)} - \frac{\cosh(\kappa z)}{\kappa^2} + \frac{z^2}{2} + \frac{1}{\kappa^2}\right)$$

With the above equation of the normal force, Equation (2.11) can now be used to determine the deflection. After substituting for *N*(*z*), Equation (2.11) assumes the form:

$$y''(z) = \frac{w}{EI}\left(\frac{bz^2}{2\kappa^2} + \frac{b}{\kappa^4} - \frac{z^2}{2} + \frac{b\sinh(\kappa H)\sinh(\kappa z)}{\kappa^4\cosh(\kappa H)} - \frac{bH\sinh(\kappa z)}{\kappa^3\cosh(\kappa H)} - \frac{b\cosh(\kappa z)}{\kappa^4}\right) \qquad (2.17)$$

The boundary conditions for the equation express that there is no translation at the top of the structure (as the origin of the coordinate system is fixed to the top)

$$y(0) = 0$$

and that the tangent to the equivalent column is vertical at the bottom (Figure 2.6/b):

$$y'(H) = 0$$

Integrating Equation (2.17) once

$$y'(z)=\frac{w}{EI}\left(\frac{bz^3}{6\kappa^2}+\frac{bz}{\kappa^4}-\frac{z^3}{6}+\frac{b\sinh(\kappa H)\cosh(\kappa z)}{\kappa^5\cosh(\kappa H)}-\frac{bH\cosh(\kappa z)}{\kappa^4\cosh(\kappa H)}-\frac{b\sinh(\kappa z)}{\kappa^5}\right)+C_1$$

and making use of the second boundary condition, constant C_1 is obtained as

$$C_1=\frac{w}{EI}\left(\frac{H^3}{6}-\frac{bH^3}{6\kappa^2}\right)$$

Integrating once again, the deflection is obtained as

$$y(z)=\frac{w}{EI}\left(\frac{bz^4}{24\kappa^2}+\frac{bz^2}{2\kappa^4}+\frac{H^3z}{6}-\frac{bH^3z}{6\kappa^2}-\frac{z^4}{24}+\right.$$

$$\left.+\frac{b\sinh(\kappa H)\sinh(\kappa z)}{\kappa^6\cosh(\kappa H)}-\frac{bH\sinh(\kappa z)}{\kappa^5\cosh(\kappa H)}-\frac{b\cosh(\kappa z)}{\kappa^6}\right)+C_2$$

Making use of the first boundary condition leads to constant C_2:

$$C_2=\frac{bw}{EI\kappa^6}$$

and the deflection emerges as

$$y(z)=\frac{w}{EI}\left[\frac{H^3z}{6}-\frac{z^4}{24}+\frac{b}{\kappa^2}\left(\frac{z^4}{24}-\frac{H^3z}{6}+\frac{z^2}{2\kappa^2}+\frac{1}{\kappa^4}\right)-\right.$$

$$\left.-\frac{b}{\kappa^6}\left(\cosh(\kappa z)+\frac{\kappa H\sinh(\kappa z)}{\cosh(\kappa H)}-\tanh(\kappa H)\sinh(\kappa z)\right)\right]$$

This is not the kind of formula that was aimed at in the beginning of the derivation following the establishment of governing differential equation (2.15). This formula is not particularly simple and it does not show what role the characteristic stiffnesses play in the behaviour of the structure under horizontal load. However, after returning to the original structural engineering notation [Equation (2.14)] and following a lengthy rearrangement, the above formula can be transformed into a much simpler, meaningful and "user-friendly" form:

$$y(z)=\frac{w}{EI_f}\left(\frac{H^3z}{6}-\frac{z^4}{24}\right)+\frac{wz^2}{2Ks^2}-\frac{wEI}{K^2s^3}\left(\frac{\cosh[\kappa(H-z)]+\kappa H\sinh(\kappa z)}{\cosh(\kappa H)}-1\right)\tag{2.18}$$

or in a shorter form:

$$y(z) = y_b(z) + y_s(z) - y_i(z) \tag{2.19}$$

These equations clearly show the two important contributors to the deflection—bending deformation y_b and shear deformation y_s—as well as the effect of interaction (y_i) between the two characteristic types of deformation:

$$y_b(z) = \frac{w}{EI_f}\left(\frac{H^3 z}{6} - \frac{z^4}{24}\right) \tag{2.20}$$

$$y_s(z) = \frac{wz^2}{2Ks^2} \tag{2.21}$$

and

$$y_i(z) = \frac{wEI}{K^2 s^3}\left(\frac{\cosh[\kappa(H-z)] + \kappa H \sinh(\kappa z)}{\cosh(\kappa H)} - 1\right) \tag{2.22}$$

In Equations (2.18) and (2.20)

$$I_f = I + I_g = I_c r + I_g \tag{2.23}$$

represents the sum of the local and the global second moments of area of the columns.

Maximum deflection develops at $z = H$:

$$y_{\max} = y(H) = \frac{wH^4}{8EI_f} + \frac{wH^2}{2Ks^2} - \frac{wEI}{K^2 s^3}\left(\frac{1 + \kappa H \sinh(\kappa H)}{\cosh(\kappa H)} - 1\right) \tag{2.24}$$

or

$$y_{\max} = y(H) = y_b(H) + y_s(H) - y_i(H) \tag{2.25}$$

where

$$y_b(H) = \frac{wH^4}{8EI_f}, \quad y_s(H) = \frac{wH^2}{2Ks^2}, \quad y_i(H) = \frac{wEI}{K^2 s^3}\left(\frac{1 + \kappa H \sinh(\kappa H)}{\cosh(\kappa H)} - 1\right) \tag{2.26}$$

are the three characteristic parts of the top deflection.

2.1.3 Extension of the results: multi-bay, multi-storey frames

Although the formulae in the previous section were derived for one-bay structures, their validity can be extended to cover multi-bay structures (Figure 2.7) as well.

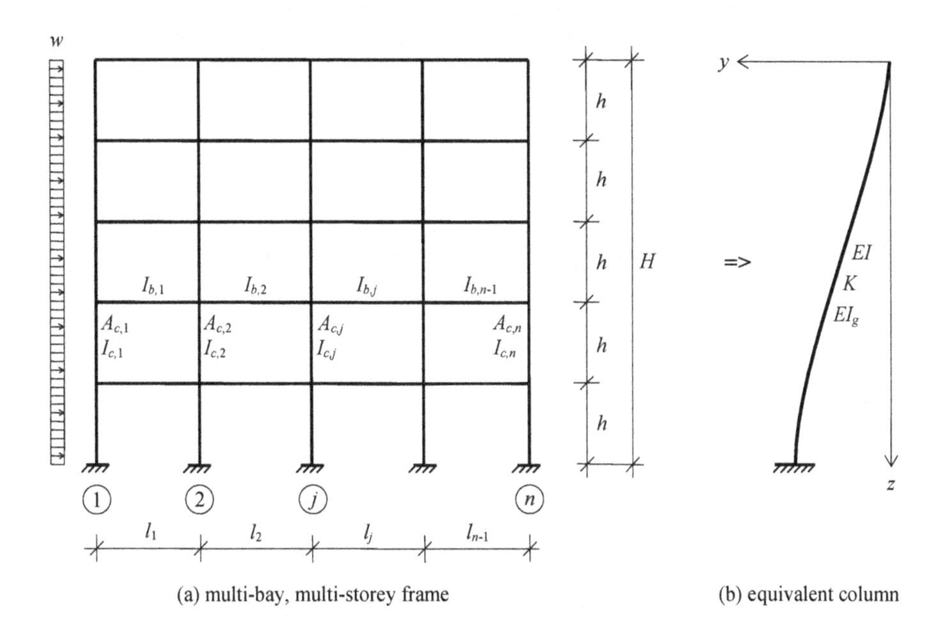

Figure 2.7 Model for multi-bay sway-frame.

The shear stiffness for the whole multi-bay structure is obtained using

$$K = \left(\frac{1}{K_b} + \frac{1}{K_c} \right)^{-1} = K_b \frac{K_c}{K_b + K_c} = K_b r \tag{2.27}$$

where the two contributors to the shear stiffness are

$$K_b = \sum_{j=1}^{n-1} \frac{12EI_{b,j}}{l_j h} \tag{2.28}$$

and

$$K_c = \sum_{j=1}^{n} \frac{12EI_{c,j}}{h^2} \tag{2.29}$$

The reduction factor is

$$r = \frac{K_c}{K_b + K_c} \tag{2.30}$$

In the above equations n is the number of columns of height H.

For the local bending stiffness ($EI = EI_c r$), the sum of the second moments of area of the columns has to be produced (and multiplied by reduction factor r):

$$I = r\sum_{j=1}^{n} I_{c,j} \tag{2.31}$$

where the summation goes from $j = 1$ to $j = n$. When the cross-sections of the columns of the frame are identical (as is often the case), the second moment of area of the columns is simply multiplied by n and r.

For the global bending stiffness (EI_g), the formula

$$I_g = \sum_{j=1}^{n} A_{c,j} t_j^2 \tag{2.32}$$

should be used, where $A_{c,j}$ is the cross-sectional area of the jth column and t_j is the distance of the jth column from the centroid of the cross-sections. See also Figures 2.9 and 2.13 for the interpretation of t_j.

It should be noted, however, that Equation (2.32) represents an approximation and its use for multi-bay frames may lead to slightly unconservative estimates in the region of 0–3% for four-bay structures and up to 6% for ten-bay structures (Kollár, 1986).

2.1.4 Discussion and special cases

To evaluate Equations (2.18) and (2.24), a comprehensive accuracy analysis was carried out—see Section 7.2.1 for details. The deflection shape and the top deflection of 216 individual frames were determined. The stiffness characteristics covered a large range and the height of the frames was varied from five to eighty storeys in eight steps. Typical deflection shapes are shown in Figure 2.8. (Note that z is measured from ground floor level upwards.) The results lead to the following observations:

(a) The effect of interaction between the bending and shear modes is always beneficial as it reduces the deflection. The range of the reduction of the top deflection was between 0% and 64%.

(b) The effect of interaction significantly becomes smaller as the height of the structures increases. For structures of height over twenty storeys, the reduction in the top deflection dropped below 20%.

(c) The effect of interaction (y_i in Figure 2.8) is roughly constant over the height of the structures.

(d) The accuracy of the procedure developed for the calculation of the top deflection is acceptable for practical application. For a group of 90 rigid sway-frames, for example, the range of difference (between the proposed method and the "exact" solution) was between -5% and 7% and the average absolute difference was 1.2% (Figure 7.6 and Table 7.2).

(e) Ignoring the effect of interaction leads to a very simple, albeit less accurate, conservative solution [with the first two terms in Equation (2.24)].

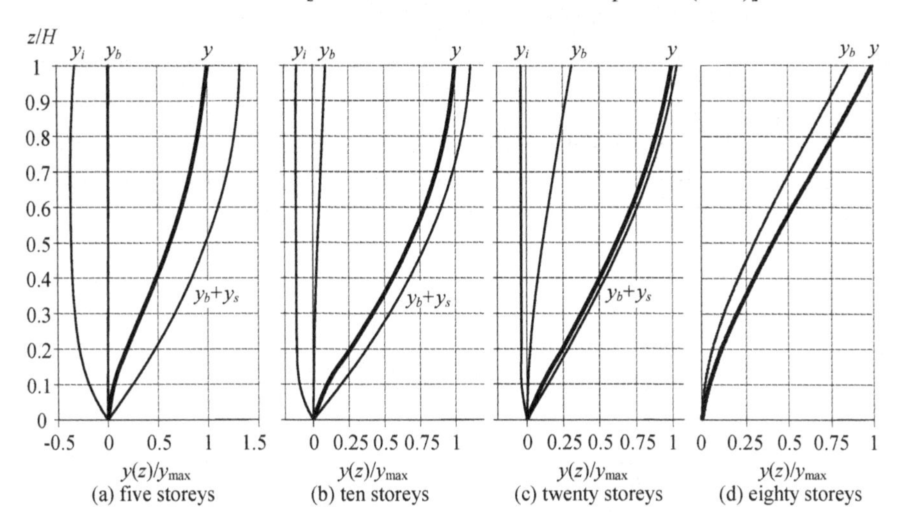

Figure 2.8 Typical deflection shapes with components y_b (bending), y_s (shear), y_i (interaction) and the overall deflection y for the five-, ten-, twenty- and eighty-storey frame F1 shown in Figure 7.1/a.

To conclude the investigation of the behaviour of frames under lateral load, some special, sometimes theoretical, cases will now be considered.

(a) Frames with dominant shear deformation

Multi-bay, low-rise frames may belong here whose resistance against global bending deformation is relatively great. Their deflection shape is dominated by global shear deformation.

This case is characterised by $I_g >> I_c$, and consequently, $a \to 0$, $b \to \infty$. Governing differential equation (2.15) cannot be used directly because of singularity but, after some rearrangement, Equation (2.12) can, which then simplifies to

$$y''(z) = \frac{w}{K}$$

where $K \cong K_b$. This differential equation, together with the boundary conditions $y(0) = 0$ and $y'(0) = 0$, lead to the deflection and the top deflection as

$$y(z) = \frac{wz^2}{2K} \tag{2.33}$$

and

$$y_{\max} = y(H) = \frac{wH^2}{2K} \tag{2.34}$$

The characteristic deflection shape is shown in Figure 2.1/a.

(b) Frames with dominant local bending deformation

In this group the connecting beams have no or negligible bending stiffness and their main task is to transmit the horizontal load and to make the full-height columns work together.

This case is characterized by $K_b \approx 0$. Consequently, the shear stiffness of the structure becomes zero ($K = 0$), which leads to $a = 0$, $b = 0$ and $\kappa = 0$. Governing differential equation (2.15) simplifies to

$$y''''(z) = -\frac{w}{EI}$$

and the solutions for the deflection and the top deflection are

$$y(z) = \frac{w}{EI_c}\left(\frac{H^3 z}{6} - \frac{z^4}{24}\right) \tag{2.35}$$

and

$$y_{\max} = y(H) = \frac{wH^4}{8EI_c} \tag{2.36}$$

where EI_c is the sum of the bending stiffnesses of the columns. This case is identified in Figure 2.1/c as one of the three characteristic types of behaviour of the frame, when the columns are linked by beams that can only pass on axial (horizontal) forces but no moments to the columns.

(c) Frames developing dominant global bending deformation

The structures in this group are tall and relatively slender (with great height/width ratio).

The second and third terms in Equations (2.18) and (2.24) tend to be by orders of magnitude smaller than the first term, and the solutions for the deflection and the top deflection effectively become

$$y(z) = \frac{w}{EI_f}\left(\frac{H^3 z}{6} - \frac{z^4}{24}\right) \tag{2.37}$$

and

$$y_{\max}=y(H)=\frac{wH^4}{8EI_f} \tag{2.38}$$

where $I_f = I_c r + I_g$. This case is illustrated in Figure 2.1/b.

(d) The columns of the structures do not undergo axial deformations

In this theoretical case, no global bending occurs.

This case is characterised by $A_{c,j}\rightarrow\infty$. As a consequence, $I_g\rightarrow\infty$, $a\rightarrow 0$, $\kappa^2\rightarrow b$ and $s\rightarrow 1$. The governing differential equation, Equation (2.15), simplifies to

$$y''''(z)-by''(z)=-\frac{w}{EI}$$

The solutions of this equation for the deflection and the top deflection are

$$y(z)=\frac{wz^2}{2K}-\frac{wEI}{K^2}\left(\frac{\cosh[\kappa(H-z)]+\kappa H\sinh(\kappa z)}{\cosh(\kappa H)}-1\right) \tag{2.39}$$

and

$$y_{\max}=y(H)=\frac{wH^2}{2K}-\frac{wEI}{K^2}\left(\frac{1+\kappa H\sinh(\kappa H)}{\cosh(\kappa H)}-1\right) \tag{2.40}$$

It is interesting to note that the above two formulae can also be originated from Equations (2.18) and (2.24) by, in addition to setting $s = 1$, dropping the first term which is associated with the bending deformation of the structure. It follows that when the columns do not develop axial deformations the structure cannot—at least not directly—"utilise" its bending stiffness. (The bending stiffness does enter the picture, but indirectly, through the last term that is responsible for the interaction between the bending and shear modes.)

2.1.5 Worked example: two-bay, ten-storey frame

The maximum deflection of the two-bay, ten-storey frame shown in Figure 2.9 is determined here, using the equations given in Sections 2.1.2 and 2.1.3. The structure is subjected to a uniformly distributed wind load of intensity $w = 5.0$ kN/m. The modulus of elasticity is $E = 25\cdot10^6$ kN/m^2. The cross-sections of both the columns and the beams are 0.4 m/0.4 m. The two bays are identical at 6.0 m and the storey height is 3.0 m. The numbers of the equations used will be given on the right-hand side in curly brackets. The frame is identical to F7 (Figure 7.1/g) used for the comprehensive accuracy analysis presented in Chapter 7.

The part of the shear stiffness which is associated with the beams is

$$K_b = \sum_{j=1}^{n-1} \frac{12EI_{b,j}}{l_j h} = 2\frac{12 \cdot 25 \cdot 10^6 \cdot 0.4 \cdot 0.4^3}{12 \cdot 6 \cdot 3} = 71111 \text{ kN} \qquad \{2.28\}$$

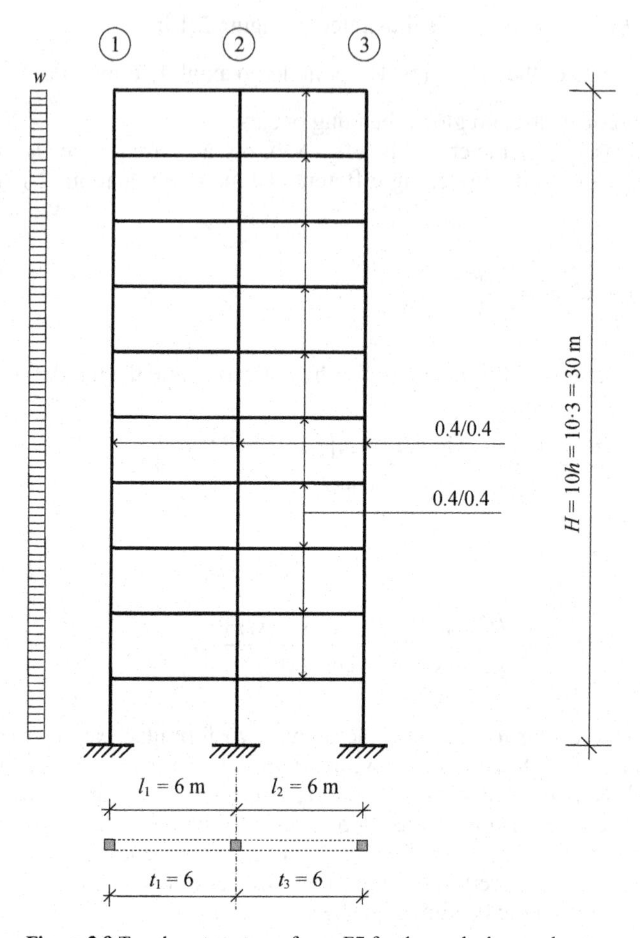

Figure 2.9 Two-bay, ten-storey frame F7 for the worked example.

The part of the shear stiffness which is associated with the columns is

$$K_c = \sum_{j=1}^{n} \frac{12EI_{c,j}}{h^2} = 3\frac{12 \cdot 25 \cdot 10^6 \cdot 0.4^4}{12 \cdot 3^2} = 213333 \text{ kN} \qquad \{2.29\}$$

The above two part stiffnesses define reduction factor r as

$$r = \frac{K_c}{K_b + K_c} = \frac{213333}{71111 + 213333} = 0.75 \qquad \{2.30\}$$

The shear stiffness of the frame can now be determined:

$$K = K_b r = K_b \frac{K_c}{K_b + K_c} = 71111 \cdot 0.75 = 53333 \text{ kN} \qquad \{2.27\}$$

For the local bending stiffness ($EI = EI_c r$), the sum of the second moments of area of the columns should be produced (and multiplied by reduction factor r). As the columns of the frame are identical, the second moment of area of one column is simply multiplied by n and r:

$$I = r\sum_{j=1}^{n} I_{c,j} = 0.75 \cdot 3 \cdot \frac{0.4^4}{12} = 0.0048 \text{ m}^4 \qquad \{2.31\}$$

The global second moment of area is

$$I_g = \sum_{j=1}^{n} A_{c,j} t_j^2 = 0.4 \cdot 0.4(6^2 + 6^2) = 11.52 \text{ m}^4 \qquad \{2.32\}$$

The total second moment of area for the bending stiffness is

$$I_f = I + I_g = I_c r + I_g = 11.5248 \text{ m}^4 \qquad \{2.23\}$$

Parameters s, κ and κH are also needed for the calculation of the maximum deflection:

$$s = 1 + \frac{I_c r}{I_g} = 1 + \frac{0.0048}{11.52} = 1.000417 \qquad \{2.14\}$$

$$\kappa = \sqrt{\frac{Ks}{EI}} = \sqrt{\frac{53333 \cdot 1.000417}{25 \cdot 10^6 \cdot 0.0048}} = 0.6668 \frac{1}{\text{m}} \qquad \text{and} \qquad \kappa H = 20$$

With the above auxiliary quantities, the maximum top deflection of the frame can now be calculated:

$$y_{\max} = y(H) = \frac{wH^4}{8EI_f} + \frac{wH^2}{2Ks^2} - \frac{wEI}{K^2 s^3}\left(\frac{1 + \kappa H \sinh(\kappa H)}{\cosh(\kappa H)} - 1\right) \qquad \{2.24\}$$

$$= \frac{5 \cdot 30^4}{8 \cdot 25 \cdot 10^6 \cdot 11.5248} + \frac{5 \cdot 30^2}{2 \cdot 53333 \cdot 1.000417^2} - \frac{5 \cdot 25 \cdot 10^6 \cdot 0.0048}{53333^2 \cdot 1.000417^3}\left(\frac{1 + 20\sinh 20}{\cosh 20} - 1\right)$$

With the three contributors to the deflection, its value is

$$y_{\max} = 0.0018 + 0.0421 - 0.0040 = 0.0399\,\text{m}$$

The Finite Element based solution, obtained using Axis VM X5 (2019), is $y_{max} = 0.0406$ m.

As in most practical cases, the effect of the interaction between the bending and shear modes is small—third term in Equation {2.24}. Neglecting the third term leads to a very simple, truly back-of-the-envelope formula with the first two terms.

2.2 FREQUENCY ANALYSIS OF RIGID SWAY-FRAMES

Because of the complexity of the problem, a number of attempts have been made to develop approximate methods for the dynamic analysis of frames. Goldberg (1973) presented several simple methods for the calculation of the fundamental frequency of (uncoupled) lateral and pure torsional vibration. The effect of the axial deformation of the vertical elements was taken into account by a correction factor in his methods. The continuous connection method enabled the development of more rigorous analysis (Rosman, 1973; Coull, 1975; Kollár, 1992). However, most approximate methods are either still too complicated for design office use or restrict the scope of analysis or neglect one or more important characteristics. Another important factor in connection with the availability of good and reliable approximate methods is the fact that their accuracy has not been satisfactorily investigated. In two excellent publications, Ellis (1980) and Jeary and Ellis (1981) reported on accuracy matters in a comprehensive manner and their findings indicated that some widely used approximate methods were of unacceptable accuracy. The method to be presented here is not only simple and gives a clear picture of the behaviour of the structure, but its accuracy has also been comprehensively investigated and found exceptionally good for practical application.

In addition to the general assumptions made in Chapter 1, it will be assumed that the mass of the structures is concentrated at floor levels (Figure 2.10).

2.2.1 Fundamental frequency

As with the deflection analysis in Section 2.1, the continuum method is used for the frequency analysis in this section. The multi-bay, multi-storey frame (Figure 2.10/a) is again characterised by its characteristic deformations and the corresponding three characteristic stiffnesses (EI, EI_g and K) (as shown and explained in Sections 2.1.1 and 2.1.3). The fundamental frequency for lateral vibration is determined using the three types of stiffness and the related vibration modes and frequencies. The three types are: local bending (=> the full-height bending of the individual columns of the frame), global bending (=> the bending of the frame as a whole unit) and shear. The deflected shape of the frame can be composed of the three deformation types and, in a similar manner, the frequency of the frame can be produced using the three "part" frequencies which are linked to the corresponding stiffnesses.

The frame can be replaced with an equivalent column for the analysis, defined by stiffnesses EI and K_e (a combination of K and EI_g) as described in the following. The structure is subjected to a uniformly distributed floor load of intensity p [kN/m] which leads to concentrated masses M_i [kg] at floor levels (Figure 2.10/b). Masses M_i are then distributed over storey-height (Figure 2.10/c) resulting in the mass density per unit height of the structure as

$$m = \frac{M_i}{h} = \frac{pL}{gh} \tag{2.41}$$

where L is the width of the frame, h is the storey height and $g = 9.81$ m/s^2 is the gravity acceleration.

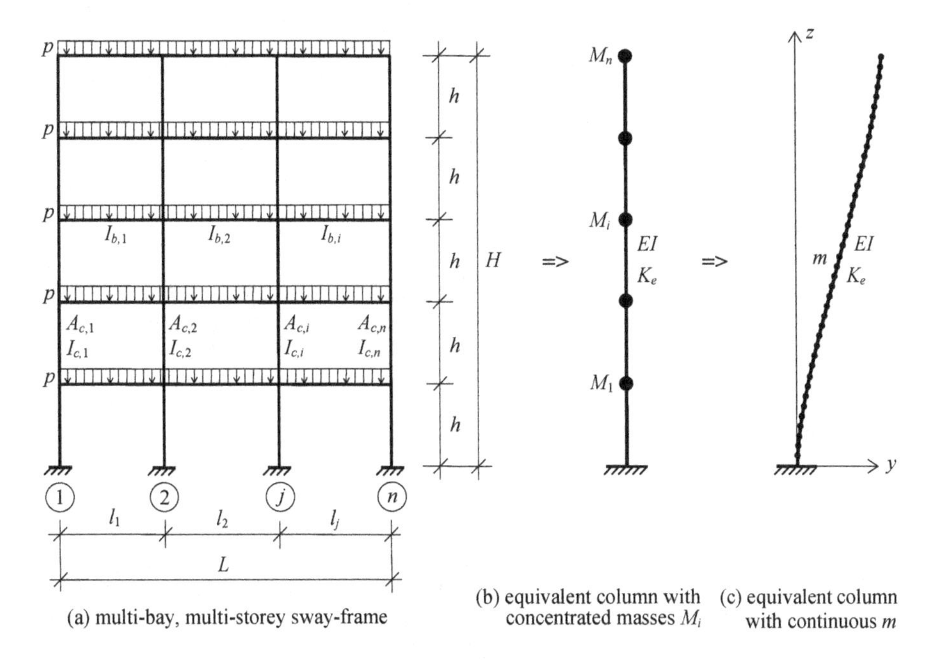

Figure 2.10 Origination of the continuum model for the frequency analysis.

The nature of the load on the structure is now different from that with the deflection analysis in Section 2.1. Wind load w in Section 2.1 was a genuinely uniformly distributed load. Mass density m has been *made* uniformly distributed. When the original concentrated floor loads M_i were arbitrarily distributed over the height (Figure 2.10/b → 2.10/c), the procedure resulted in a detrimental approximation as the centroid of the total load shifted downwards. This approximation can easily be accounted for by using the Dunkerley theorem (Zalka, 2000) and introducing mass distribution factor r_f which will be attached to the relevant stiffnesses when the frequencies will be calculated later on. Its value is given by

$$r_f = \sqrt{\frac{n}{n+2.06}}$$

where n ($n > 2$) is the number of storeys. Values for r_f are given in Figure 2.11 for frames up to twenty storeys high. Table 4.1 in Chapter 4 can be used for more accurate values and/or for higher frames.

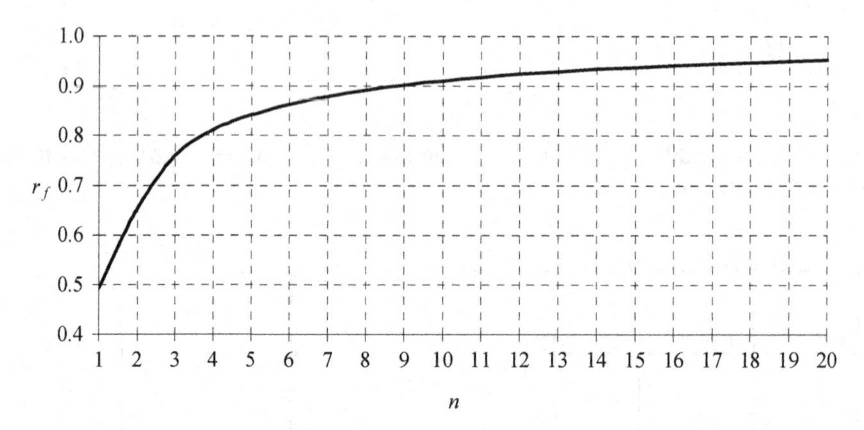

Figure 2.11 Mass distribution factor r_f as a function of n (the number of storeys).

Vibration in shear (Figure 2.1/a) is associated with the shear stiffness of the frame. Based on the classical formula of a cantilever with uniformly distributed mass and shear stiffness (Vértes, 1985), and amended by r_f, the fundamental frequency of the frame due to shear deformation can be calculated from

$$f_{s'}^2 = \frac{1}{(4H)^2} \frac{r_f^2 K}{m} \tag{2.42}$$

In the above formula K is the shear stiffness calculated using Equations (2.27), (2.28), (2.29) and (2.30). Mass distribution factor r_f is introduced into the formula to allow for the fact that the mass of the structure is concentrated at floor levels (M_i in Figure 2.10/b) and is not uniformly distributed over the height (as assumed for the derivation of the classical formula).

The full-height global bending vibration of the frame as a whole unit represents pure bending type deformation (Figure 2.1/b). In this case, the columns act as longitudinal fibres (in tension and compression) and the role of the beams is to transfer shear so as to make the columns work together in this fashion. The bending stiffness associated with this bending deformation is the global bending stiffness (EI_g) defined by Equation (2.32). The fundamental frequency that belongs to this global bending deformation is obtained using Timoshenko's (1928) classical formula, which is again amended with factor r_f, as

$$f_g^2 = \frac{0.313 r_f^2 EI_g}{H^4 m} \tag{2.43}$$

Although frames are routinely associated with shear-type deformation, reality is somewhat more complicated. As Figure 2.8 demonstrates, and the application of any FE package can confirm, as a function of height, a frame with the same (beam/column) stiffness characteristics may assume different types of deformation. Low frames tend to show a predominantly shear-type vibration mode, in the case of medium-rise frames the vibration shape can be a mixture of bending and shear-type deformations, and tall, “slender” structures normally vibrate in a predominantly bending mode. The reason for this type of behaviour lies in the fact that there is an interaction between sway in shear and in global bending. Low and/or wide (multi-bay) frames tend to undergo shear deformation while as the height of the frame increases, the effect of the axial deformation of the columns becomes more and more important. The axial deformation of the columns can be interpreted as a “compromising” factor, as far as the shear stiffness is concerned. Because of the lengthening and shortening of the columns, there is less and less “scope” for the structure to develop shear deformation. As indeed is the case with narrow and very tall frames; very often they do not show any shear deformation at all.

This phenomenon can be easily taken into account by applying the Föppl-Papkovich theorem (Tarnai, 1999) to the vibration mode in shear (subscript: s') and the vibration mode in full-height global bending (subscript: g) and then introducing the *effective* shear stiffness as follows.

The application of the Föppl-Papkovich theorem leads to

$$\frac{1}{f_s^2} = \frac{1}{f_{s'}^2} + \frac{1}{f_g^2}$$

where f_s is the natural frequency which is associated with the effective shear stiffness, i.e. with the shear stiffness that remains after taking into account the effect of the axial deformation of the full-height columns.

In rearranging the above equation and substituting for $f_{s'}$ and f_g [using Equations (2.42) and (2.43)], the natural frequency can be expressed as a function of the effective shear stiffness:

$$f_s^2 = f_{s'}^2 \frac{f_g^2}{f_{s'}^2 + f_g^2} = \frac{1}{(4H)^2} \frac{r_f^2 K_e}{m} \tag{2.44}$$

In the above equation K_e is the effective shear stiffness, according to

$$K_e = s_f^2 K \tag{2.45}$$

and

$$s_f = \sqrt{\frac{f_g^2}{f_{s'}^2 + f_g^2}} = \sqrt{\frac{K_e}{K}} \tag{2.46}$$

is the effectiveness factor.

Finally, the frame may develop bending vibration in a different manner. The full-height bending vibration of the individual columns of the frame—also called local bending vibration—also represents pure bending type deformation (Figure 2.1/c). The characteristic stiffness is defined by EI given by Equation (2.31). With the columns of the frame built in at ground floor level, the fundamental frequency which is associated with the local bending stiffness is again obtained using Timoshenko's formula for cantilevers under uniformly distributed mass:

$$f_b^2 = \frac{0.313 r_f^2 EI}{H^4 m} \tag{2.47}$$

The frame can now be characterised by its local bending stiffness and its effective shear stiffness (and the related frequencies). It follows that the complex behaviour of a frame in lateral vibration can now be analysed by using an equivalent column with stiffnesses EI and K_e (Figure 2.10/c).

Before the governing differential equation of the problem is presented and solved, an approximate solution is given below. This approximate solution is a lower bound to the exact solution, which can also be used when the governing differential equation is solved using numerical methods. Applying the Southwell theorem to the squares of the natural frequencies that belong to the local bending and shear stiffnesses—Equations (2.44) and (2.47)—a very simple formula can be given for the calculation of the fundamental frequency of the frame:

$$f_{\text{Southwell}}^2 = f_b^2 + f_s^2 = \frac{0.313 r_f^2 EI}{H^4 m} + \frac{1}{(4H)^2} \frac{r_f^2 K_e}{m} \tag{2.48}$$

The governing differential equation of the equivalent column is obtained by examining the equilibrium of its elementary section (Zalka, 1994). This leads to

$$r_f^2 EIv''''(z,t) - r_f^2 K_e v''(z,t) + m\ddot{v}(z,t) = 0$$

where $v(z,t)$ denotes horizontal motion in direction y. Primes and dots mark differentiation by z and t (time), respectively.

The solution can be sought in the product form

$$v(z,t) = v_1(z) v_2(t)$$

where $v_1(z)$ only depends on place and $v_2(t)$ only depends on time.

Separating the variables and eliminating the time dependent functions lead to the boundary value problem

$$r_f^2 EIv_1''''(z) - r_f^2 K_e v_1''(z) - \omega^2 m v_1(z) = 0 \tag{2.49}$$

where ω is the circular frequency.

The origin of the coordinate system is set at the lower built-in end of the equivalent column (Figure 2.10/c). The four boundary conditions needed for the solution are:

$$v_1(0) = 0\,, \qquad\qquad v_1'(0) = 0$$

and

$$v_1''(H) = 0\,, \qquad\qquad EIv_1'''(H) - K_e v_1'(H) = 0$$

The solution can be written in the form

$$v_1 = c_1 \sinh(\alpha z) + c_2 \cosh(\alpha z) + c_3 \sin(\beta z) + c_4 \cos(\beta z)$$

where

$$\alpha = \sqrt{\frac{K_e + \sqrt{K_e^2 + 4EI\omega^2 m}}{2EI}} \qquad \text{and} \qquad \beta = \sqrt{\frac{-K_e + \sqrt{K_e^2 + 4EI\omega^2 m}}{2EI}}$$

are auxiliary parameters.

The circular frequency is given by

$$\omega = \frac{2\pi\eta}{H^2}\sqrt{\frac{EIr_f^2}{m}} \tag{2.50}$$

where η is the frequency parameter, yet unknown.

Making use of the boundary conditions leads to the frequency equation

$$\frac{(\alpha H)^4 + (\beta H)^4}{(\alpha H)^2(\beta H)^2}\cosh(\alpha H)\cos(\beta H) + \frac{(\alpha H)^2 - (\beta H)^2}{\alpha H \beta H}\sinh(\alpha H)\sin(\beta H) + 2 = 0$$

whose first root furnishes the frequency parameter that is needed for the determination of the fundamental frequency. This parameter is the function of two stiffnesses, K_e and EI, and it is therefore advantageous to use their ratio. This is done by introducing the non-dimensional parameter

$$k = H\sqrt{\frac{K_e}{EI}} \tag{2.51}$$

The frequency equation can now be solved using the above non-dimensional parameter and the two constants

$$\alpha H = \sqrt{\frac{k^2 + \sqrt{k^4 + (4\pi\eta)^2}}{2}} \qquad \text{and} \qquad \beta H = \sqrt{\frac{-k^2 + \sqrt{k^4 + (4\pi\eta)^2}}{2}}$$

The solution is best used in practice if the values of frequency parameter η are tabulated or given by a graph. Figure 2.12 covers the range $0 \le k \le 10$. Table 4.2 in Chapter 4 can also be used if a more accurate value of η is needed or a wider range of k should be covered. Values of parameter η for the second and third frequencies are tabulated in (Zalka, 2000).

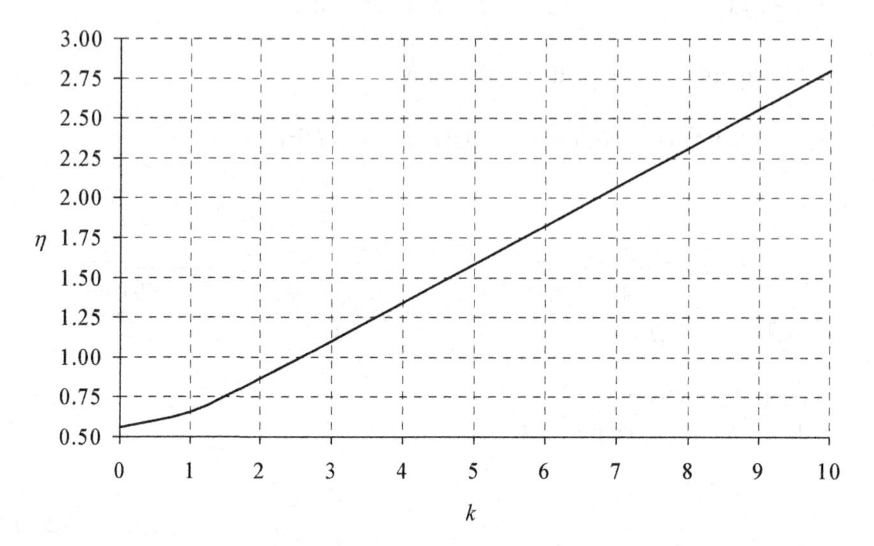

Figure 2.12 Frequency parameter η as a function of non-dimensional parameter k.

Knowing frequency parameter η, Equation (2.50) leads to the natural frequency as

$$f^2 = \left(\frac{\omega}{2\pi}\right)^2 = \frac{\eta^2}{H^4}\frac{r_f^2 EI}{m}$$

Before this solution is used for the lateral vibration analysis, however, some modification is needed. The above formula combines three phenomena in one: the value of the frequency of a frame depends on the contribution of the bending stiffness, the contribution of the shear stiffness and it is also influenced by the interaction between the bending and shear modes. (The nature of the behaviour is exactly the same as with the deflection analysis in Section 2.1.) The three items have to be separated as the third item—the one that is responsible for the interaction—has to be amended because interaction can only take place in line with the effectiveness of the shear stiffness.

Making use of Equations (2.44), (2.47) and (2.51), the separation of the three characteristic items can easily be carried out. The first step is expressing the part frequency that belongs to the shear stiffness of the frame by the part frequency that belongs to the bending stiffness of the frame. This is done by replacing K_e and EI in Equation (2.51) of parameter k by the relevant part frequencies using Equations (2.44) and (2.47):

$$k^2 = H^2 \frac{K_e}{EI} = 5\frac{f_s^2}{f_b^2} \quad \rightarrow \quad f_s^2 = \frac{k^2}{5} f_b^2$$

Next, the equation of the natural frequency is rearranged by involving Equation (2.47) of f_b and by adding to it the term

$$\left(f_s^2 - \frac{k^2}{5} f_b^2 + f_b^2 - f_b^2 \right)$$

of zero value:

$$f^2 = \frac{\eta^2}{H^4} \frac{EIr_f^2}{m} = \frac{\eta^2}{0.313} f_b^2 + f_s^2 - \frac{k^2}{5} f_b^2 + f_b^2 - f_b^2$$

Finally, the two part frequencies f_b and f_s and the term responsible for the interaction can now be separated:

$$f^2 = f_b^2 + f_s^2 + \left(\frac{\eta^2}{0.313} - \frac{k^2}{5} - 1 \right) f_b^2$$

The third term in the above equation is responsible for the interaction between the bending and shear modes of vibration. This is the term that has to be adjusted as interaction can only occur in line with the effectiveness of the shear stiffness. As the effectiveness of the shear stiffness is normally smaller than 100% [c.f. Equation (2.46) where $s_f \leq 1$ holds], the effectiveness factor should be applied to the part which is responsible for the interaction. This amendment leads to the formula for the lateral vibration as

$$f = \sqrt{f_b^2 + f_s^2 + \left(\frac{\eta^2}{0.313} - \frac{k^2}{5} - 1 \right) s_f f_b^2} \tag{2.52}$$

In the right-hand side of the above equation, the first two terms stand for the lateral frequency associated with bending and shear deformations, respectively, while the third term represents the effect of the interaction between the bending and shear modes. Values for η are available in Figure 2.12 and in Table 4.2.

2.2.2 Discussion

To evaluate the accuracy of Equation (2.52), a comprehensive accuracy analysis was carried out—see Section 7.2.2 for details. The fundamental frequency of 216 frames ranging in height from five to eighty storeys was determined and compared to the "exact" (FE) solution. The results lead to the following observations:

(a) As is the case with the deflection analysis, the interaction between the bending and shear modes is always beneficial. Bearing in mind that $(\eta^2/0.313 - k^2/5) \geq 1$ always holds, the evaluation of the third term in Equation (2.52) demonstrates that the effect of the interaction increases the value of the lateral frequency of the frame. According to the data given in Table 4.2, the maximum increase (at $k = 3.2$) is 62%.
(b) The effect of interaction significantly becomes smaller as the height of the frame increases. For structures of height over twenty storeys, the increase dropped below 20% in the test cases. The characteristic deflection types shown in Figure 2.8 are still relevant as lateral vibration shapes.
(c) The effect of interaction is roughly constant over the height of the structures.
(d) The accuracy of the procedure developed for the calculation of the fundamental frequency is acceptable for practical application. In the case of a group of 90 rigid sway-frames, for example, the range of difference (between the proposed method and the "exact" solution) was between -3% and 7% and the average absolute difference was 1.6%—see Figure 7.9 and Table 7.5.
(e) Ignoring the effect of interaction leads to a very simple, albeit less accurate, conservative solution, the Southwell formula [Equation (2.48)].

2.2.3 Worked example: three-bay, twenty-five storey frame

The fundamental frequency of the three-bay, twenty-five storey frame shown in Figure 2.13 is determined in this section.

The modulus of elasticity is $E = 25 \cdot 10^6$ kN/m^2. The cross-sections of the columns and beams are 0.4 m/0.4 m and 0.4 m/0.7 m, respectively, where the first number stands for the width (perpendicular to the plane of the structure) and the second number is the depth (in-plane size of the member). The three bays are identical at 6.0 m and the storey height is 3.0 m. The numbers of the equations used for the calculation will be given on the right-hand side in curly brackets. The frame is one of the frames used for the accuracy analysis in Chapter 7 (frame F6 in Figure 7.1/f).

The structure is subjected to a uniformly distributed vertical load of intensity $p = 50$ kN/m, representing the total load on a storey. This leads to the mass density per unit length

$$m = \frac{pL}{gh} = \frac{50 \cdot 18}{9.81 \cdot 3} = 30.58 \text{ kg/m} \qquad \{2.41\}$$

The part of the shear stiffness which is associated with the beams is

$$K_b = \sum_{j=1}^{n-1} \frac{12EI_{b,j}}{l_j h} = 3\frac{12 \cdot 25 \cdot 10^6 \cdot 0.4 \cdot 0.7^3}{12 \cdot 6 \cdot 3} = 571667 \text{ kN} \qquad \{2.28\}$$

The part of the shear stiffness which is associated with the columns is

$$K_c = \sum_{j=1}^{n} \frac{12EI_{c,j}}{h^2} = 4\frac{12 \cdot 25 \cdot 10^6 \cdot 0.4^4}{12 \cdot 3^2} = 284444\,\text{kN} \qquad \{2.29\}$$

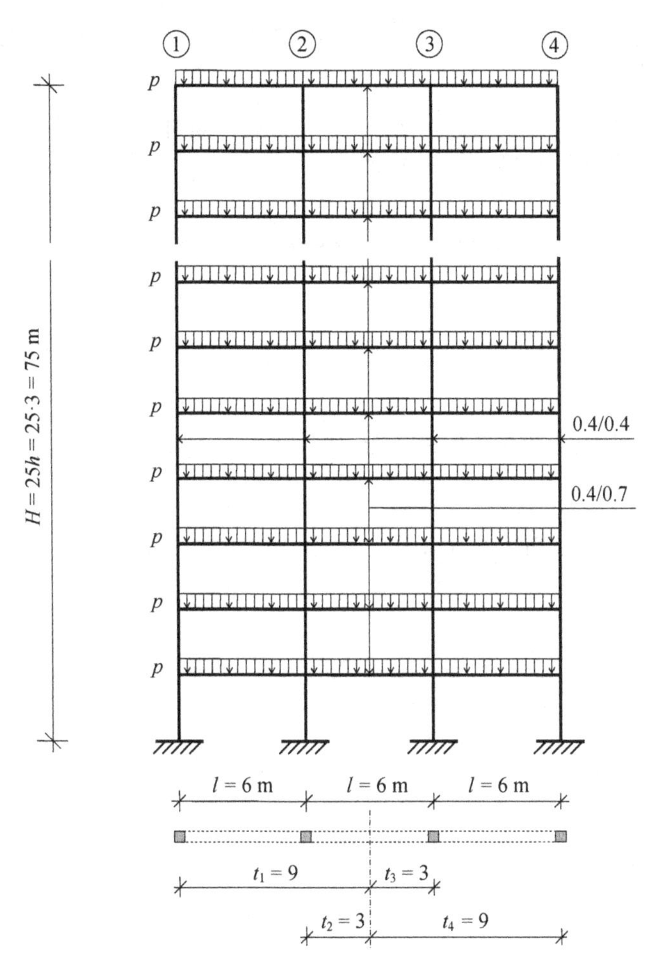

Figure 2.13 Three-bay, twenty-five storey frame F6 for the worked example.

The above two part stiffnesses define reduction factor r as

$$r = \frac{K_c}{K_b + K_c} = \frac{284444}{571667 + 284444} = 0.33225 \qquad \{2.30\}$$

The "original" shear stiffness of the frame can now be determined:

$$K = K_b r = K_b \frac{K_c}{K_b + K_c} = 571667 \cdot 0.33225 = 189936\,\text{kN} \qquad \{2.27\}$$

With the above shear stiffness, the square of the fundamental frequency of the frame due to shear deformation can be calculated as

$$f_{s'}^2 = \frac{1}{(4H)^2}\frac{r_f^2 K}{m} = \frac{0.961^2 \cdot 189936}{(4 \cdot 75)^2 \cdot 30.58} = 0.06373 \text{ Hz}^2 \qquad \{2.42\}$$

where mass distribution factor $r_f = 0.961$ was obtained from Table 4.1.

The global second moment of area is

$$I_g = \sum_{j=1}^{n} A_{c,j} t_j^2 = 0.4 \cdot 0.4 \cdot 2(9^2 + 3^2) = 28.8 \text{ m}^4 \qquad \{2.32\}$$

The square of the fundamental frequency that belongs to this global second moment of area is

$$f_g^2 = \frac{0.313 r_f^2 E I_g}{H^4 m} = \frac{0.313 \cdot 0.961^2 \cdot 25 \cdot 10^6 \cdot 28.8}{75^4 \cdot 30.58} = 0.2151 \text{ Hz}^2 \qquad \{2.43\}$$

The effectiveness factor shows the extent the global bending deformation erodes the shear stiffness:

$$s_f = \sqrt{\frac{f_g^2}{f_{s'}^2 + f_g^2}} = \sqrt{\frac{0.2151}{0.06373 + 0.2151}} = \sqrt{0.7714} = 0.8783 \qquad \{2.46\}$$

With the effectiveness factor, the effective shear stiffness is

$$K_e = s_f^2 K = 0.7714 \cdot 189936 = 146517 \text{ kN} \qquad \{2.45\}$$

The square of the fundamental frequency that belongs to the effective shear stiffness can now be calculated:

$$f_s^2 = \frac{1}{(4H)^2}\frac{r_f^2 K_e}{m} = \frac{1}{(4 \cdot 75)^2}\frac{0.961^2 \cdot 146517}{30.58} = 0.04917 \text{ Hz}^2 \qquad \{2.44\}$$

For the local bending stiffness, the sum of the second moments of area of the columns should be produced. As all the columns of the frame are identical, the second moment of area of one column is simply multiplied by n and r:

$$I = r\sum_{j=1}^{n} I_{c,j} = 0.33225 \cdot 4 \cdot \frac{0.4 \cdot 0.4^3}{12} = 0.002835 \text{ m}^4 \qquad \{2.31\}$$

The fundamental frequency which is associated with the local bending stiffness is defined by

$$f_b^2 = \frac{0.313 r_f^2 EI}{H^4 m} = \frac{0.313 \cdot 0.961^2 \cdot 25 \cdot 10^6 \cdot 0.002835}{75^4 \cdot 30.58} = 0.0000212\ \text{Hz}^2 \qquad \{2.47\}$$

As a function of the non-dimensional parameter

$$k = H\sqrt{\frac{K_e}{EI}} = 75\sqrt{\frac{146517}{25 \cdot 10^6 \cdot 0.002835}} = 107.83 \qquad \{2.51\}$$

the frequency parameter is obtained using Table 4.2 as

$$\eta = 25.26 + \frac{50.25 - 25.26}{200 - 100}(107.83 - 100) = 27.22 \qquad \{\text{Table 4.2}\}$$

Finally, the fundamental frequency is

$$f = \sqrt{f_b^2 + f_s^2 + \left(\frac{\eta^2}{0.313} - \frac{k^2}{5} - 1\right) s_f f_b^2} \qquad \{2.52\}$$

$$= \sqrt{0.0000212 + 0.04917 + \left(\frac{27.22^2}{0.313} - \frac{107.83^2}{5} - 1\right) 0.8783 \cdot 0.0000212} = 0.2235\ \text{Hz}$$

The Finite Element based solution, obtained using Axis VM X5 (2019), is $f = 0.2226$ Hz.

2.3 STABILITY ANALYSIS OF RIGID SWAY-FRAMES

If the dynamic analysis of complex bar structures is said to be complex, then the stability analysis certainly presents an even greater challenge as numerical difficulties may further aggravate the situation in the course of the solution of the eigenvalue problem. The determination of the critical load of even a small frame may be a formidable task using conventional methods.

It would take many pages to list all the approximate methods that are worth mentioning as the field has been more than well cultivated and it would be unjust to choose only one or two, so no retrospective references are given here.

The best way, perhaps, towards a relatively simple solution of good accuracy is the application of the continuum method. If the structure is considered a continuous medium, as shown in the previous sections, the analysis can be carried out in a relatively simple way. In doing so, a closed-form solution can be produced for the critical load, which can then directly be used in practical structural design—see Chapter 6 on the global critical load ratio.

Because of the complexity of the problem, two methods will be presented for the stability analysis. The first one is a comprehensive method which is of general validity—see Section 2.3.1. The second one is a much simpler method; however, its applicability is limited to special cases—see Sections 2.4.1 and 2.4.6.

In addition to the general assumptions listed in Chapter 1, it will be assumed that

(a) the frames are subjected to uniformly distributed vertical load at floor levels (Figure 2.14/a).
(b) the critical load defines the bifurcation point.

2.3.1 A comprehensive method for the stability analysis

Investigating the stability of sandwich columns, Hegedűs and Kollár (1984) derived the governing differential equation of a sandwich column with thick faces:

$$\frac{B_0 B_l}{S}\varphi''''(z) - (B_0 + B_l)\varphi''(z) + N(z)\left(\frac{B_0}{S}\varphi''(z) - \varphi(z)\right) = 0$$

where $\varphi(z)$ is the rotation of the normal to the cross-section of the sandwich column, $N(z)$ is the axial load and B_0, B_l and S are the global bending, local bending and shear stiffnesses of the sandwich column.

For a sandwich column with a free upper end and a built-in lower end and using a coordinate system whose origin is fixed at the upper end, the boundary conditions are

$$\varphi(H) = 0, \qquad \varphi'(0) = 0$$

$$\varphi''(H) = 0, \qquad \varphi'''(0) = 0$$

Hegedűs and Kollár (1984 and 1999) solved the above differential equation for different load cases. The solution for the uniformly distributed axial load [when $N(z) = qz$ holds and q is the intensity of the vertical load] assumes the form

$$N_{cr} = qH = c_1 \frac{B_0 + B_l}{H^2}$$

where coefficient c_1 is obtained using a table as a function of $B_l/(B_0 + B_l)$ and $SH^2/(B_0 + B_l)$.

With the above stiffnesses attached to the sandwich column, the sandwich column is an ideal choice as the equivalent column of multi-storey frames. With some modification, the above simple formula can also be used for determining the global critical load of multi-storey, multi-bay frames. First, the stiffnesses that correspond to those of the sandwich column should be identified. These stiffnesses are obtained by distributing the global bending, local bending and shear stiffnesses of the frame over the height of the structure.

This procedure is presented in the following, using the terminology common in structural engineering. Most of the characteristics are shown in Figure 2.14. The stiffnesses are very similar to those introduced in Section 2.1.

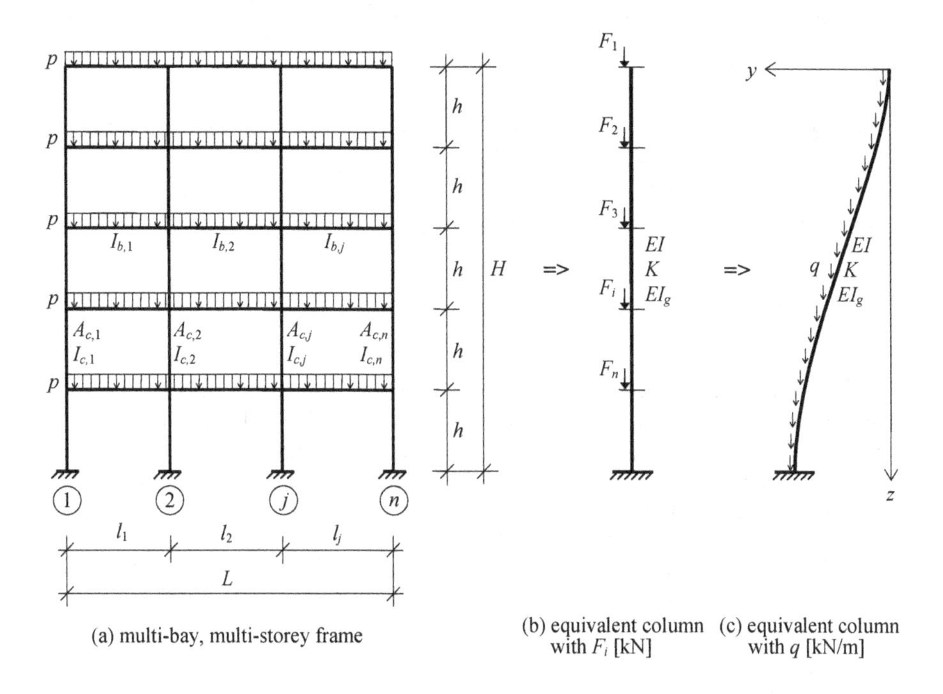

Figure 2.14 Origination of the equivalent column for the stability analysis.

The shear stiffness of the frame (K) is composed using two parts. The first part is associated with the beams of the frame as

$$K_b = \sum_{j=1}^{n-1} \frac{12EI_{b,j}}{l_j h} \tag{2.53}$$

with

- E the modulus of elasticity
- $I_{b,j}$ the second moment of area of the jth beam
- h the storey height
- l_j the width of the jth bay
- n the number of columns

The second part of the shear stiffness is associated with the columns and the local sway of the frame between two storeys:

$$K_c = \sum_{j=1}^{n} \frac{\pi^2 EI_{c,j}}{h^2} \tag{2.54}$$

In the above equation $I_{c,j}$ is the second moment of area of the *j*th column.

With the two components, the shear stiffness of the structure assumes the form

$$K = K_b r = K_b \frac{K_c}{K_b + K_c} \tag{2.55}$$

where the reduction factor r is also introduced as

$$r = \frac{K_c}{K_b + K_c} \tag{2.56}$$

Note the similarity to—and the difference from—the shear stiffness that was used for the deflection and frequency analyses in Section 2.1 and Section 2.2, respectively. For more detailed explanation regarding the components of the shear stiffness, see Chapter 5 where the stability analysis of whole three-dimensional bracing systems is discussed.

The global bending stiffness (EI_g) is associated with the full-height bending of the frame when the columns act as longitudinal fibres of a solid body in bending. It is calculated in the same way as described in Section 2.1, with

$$I_g = \sum_{j=1}^{n} A_{c,j} t_j^2 \tag{2.57}$$

where $A_{c,j}$ is the cross-sectional area of the *j*th column and t_j is the distance of the *j*th column from the centroid of the cross-sections.

The local bending stiffness (EI) of the frame is associated with the full-height bending of the individual columns. Again, it is obtained in the same way as explained in Section 2.1, with

$$I = r \sum_{j=1}^{n} I_{c,j} \tag{2.58}$$

where $I_{c,j}$ is the second moment of area of the *j*th column and r is the reduction factor [Equation (2.56)].

Having identified the stiffnesses for the use of the sandwich solution above, the way the frame is loaded should now be considered. (The situation is similar to that with the frequency analysis in Section 2.2.) The sandwich solution was produced for a cantilever subjected to a uniformly distributed axial load. The load of multi-storey frames, however, is not uniformly distributed over the height but it consists of floor loads (p in Figure 2.14/a). When the frame is modelled for the continuum method by an equivalent column, the floor load can first be considered as a system of concentrated forces at floor levels (F_i in Figure 2.14/b). This load system can then be distributed over the height of the column (q in Figure 2.14/c). This procedure represents an approximation and this approximation is unconservative as the distribution of the load occurs *downwards* at each storey and the centroid of the total load also moves downwards. The lower the frame, the greater the approximation. For a four-storey structure, for example, this approximation can lead to a critical load that is up to 40% greater than the exact one; therefore, this phenomenon cannot be ignored.

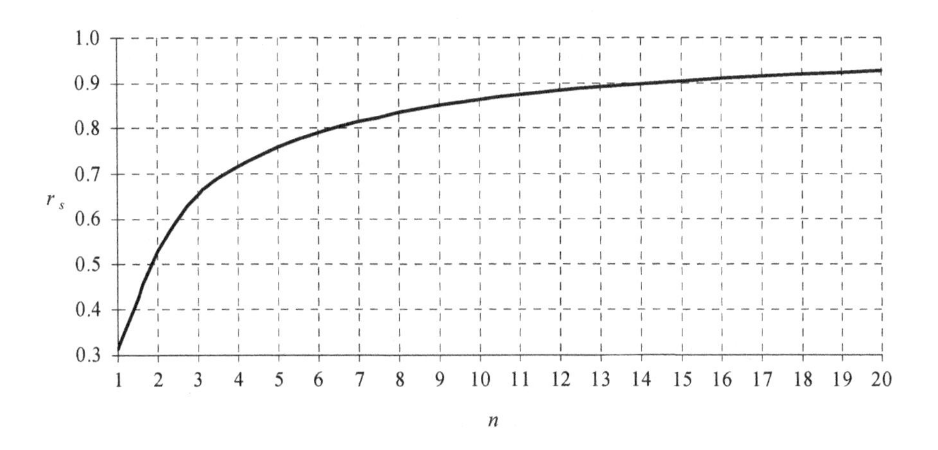

Figure 2.15 Load distribution factor r_s as a function of n (number of storeys).

This unfavourable phenomenon, however, can easily be taken into account by using Dunkerley's summation theorem and introducing a load distribution factor into the formula of the critical load. A short derivation (Zalka, 2000) produces the load distribution factor as

$$r_s = \frac{n}{n+1.588} \tag{2.59}$$

where n is the number of storeys. [Equation (2.59) can be used for $n > 2$.] Values for load distribution factor r_s are given in Figure 2.15 for frames up to twenty storeys high. Equation (2.59) and Table 5.1 can be used for higher frames.

The sandwich solution (Hegedűs and Kollár, 1984 and 1999) can now be applied and—using structural engineering notation—the critical load of the frame can be produced as

$$N_{cr} = qH = c_1 \frac{EI_f r_s}{H^2} \tag{2.60}$$

In the above equation EI_f is the total bending stiffness characterized by the sum of the local and global second moments of area as

$$I_f = I + I_g \tag{2.61}$$

Table 2.1 Values for coefficient c_1.

$\frac{KH^2}{EI_f r_s}$	$\frac{I}{I_f}$													
	0	0.001	0.005	0.01	0.05	0.1	0.2	0.3	0.4	0.5	0.6	0.7	0.8	1
0.00	0.000	0.0078	0.039	0.078	0.392	0.784	1.567	2.351	3.135	3.918	4.702	5.486	6.269	7.837
0.05	0.050	0.099	0.161	0.211	0.535	0.928	1.712	2.496	3.279	4.062	4.844	5.626	6.405	7.837
0.1	0.100	0.171	0.255	0.320	0.668	1.064	1.850	2.632	3.414	4.195	4.974	5.750	6.519	7.837
0.2	0.200	0.304	0.412	0.500	0.904	1.314	2.102	2.882	3.658	4.432	5.219	5.957	6.698	7.837
0.5	0.500	0.665	0.815	0.933	1.465	1.917	2.717	3.486	4.238	5.025	5.691	6.378	7.015	7.837
1	1.000	1.222	1.403	1.536	2.142	2.642	3.449	4.185	4.887	5.551	6.179	6.757	7.265	7.837
2	2.000	2.289	2.483	2.574	3.094	3.589	4.366	5.026	5.618	6.178	6.679	7.111	7.473	7.837
5	4.211	4.364	4.475	4.524	4.858	5.057	5.637	6.117	6.532	6.892	7.202	7.458	7.655	7.837
10	5.597	5.600	5.626	5.655	5.861	6.080	6.457	6.773	7.052	7.279	7.466	7.620	7.736	7.837
20	6.570	6.572	6.584	6.599	6.706	6.828	7.045	7.230	7.388	7.522	7.632	7.719	7.783	7.837
50	7.287	7.288	7.292	7.298	7.344	7.395	7.485	7.574	7.641	7.700	7.749	7.787	7.815	7.837
100	7.554	7.555	7.557	7.560	7.583	7.609	7.657	7.700	7.736	7.767	7.792	7.812	7.826	7.837
∞	7.837	7.837	7.837	7.837	7.837	7.837	7.837	7.837	7.837	7.837	7.837	7.837	7.837	7.837

Values for the critical load factor c_1 are given in Table 2.1 as a function of

$$\frac{I}{I_f} \quad \text{and} \quad \frac{KH^2}{EI_f r_s} \tag{2.62}$$

where K is the shear stiffness.

To evaluate the accuracy of Equation (2.60), a comprehensive accuracy analysis was carried out. Details are to be found in Section 7.2.3 where the critical load of 216 frames ranging in height from five to eighty storeys determined using Equation (2.60) is compared to the "exact" (FE) results. The accuracy analysis shows that Equation (2.60) can safely be used in structural engineering practice.

2.3.2 Worked example: two-bay, twenty-five storey frame

The critical load of frame F5 shown in Figure 2.16 is determined in this section. The two-bay, twenty-five storey structure is subjected to uniformly distributed load on the beams. The modulus of elasticity is $E = 25{\cdot}10^6$ kN/m². The cross-sections of the columns and beams are 0.4 m/0.7 m and 0.4 m/0.4 m, respectively, where the first number stands for the width (perpendicular to the plane of the structure) and the second number is the depth (in-plane size of the member). The two bays are identical at 6.0 m and the storey height is $h = 3.0$ m. Frame F5 is part of the group of frames used for a comprehensive accuracy analysis in Chapter 7.

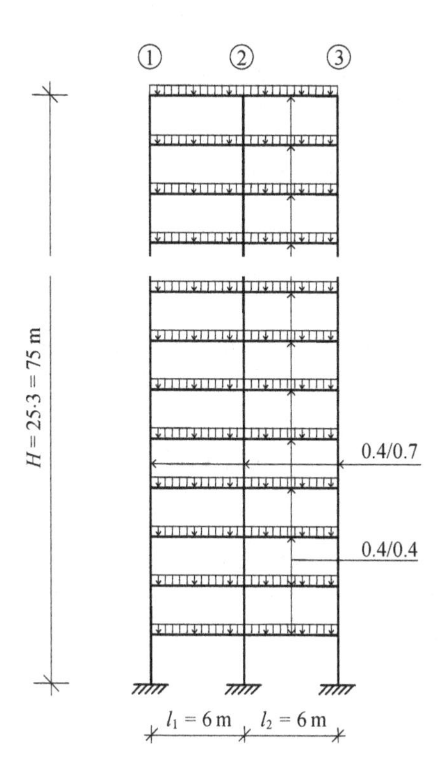

Figure 2.16 Two-bay, twenty-five storey frame F5 for the worked example.

The shear stiffness of a frame is composed using two parts. The first part is associated with the beams of the frame:

$$K_b = \sum_{j=1}^{n-1} \frac{12EI_{b,j}}{l_j h} = 2\frac{12 \cdot 25 \cdot 10^6 \cdot 0.4 \cdot 0.4^3}{12 \cdot 6 \cdot 3} = 71111\,\text{kN} \qquad \{2.53\}$$

The second part of the shear stiffness is associated with the columns:

$$K_c = \sum_{j=1}^{n} \frac{\pi^2 EI_{c,j}}{h^2} = 3\frac{\pi^2 \cdot 25 \cdot 10^6 \cdot 0.4 \cdot 0.7^3}{12 \cdot 3^2} = 940354\,\text{kN} \qquad \{2.54\}$$

With the two components, the shear stiffness is

$$K = K_b r = K_b \frac{K_c}{K_b + K_c} = 71111\frac{940354}{71111 + 940354} = 66112\,\text{kN} \qquad \{2.55\}$$

where the reduction factor is

$$r = \frac{K_c}{K_b + K_c} = \frac{940354}{71111 + 940354} = 0.9297 \qquad \{2.56\}$$

The global second moment of area is

$$I_g = \sum_{1}^{n} A_{c,i} t_i^2 = 0.4 \cdot 0.7(6^2 + 6^2) = 20.16\,\text{m}^4 \qquad \{2.57\}$$

The local second moment of area of the frame, amended by r, making use of the fact that the three columns have the same cross-section, is

$$I = r\sum_{j=1}^{n} I_{c,j} = 0.9297 \cdot 3\frac{0.4 \cdot 0.7^3}{12} = 0.0319\,\text{m}^4 \qquad \{2.58\}$$

Load distribution factor r_s is obtained from Equation (2.59) as $r_s = 0.940$.
Parameter c_1 is needed for the calculation of the critical load. As a function of the two parameters (2.62)

$$\frac{I}{I_f} = \frac{0.0319}{0.0319 + 20.16} = 0.00158, \qquad \frac{KH^2}{EI_f r_s} = \frac{66112 \cdot 75^2}{25 \cdot 10^6 \cdot 20.1919 \cdot 0.94} = 0.7837$$

its value is obtained from Table 2.1.
With the relevant values from Table 2.1

	0.001	0.005
0.5	0.665	0.815
1.0	1.222	1.403

parameter c_1 is obtained either as a quick estimate (between 0.665 and 1.403), $c_1 = 1.0$, for example, or after three interpolations

$$1:\quad 0.665+\frac{0.815-0.665}{0.005-0.001}(0.00158-0.001)=0.68675$$

$$2:\quad 1.222+\frac{1.403-1.222}{0.005-0.001}(0.00158-0.001)=1.248245$$

$$3:\quad c_1=0.68675+\frac{1.248245-0.68675}{1-0.5}(0.7837-0.5)=1.0053$$

The critical load of the frame can now be calculated:

$$N_{cr}=c_1\frac{EI_f r_s}{H^2}=1.0053\frac{25\cdot 10^3\cdot 20.1919\cdot 0.94}{75^2}=84.8\text{ MN} \qquad \{2.60\}$$

The Finite Element based solution, obtained using Axis VM X5 (2019), is N_{cr} = 85.6 MN.

2.4 OTHER TYPES OF FRAME

The investigations in the previous sections centred on regular sway-frames on fixed supports. These frames are ideal candidates for using the continuum method as their geometrical and stiffness characteristics do not vary over the height. However, the methods presented for the deflection, frequency and stability analyses may also be used for other types of frame; sometimes unaltered, sometimes with some modification and sometimes even in a simpler form. Some such cases will be discussed later in this section but first a very simple method will be introduced that can be used for the stability analysis of certain types of frame. As the detailed accuracy analysis presented in Chapter 7 shows, compromising on regularity has its price: accuracy and/or applicability suffer. However, even with reduced accuracy, the use of the procedures presented in the following may well be worth considering, especially for the stability analysis, where safety margins should be kept much wider. See Chapter 6 for more details concerning the global safety factor.

2.4.1 A simple method for the stability analysis

The comprehensive method, based on the sandwich column with thick faces, presented in Section 2.3.1, is a method of general validity and is ideally used when each of the three characteristic stiffnesses—shear, global bending and local bending—may have a sizable effect on the behaviour of the structure. There are special cases, however, when, compared to the other two stiffnesses, the contribution of one of the three stiffnesses is very small (or even non-existent) and can safely be ignored. Frames on pinned support—to be discussed in Section 2.4.2—fall into this category. With frames with cross-bracing—to be discussed in

Section 2.4.4—the effect of the local bending stiffness tends to become very small compared to that of the other two stiffnesses.

Neglecting the local bending stiffness makes it possible to create and use a very simple model for the equivalent column. In this procedure the multi-storey frame (Figure 2.17/a) is replaced by an equivalent column which has global bending and shear stiffnesses (Figure 2.17/b). This model—also called the sandwich model with thin faces—is used in this section.

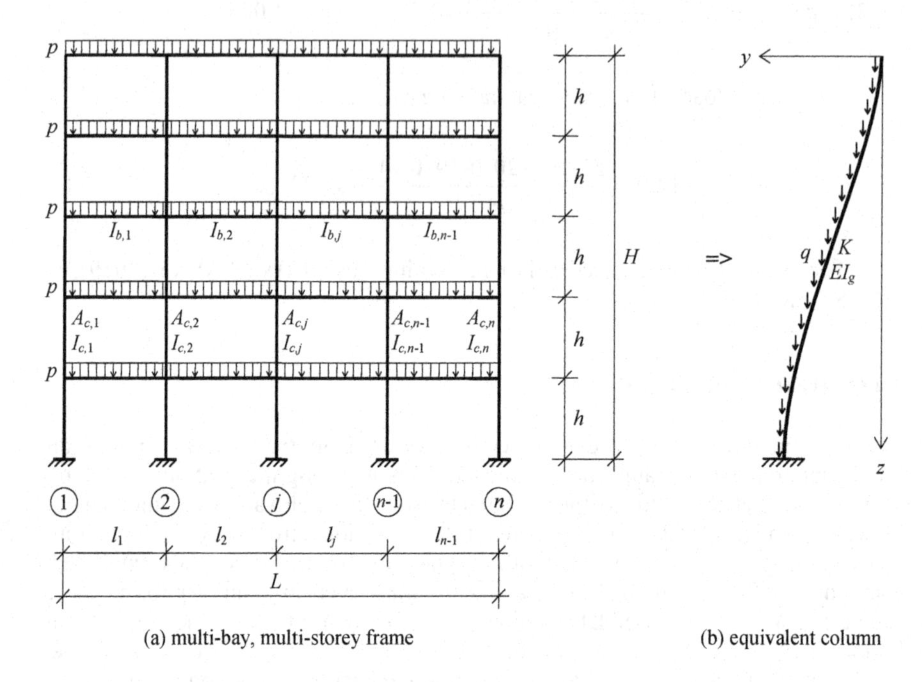

Figure 2.17 Origination of the equivalent sandwich column with thin faces.

The analysis of the equilibrium of an elementary section of the column leads to the differential equation

$$y''''(z)\left(1-\frac{zq}{K}\right)-\frac{3q}{K}y'''(z)+\frac{q}{r_s EI_g}\left[zy''(z)+y'(z)\right]=0$$

where critical load intensity q is the eigenvalue of the problem, r_s is the load distribution factor, EI_g is the global bending stiffness of the frame and K is the shear stiffness (Zalka, 1979). By definition, shear stiffness K is also the shear critical load of the frame. The fact that the load of the original frame is not uniformly distributed over the height of the structure (but it is uniformly distributed floor load) has already been taken into account by introducing load distribution factor r_s in the differential equation.

The boundary conditions are

$$y(0) = 0\,, \qquad y'(H)\left(1 - \frac{qH}{K}\right) = 0$$

$$y''(0) - \frac{q}{K}y'(0) = 0\,, \qquad y'''(0) - \frac{2q}{K}y''(0) = 0$$

Table 2.2 Critical load parameter α_s as a function of β_s.

β_s	α_s	β_s	α_s	β_s	α_s	β_s	α_s
0.0	1.0000	2.4	0.3457	6.0	0.1543	55	0.01803
0.1	1.0000	2.5	0.3342	6.5	0.1433	60	0.01654
0.2	1.0000	2.6	0.3235	7.0	0.1337	65	0.01527
0.3	1.0000	2.7	0.3134	7.5	0.1253	70	0.01419
0.4	0.9972	2.8	0.3039	8.0	0.1179	80	0.01243
0.5	0.9325	2.9	0.2950	8.5	0.1114	90	0.01105
0.6	0.8663	3.0	0.2866	9.0	0.1055	100	0.00995
0.7	0.8051	3.1	0.2787	10	0.09544	110	0.00905
0.8	0.7501	3.2	0.2711	11	0.08713	120	0.00830
0.9	0.7011	3.3	0.2640	12	0.08015	130	0.00767
1.0	0.6575	3.4	0.2572	13	0.07420	140	0.00712
1.1	0.6186	3.5	0.2508	14	0.06908	150	0.00665
1.2	0.5838	3.6	0.2447	15	0.06462	160	0.00623
1.3	0.5526	3.7	0.2389	16	0.06069	170	0.00587
1.4	0.5243	3.8	0.2333	17	0.05722	180	0.00554
1.5	0.4988	3.9	0.2280	18	0.05413	190	0.00525
1.6	0.4755	4.0	0.2230	19	0.05135	200	0.00499
1.7	0.4543	4.1	0.2181	20	0.04884	220	0.00454
1.8	0.4349	4.2	0.2135	25	0.03926	240	0.00416
1.9	0.4170	4.3	0.2090	30	0.03282	260	0.00384
2.0	0.4005	4.4	0.2047	35	0.02819	280	0.00357
2.1	0.3852	4.5	0.2006	40	0.02471	300	0.00333
2.2	0.3711	5.0	0.1824	45	0.02199	350	0.00285
2.3	0.3579	5.5	0.1672	50	0.01981	>350	$1/\beta_s$

After introducing non-dimensional parameters α_s and β_s, the solution of the above fourth-order differential equation can easily be produced using the generalized power series method (Zalka, 2000) and the critical load is obtained from the simple formula

$$N_{cr} = \alpha_s K \tag{2.63}$$

Values for critical load parameter α_s are given in Table 2.2 and in Figure 2.18 as a function of part critical load ratio

$$\beta_s = \frac{K}{N_g} \tag{2.64}$$

In Equation (2.64) N_g is the part critical load characterizing the full-height global buckling of the frame as a whole:

$$N_g = \frac{7.837 r_s E I_g}{H^2} \tag{2.65}$$

In the above equation I_g is the global bending second moment of area defined by Equation (2.32) and r_s is the load distribution factor whose values are given in Figure (2.15) and in Table 5.1.

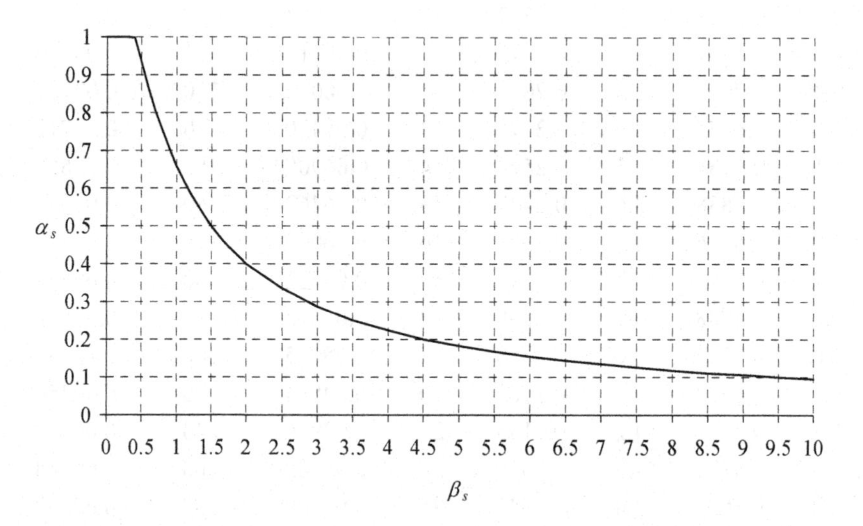

Figure 2.18 Critical load parameter α_s as a function of β_s.

As Equations (2.63) and (2.64) show, the value of the critical load depends on the two part critical loads K and N_g, and its value increases as the value of the shear critical load (K) and that of the global bending critical load (N_g) increases. However, it is important to know how these part critical loads compare and influence the value of the critical load.

Based on the Föppl-Papkovich summation theorem (Tarnai, 1999), applied to this situation, the critical load can be approximated as

$$N_{cr} = \left(\frac{1}{K} + \frac{1}{N_g} \right)^{-1}$$

Figure 2.19 demonstrates that the most favourable situation arises when the two part critical loads are equal (Figure 2.19/c). In this case, the critical load of the frame is maximum and its value increases in direct proportion with the increase of the part critical loads, i.e., doubling the part critical loads leads to a critical load which is twice as much as the original critical load. Figure 2.19 also demonstrates that, for unequal part critical loads, there is not much sense in increasing the greater part critical load, as the overall critical load is always governed by the value of the smaller part critical load (Figures 2.19/a and 2.19/b).

As for the applicability of the thin sandwich model, the two cases mentioned in the beginning of this section (and to be discussed in detail in Sections 2.4.2 and 2.4.4), are only the obvious choices. As a detailed accuracy analysis presented in Section 7.2.3 demonstrates, Equation (2.63) can also be used for the determination of the critical load of other types of sway-frames and coupled shear walls, if the contribution of the local bending stiffness to the critical load is "small enough" and/or the contribution of the shear stiffness is "great enough".

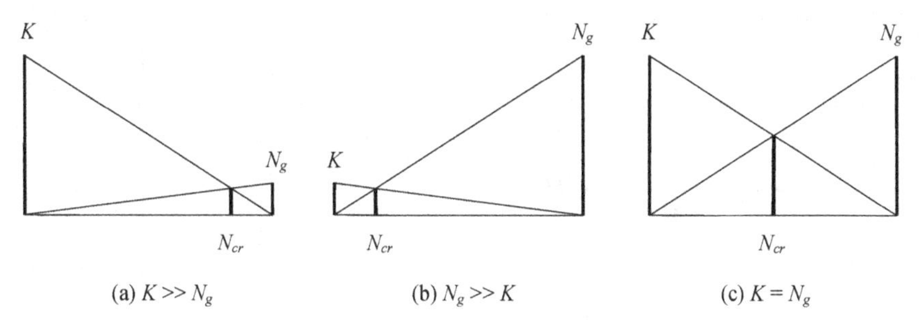

Figure 2.19 The effect of the relative values of the part critical loads on the critical load.

According to the data of the accuracy analysis in Section (7.2.3), the loose terms "small enough" and "great enough" can be defined more precisely by introducing two conditions:

$$\frac{I}{I_f} \leq 0.0003 \qquad \text{and} \qquad \frac{KH^2}{EI_f r_s} \geq 3 \tag{2.66}$$

If either of the above conditions is met, Equation (2.63) can be used. Its accuracy is satisfactory for practical application.

2.4.2 Frames on pinned support. Stability analysis

With frames on pinned supports (Figure 2.20/a), the ground floor section of the frame is "softer" than the sections above (Figure 2.20/b). Two of the basic assumptions listed in the Introduction are not fulfilled: the supports of the frames are not fixed (but pinned) and, as a consequence of the pinned supports, the shear stiffness is not constant (but varies) over the height of the structure.

These frames have full-height columns that, because of the pinned support, would not be stable by themselves and therefore the critical load that would belong to their local bending stiffness is zero. It follows that the model with K and EI_g (introduced in the previous section) can be used to produce a simple and good estimate of the global critical load of the frame.

The critical load is obtained using Equation (2.63), as

$$N_{cr} = \alpha_s K$$

where values for α_s are given in Figure 2.18 and in Table 2.2 as a function of part critical load ratio β_s [Equation (2.64)]. The shear stiffness of the frame, K, originates from two sources. The global part of the shear stiffness (K_b) depends on the stiffness of the beams and its value is not affected by the type of support of the columns. It follows that Equation (2.53) originally derived for frames on fixed supports can also be used here.

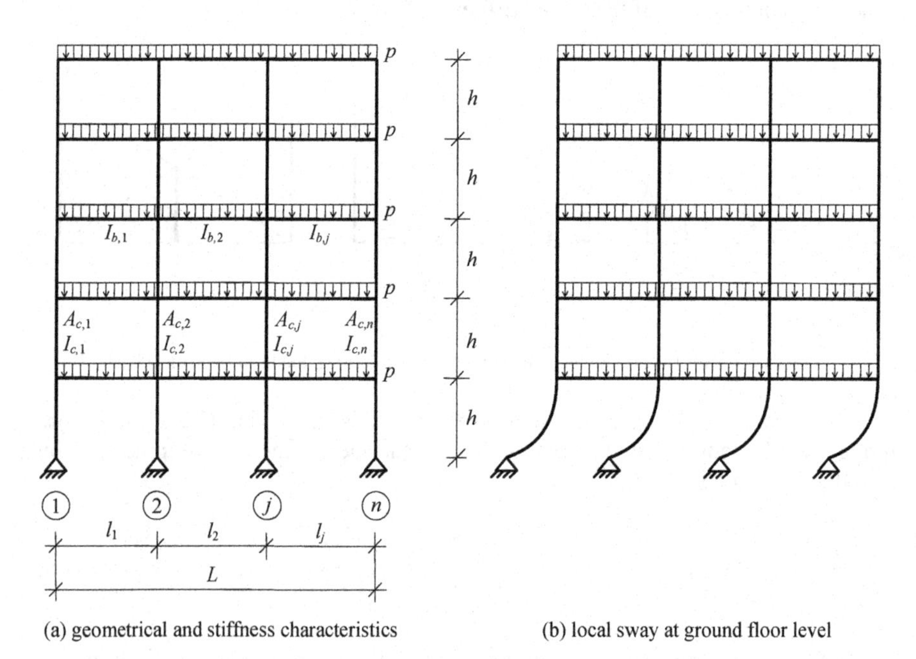

Figure 2.20 Frames on pinned supports.

The other—local—component of the shear stiffness is associated with the columns of the frame. It is determined by the local sway of the frame between two storeys. With frames on pinned supports, the vulnerability of the storey sections against sway is not the same over the height of the structure. The most vulnerable level is ground floor level (Figure 2.20/b). The local shear stiffness that is needed for the establishment of the continuum model belongs to sway buckling between ground floor and first floor level (Zalka and Armer, 1992). The local shear stiffness is therefore defined by

$$K_c = \sum_{j=1}^{n} \frac{\pi^2 EI_{c,j}}{(2h)^2} = \sum_{j=1}^{n} \frac{\pi^2 EI_{c,j}}{4h^2} \tag{2.67}$$

With this local shear stiffness, the shear stiffness of the frame on pinned supports is given by Equation (2.55). Critical load parameter α_s in Equation (2.63) is again given in Figure 2.18 and in Table 2.2 as a function of part critical load ratio β_s, defined by Equation (2.64), where the value of N_g does not depend on the type of support of the frame; therefore Equation (2.65) can be used.

2.4.3 Frames with longer columns at ground floor level. Stability analysis

Functional demands often lead to the construction of frames whose columns on the ground floor level are longer than those above—see Figure 2.21 where $h^* > h$ holds. Two cases will be considered: the frames have pinned supports (Figure 2.21/a) and the frames have fixed supports (Figure 2.21/b).

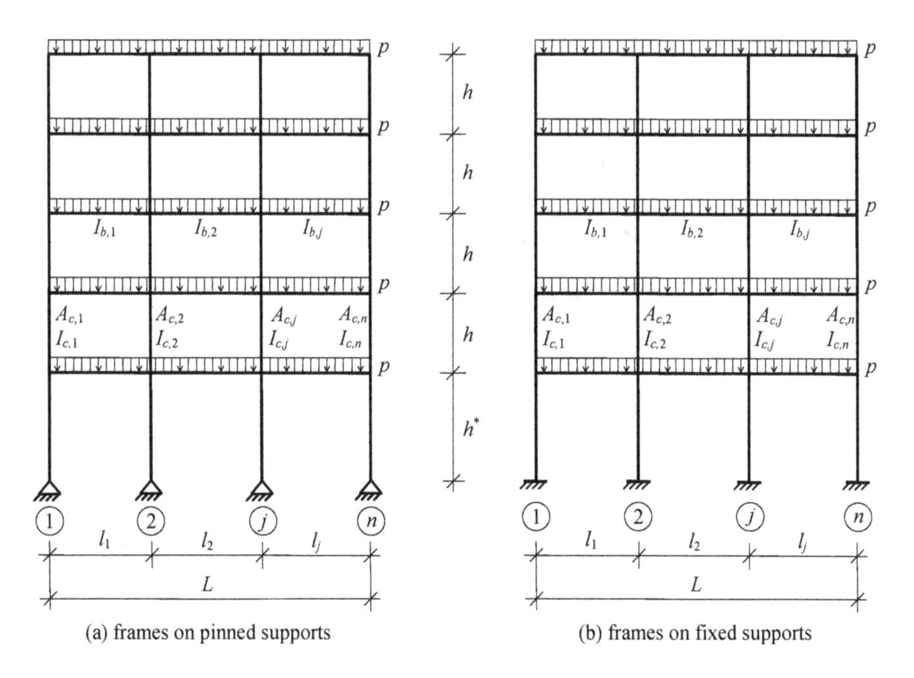

Figure 2.21 Frames with longer columns at ground floor level.

With frames on pinned support (Figure 2.21/a), the behaviour of the structure is similar to that in the previous section. However, the situation is further aggravated by the longer ground floor level columns. Of the basic assumptions listed in the Introduction, three are now unfulfilled: the supports are not fixed, the storey heights are not identical and the shear stiffness is not constant over the height of the structure.

When the frames have fixed supports (Figure 2.21/b), two assumptions are still not fulfilled: the storey heights are not identical and the shear stiffness is not constant over the height of the structure.

In both cases, the characteristic feature is that the first storey section is "softer" than the sections above.

When the frame has columns that are longer at ground floor level than those above, then the corresponding equations can still be used if the value of the local shear stiffness is determined according to the greater storey height, i.e., Equations (2.54) or (2.67) should be used for the stability analysis but with storey height h^* (Figure 2.21) instead of h.

2.4.4 Frames with cross-bracing

In some practical circumstances the structural performance of sway-frames may need to be enhanced, but it is not feasible to use stronger members or there is no space to increase the number of the bays to create a larger structure. In such situations, this aim can be achieved by building in some cross-bracing in the form of diagonal struts. Cross-bracing can considerably reduce lateral movements as it increases the lateral stiffness of the frame.

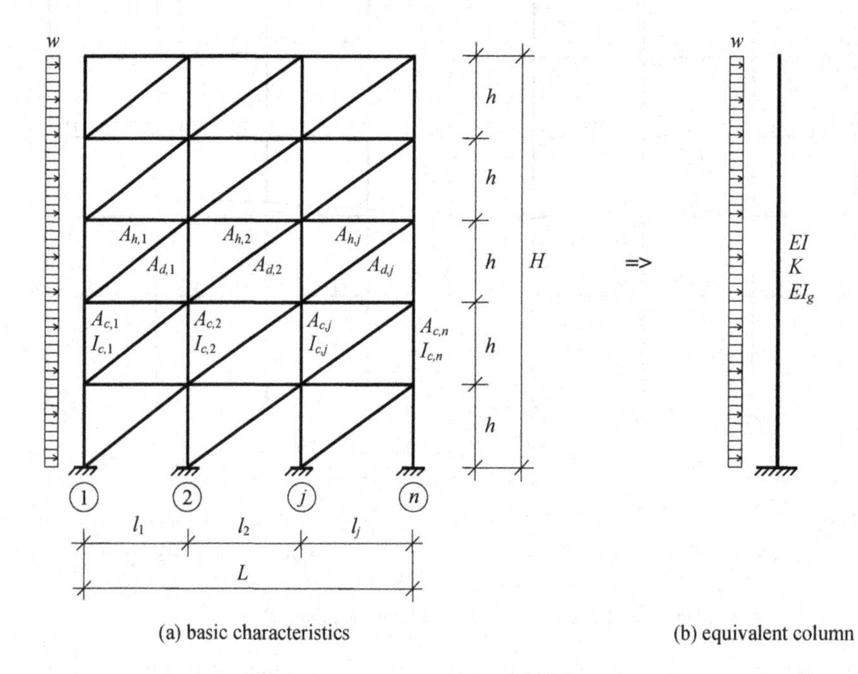

(a) basic characteristics

(b) equivalent column

Figure 2.22 Frame with single cross-bracing.

The equivalent columns introduced in Sections 2.1, 2.2 and 2.3 can be used to investigate frames with cross-bracing (Figure 2.22) and the formulae derived for the deflection, the fundamental frequency and the critical load of rigid sway-frames can still be used if certain considerations are made concerning the three characteristic stiffnesses.

The main difference between sway-frames and braced frames lies with the shear stiffness. It is obvious that the formulae used in connection with sway-frames cannot be used with braced frames. However, knowing the lateral displacement of a one-storey, one-bay section of a braced frame, it is easy to derive simple formulae for the shear stiffness of different types of braced frames (Zalka, 2000).

The shear stiffness of six characteristic cross-bracing arrangements is collected in Tables 2.3 and 2.4, and they can be used in the relevant equations.

Table 2.3 Shear stiffness K for single, double and continuous cross-bracing arrangements.

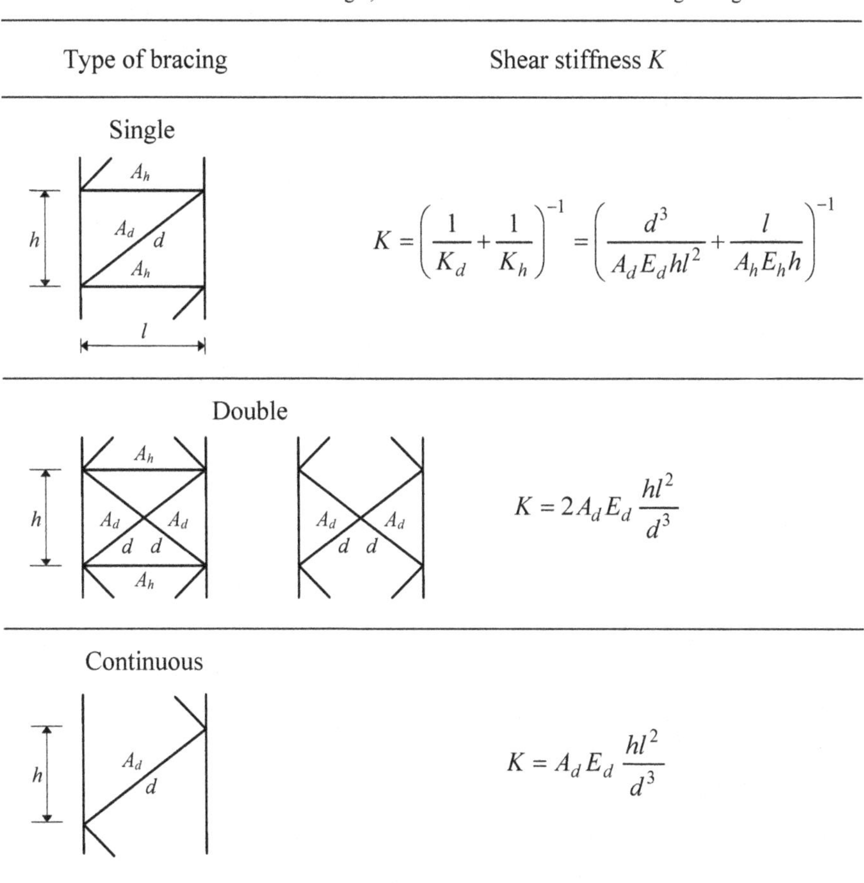

Type of bracing	Shear stiffness K
Single	$K=\left(\frac{1}{K_d}+\frac{1}{K_h}\right)^{-1}=\left(\frac{d^3}{A_d E_d h l^2}+\frac{l}{A_h E_h h}\right)^{-1}$
Double	$K=2A_d E_d \frac{hl^2}{d^3}$
Continuous	$K=A_d E_d \frac{hl^2}{d^3}$

The six bracing arrangements are divided into two groups as their behaviour in bending is slightly different. The situation is simple with single, double and continuous cross-bracing arrangements (Table 2.3). In these cases the diagonals are "full-length" diagonals, i.e. they extend over a full bay. The local bending stiffness is calculated directly as the sum of the bending stiffnesses of the columns [Equation (2.31)], but with $r = 1$, i.e. $EI = EI_c$. The calculation of the global bending stiffness EI_g is identical to that of the rigid sway-frames, i.e., according to Equation (2.32).

Table 2.4 Shear stiffness K for K-bracing and knee-bracing arrangements.

Type of bracing	Shear stiffness K
K-bracing 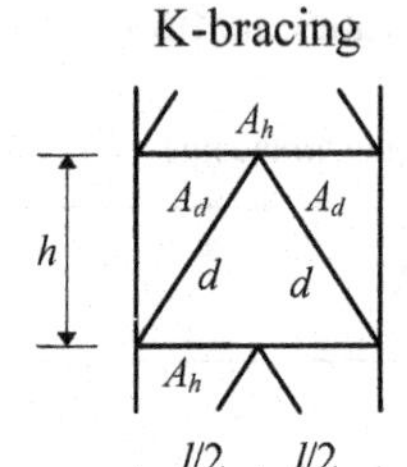	$K = \left(\frac{2d^3}{A_d E_d h l^2} + \frac{l}{4 A_h E_h h} \right)^{-1}$
K-bracing 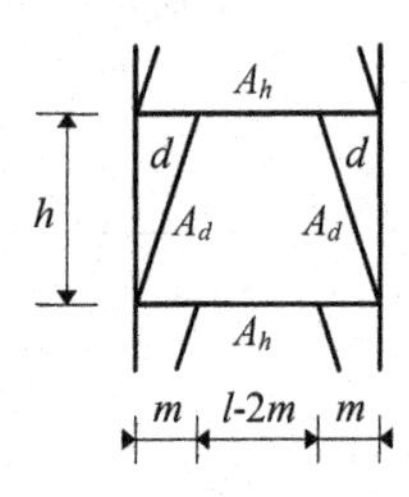	$K = \left(\frac{d^3}{2 A_d E_d h m^2} + \frac{m}{2 A_h E_h h} + \frac{h(l-2m)^2}{12 I_h E_h l} \right)^{-1}$
Knee-bracing 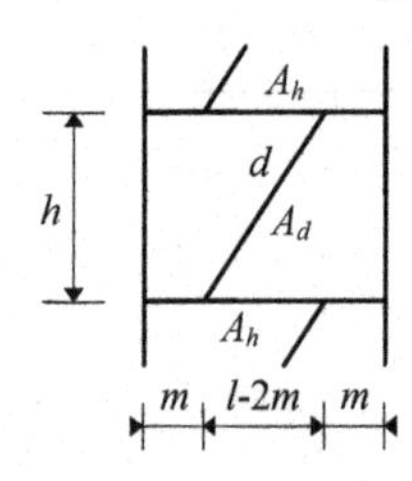	$K = \left(\frac{d^3}{A_d E_d h(l-2m)^2} + \frac{l-2m}{A_h E_h h} + \frac{h m^2}{12 I_h E_h l} \right)^{-1}$

The situation with K-bracing and knee-bracing (Table 2.4) is slightly more complicated. The diagonals of these arrangements are "short" diagonals—compared to the "full-length" diagonals—as they do not extend over a full bay. When the frame undergoes bending, these diagonals have some contribution to both the local and the global bending stiffnesses. The problem is that the extent of these contributions cannot be characterised by simple formulae.

However, a simple solution offers itself, as in practical cases the relative magnitude of the effect of the global and local bending stiffnesses is known: the contribution of the global bending stiffness is normally by an order of magnitude greater than that of the local bending stiffness. If the effect of the contribution of the diagonals to the global bending stiffness is ignored altogether and their contribution to the local bending stiffness is fully accepted, then the approximation tends to be on the safe side. It follows that the calculation of the global bending stiffness EI_g is identical to that of the rigid sway-frames, i.e. according to Equation (2.32). For the calculation of the local bending stiffness, Equation (2.31) is amended by adding the second moments of area of the diagonals, i.e. $I = I_c + I_d$, where I_d is the second moment of area of the diagonals.

With regard to the stability analysis, both procedures presented in Sections 2.3.1 and 2.4.1 can be used. The method described in Section 2.3.1 is more accurate while the method discussed in Section 2.4.1 is much simpler. See Chapter 7 for details regarding their accuracy and reliability.

2.4.5 Infilled frames

Frames filled with masonry walls (Figure 2.23/a) have increased resistance to lateral movement. Theoretical investigations (Polyakov, 1956; Mainstone and Weeks, 1972; Madan *et al.*, 1997) and experimental evidence (Mainstone and Weeks, 1972; Riddington and Stafford Smith, 1977) show that the complex behaviour of the composite structure can be handled in a relatively simple way.

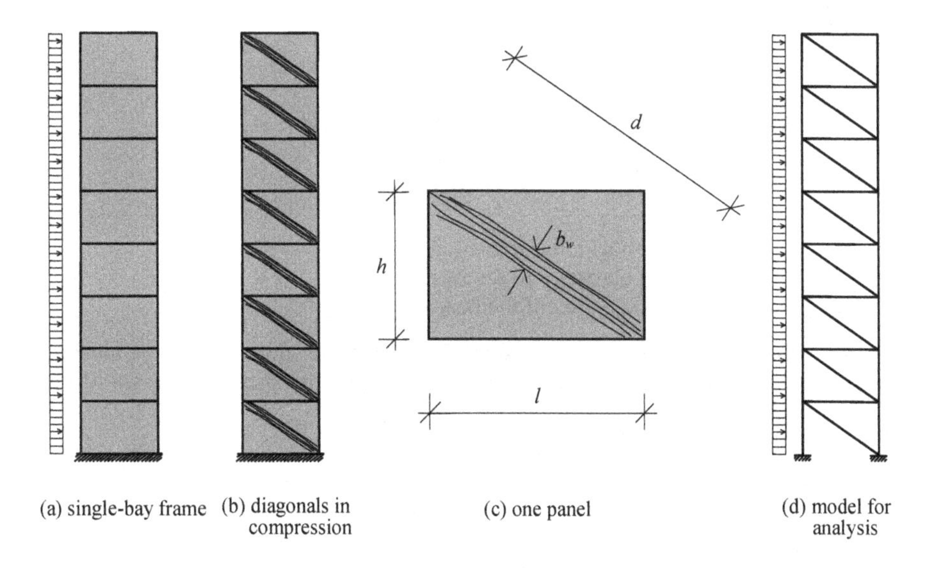

Figure 2.23 Model for single-bay infilled frames.

The contribution of the masonry infill panel to the response of the infilled frame can be modelled by replacing the panel with two equivalent struts: one in tension and the other in compression. The tensile strength of masonry is negligible and can therefore be safely neglected, leading to the structure shown in Figure 2.23/b. As for the diagonal in compression, the cross-section is

$$A_d = tb_w \tag{2.68}$$

where

t is the thickness of the masonry wall
b_w is the effective width of the diagonal strip (Figure 2.23/c)

Experimental evidence shows that the value of the effective width varies in a relatively wide range (Mainstone and Weeks, 1972; Riddington and Stafford Smith, 1977; Achyutha *et al.*, 1994): $0.10 \leq b_w/d \leq 0.40$, with d being the length of the diagonal strut. Design charts have been made available offering values for the effective width, as a function of stiffness parameters and panel proportions (Stafford Smith, 1966; Stafford Smith and Carter, 1969).

Alternatively, the value of the effective width can be approximated by

$$b_w = 0.15d \tag{2.69}$$

which normally represents a conservative estimate.

This leads to a frame with single bracing as a possible model for the structural analysis (Figure 2.23/d), whose shear stiffness, based on the first equation in Table 2.3, is composed of two parts, depending on the diagonals (K_d) and the horizontal beams (K_h), as

$$K = \left(\frac{1}{K_d} + \frac{1}{K_h} \right)^{-1} = \left(\frac{d^3}{A_d E_d h l^2} + \frac{l}{A_h E_h h} \right)^{-1} \tag{2.70}$$

with

E_d the modulus of elasticity of the masonry wall
E_h the modulus of elasticity of the beams
A_h the cross-sectional area of the beams

Equation (2.70) stands for single-bay infilled frames. For multi-bay structures, the shear stiffness is obtained by adding up the shear stiffnesses of the bays:

$$K = \sum_{j=1}^{n-1} K_j \tag{2.71}$$

where n is the number of columns and K_j refers to the shear stiffness of the jth bay.

As for increasing the value of shear stiffness K, the following rules apply. The value of the shear stiffness depends on two parts: the shear stiffness of the diagonal struts—the first term between the brackets in Equation (2.70)—and the shear stiffness of the beams of the frame (the second term). The bigger K_d and K_h, the bigger the overall shear stiffness K. The overall shear stiffness is optimized when K_d and K_h are equal. (What was said about the part critical loads in Section 2.4.1—with regard to K and N_g—is also valid here in relation to K_d and K_h.)

The value of K_d is directly proportional to the cross-sectional area A_d and the modulus of elasticity E_d of the diagonals. As for the geometry of the structure, maximum shear stiffness associated with the diagonals is achieved at

$$\frac{h}{l} = 0.708$$

where h is storey-height and l is the size of the bay of the frame (Figure 2.24).

The situation is simpler with the second part of the shear stiffness. Part stiffness K_h represents the contribution of the beams. Its value is in direct proportion to the cross-sectional area A_h, the modulus of elasticity E_h of the beams and the height of the storeys h, and is in inverse proportion to the length of the beams l.

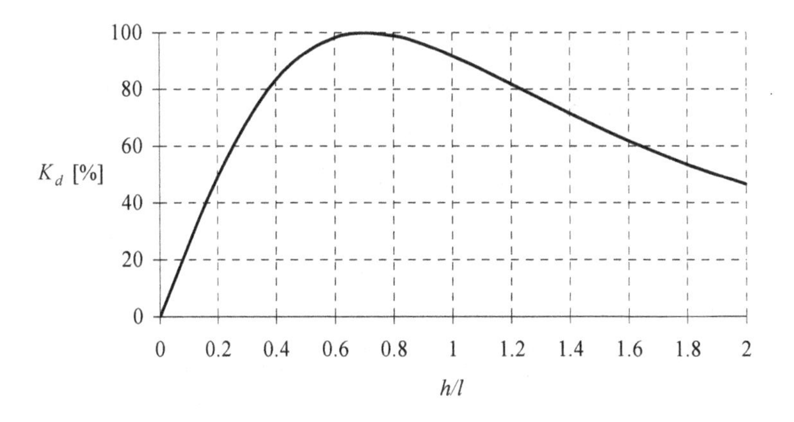

Figure 2.24 Optimum geometrical arrangement for part shear stiffness K_d associated with the diagonals.

2.4.6 Worked example: fifteen-storey frame with cross-bracing

The critical load of frame F12 shown in Figure 2.25 is determined in this section. The two-bay, fifteen-storey steel structure is subjected to uniformly distributed load on the beams. The geometrical and stiffness characteristics of the beams, columns and cross-bracing are collected in Table 2.5. The modulus of elasticity for each member is $E = 2 \cdot 10^8$ kN/m^2. The size of the bay is $l = 6$ m and the storey height is $h = 3$ m.

The critical load can be determined by using the methods presented in Section 2.3.1 and Section 2.4.1. Both methods need the value of the shear stiffness:

$$K=(n-1)\left(\frac{d^3}{A_dE_dhl^2}+\frac{l}{A_hE_hh}\right)^{-1} \qquad \{\text{Table 2.3}\}$$

$$=2\left(\frac{6.708^3}{3.75\cdot10^{-3}\cdot2\cdot10^8\cdot3\cdot6^2}+\frac{6}{5.73\cdot10^{-3}\cdot2\cdot10^8\cdot3}\right)^{-1}=365521\,\text{kN}$$

Table 2.5 Cross-sectional characteristics for frame F12.

Characteristics	Columns	Beams	Diagonals
Cross-section	305×305UC137	356×171×45UB	250/15
Area [m^2]	$1.744\cdot10^{-2}$	$5.73\cdot10^{-3}$	$3.75\cdot10^{-3}$
Second moment of area [m^4]	$3.281\cdot10^{-4}$	$1.207\cdot10^{-4}$	$1.953\cdot10^{-5}$

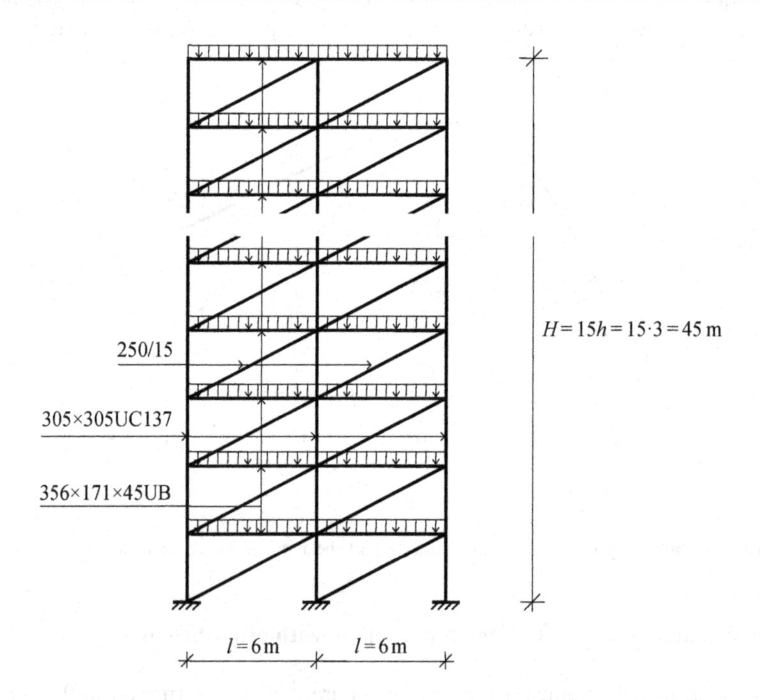

Figure 2.25 Fifteen-storey frame F12 with single cross-bracing for the worked example.

The global second moment of area is also needed by both methods:

$$I_g=\sum_{j=1}^{n}A_{c,j}t_j^2=1.744\cdot10^{-2}\cdot6^2\cdot2=1.2557\text{ m}^4 \qquad \{2.57\}$$

Load distribution factor r_s is obtained from Equation (2.59) as $r_s = 0.904$.

The value of the critical load is determined first by using the comprehensive method introduced in Section 2.3.1.

The local second moment of area (with $r = 1$) is

$$I = r\sum_{j=1}^{n} I_{c,j} = 3 \cdot 3.281 \cdot 10^{-4} = 9.843 \cdot 10^{-4}\ \text{m}^4 \qquad \{2.58\}$$

Parameter c_1 is needed for the calculation of the critical load. Its value is obtained from Table 2.1 as a function of the two parameters by Equation (2.62):

$$\frac{I}{I_f} = \frac{0.0009843}{0.0009843 + 1.2557} = 0.000783, \qquad \frac{KH^2}{EI_f r_s} = \frac{365521 \cdot 45^2}{2 \cdot 10^8 \cdot 1.2567 \cdot 0.904} = 3.2577$$

With the relevant values from Table 2.1

	0.000	0.001
2	2.000	2.289
5	4.211	4.364

parameter c_1 is obtained after three interpolations:

$$1: \quad 2.000 + \frac{2.289 - 2.000}{0.001 - 0.000}(0.000783 - 0.000) = 2.2263$$

$$2: \quad 4.211 + \frac{4.364 - 4.211}{0.001 - 0.000}(0.000783 - 0.000) = 4.3308$$

$$3: \quad c_1 = 2.2263 + \frac{4.3308 - 2.2263}{5 - 2}(3.2577 - 2.0) = 3.108$$

The critical load of the frame can now be calculated:

$$N_{cr} = c_1 \frac{EI_f r_s}{H^2} = 3.108 \frac{2 \cdot 10^5 \cdot 1.2567 \cdot 0.904}{45^2} = 348.7\ \text{MN} \qquad \{2.60\}$$

The value of ratio $KH^2/EI_f r_s$ [Equation (2.62)] is greater than 3, and condition (2.66) is fulfilled. In such cases, the simple method introduced in Section 2.4.1 is recommended and can also be used for the determination of the critical load. This procedure is shown below.

The global bending critical load is needed for the calculation:

$$N_g = \frac{7.837 r_s EI_g}{H^2} = \frac{7.837 \cdot 0.904 \cdot 2 \cdot 10^8 \cdot 1.2557}{45^2} = 878636 \text{ kN} \qquad \{2.65\}$$

As a function of

$$\beta_s = \frac{K}{N_g} = \frac{365521}{878636} = 0.416 \qquad \{2.64\}$$

the critical load parameter is obtained from Table 2.2 as

$$\alpha_s = 0.9972 - \frac{0.9972 - 0.9325}{0.5 - 0.4}(\beta_s - 0.4) = 0.9868 \qquad \{\text{Table 2.2}\}$$

Finally, the critical load of the frame is

$$N_{cr} = \alpha_s K = 0.9868 \cdot 365.521 = 360.7 \text{ MN} \qquad \{2.63\}$$

The Finite Element based solution, obtained using Axis VM X5 (2019), is $N_{cr} = 372$ MN.

2.5 COUPLED SHEAR WALLS

Coupled shear walls can be treated as special frames if two phenomena are taken into account. Both phenomena concern the part of the shear stiffness that relates to the connecting beams and both can be taken into account by modifying Equation (2.2) of the stiffness of the beams (distributed over the height). With these two modifications, the formulae derived for the maximum deflection, fundamental frequency and critical load of rigid sway-frames can be used. These two modifications will be presented in the next section and then a worked example will illustrate how the maximum deflection, the fundamental frequency and the critical load of a system of multi-storey coupled shear walls are calculated.

2.5.1 The modified frame model

Compared to frames, coupled shear walls normally have relatively wide columns (wall sections) and beams with relatively great depth (Figure 2.26).

Consequently, when the relative vertical displacement of the beams at contraflexure is established, two items have to be included [that were originally not included in $\Delta_3(z)$ in Figure 2.5 in Section 2.1.2]. First, the straight-line sections of the beams have to be considered ($s_1/2$ and $s_2/2$ in Figure 2.26) and, second, the shear deformation of the beams also has to be taken into account.

These amendments can be made in a relatively simple way (Zalka and Armer, 1992) and the resulting relative displacement at the contraflexure point of the beams (Figure 2.26) is obtained as

$$\Delta_3(z) = -\frac{q(z)l^2}{K^*}$$

where K^* is the modified shear stiffness as

$$K^* = K_b^* r^* = K_b^* \frac{K_c}{K_b^* + K_c} \tag{2.72}$$

with

$$K_b^* = \frac{6EI_b\left((l^* + s_1)^2 + (l^* + s_2)^2\right)}{l^{*3}h\left(1 + 12\frac{\rho EI_b}{l^{*2}GA_b}\right)} \tag{2.73}$$

for one bay. In the above equation

E	is the modulus of elasticity of the structure
I_b	is the second moment of area of the beams
A_b	is the cross-sectional area of the beams
l^*	is the distance between the two wall sections (Figure 2.26)
s_1, s_2	are the width of the wall sections of the coupled shear walls
G	is the modulus of elasticity in shear of the beams
ρ	is a constant depending on the shape of the cross-section of the beams ($\rho = 1.2$ for rectangular cross-sections)
h	is the storey height

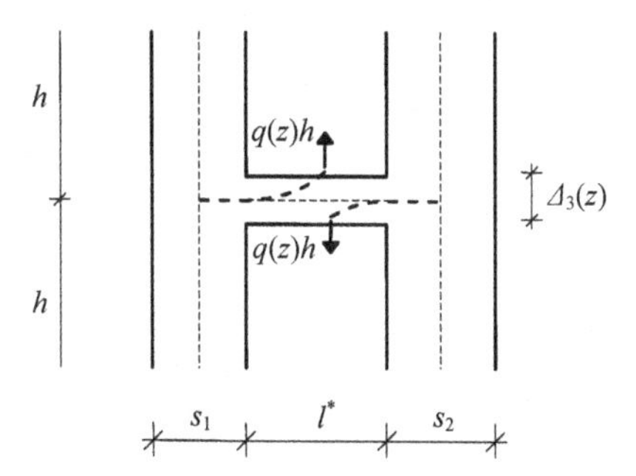

Figure 2.26 Vertical displacement Δ_3 at contraflexure point due to the bending of the connecting beam.

It follows that all the formulae for the lateral deflection, fundamental frequency and the critical load can be used if, according to Equation (2.72), K_b and r are replaced by K_b^* and r^* for coupled shear walls. For multi-bay coupled shear walls the shear stiffnesses of the bays should be added up using Equation (2.71), where n is now the number of wall sections.

2.5.2 Worked example: three-bay, thirty-storey coupled shear walls

The maximum deflection, the fundamental frequency and the critical load of coupled shear walls CSW3 (Figure 2.27) are determined in this section, using the equations given in the relevant sections.

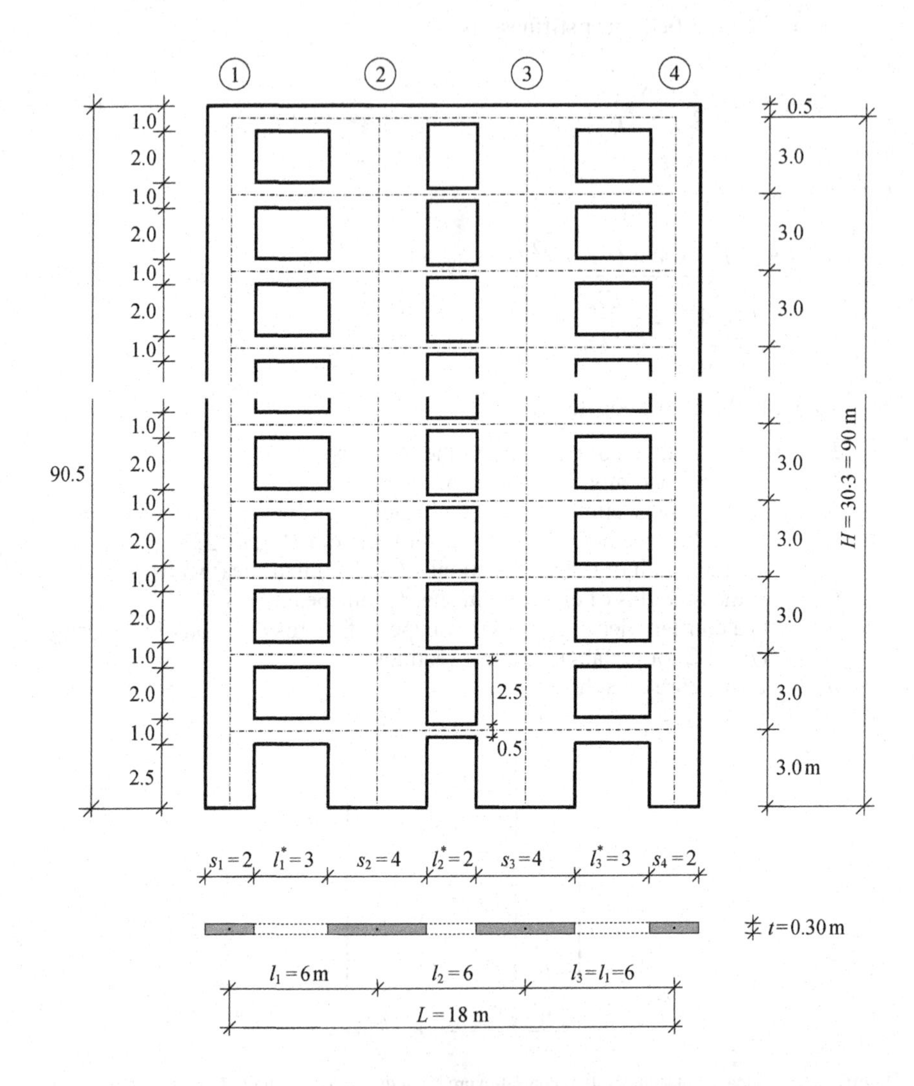

Figure 2.27 Thirty-storey coupled shear walls CSW3 for the worked example.

The structure is thirty-storey high and consists of four shear walls that are connected by beams at every floor level. The story height is $h = 3.0$ m and the total height of the structure is 90.50 m. The wall thickness is $t = 0.30$ m. The cross-section of the beams is 0.3 m/1.0 m in the first and third bay and 0.3 m/0.5 m in the second bay. The cross-sectional area and the second moment of area of the beams are $A_{b,1} = 0.3\ \text{m}^2$ and $I_{b,1} = 0.025\ \text{m}^4$, and $A_{b,2} = 0.15\ \text{m}^2$ and $I_{b,2} = 0.003125\ \text{m}^4$, respectively. The modulus of elasticity of the structure is $E = 30 \cdot 10^6\ \text{kN/m}^2$. The modulus of elasticity in shear is $G = 12.5 \cdot 10^6\ \text{kN/m}^2$.

The basic stiffness characteristics are calculated first.

Taking into consideration that the two sets of beams are of identical characteristics, the part of the shear stiffness of the coupled shear walls that is associated with the beams is

$$K_b^* = \frac{6EI_{b,1}\left((l_1^* + s_1)^2 + (l_1^* + s_2)^2\right)}{l_1^{*3} h \left(1 + 12\dfrac{\rho EI_{b,1}}{l_1^{*2} GA_{b,1}}\right)} 2 + \frac{6EI_{b,2}\left((l_2^* + s_2)^2 + (l_2^* + s_3)^2\right)}{l_2^{*3} h \left(1 + 12\dfrac{\rho EI_{b,2}}{l_2^{*2} GA_{b,2}}\right)} \qquad \{2.73;2.71\}$$

$$= \frac{6 \cdot 3 \cdot 10^4 \cdot 0.025\left((3+2)^2 + (3+4)^2\right)}{3^3 \cdot 3\left(1 + 12\dfrac{1.2 \cdot 3 \cdot 10^4 \cdot 0.025}{3^2 \cdot 1.25 \cdot 10^4 \cdot 0.3}\right)} 2 + \frac{6 \cdot 3 \cdot 10^4 \cdot 0.003125(2+4)^2 2}{2^3 \cdot 3\left(1 + 12\dfrac{1.2 \cdot 3 \cdot 10^4 \cdot 0.003125}{2^2 \cdot 1.25 \cdot 10^4 \cdot 0.15}\right)} = 7659\ \text{MN}$$

The part of shear stiffness that depends on the wall sections is

$$K_c = \sum_{j=1}^{n} \frac{12EI_{c,j}}{h^2} = \frac{12 \cdot 3 \cdot 10^4 \cdot 0.3(2^3 + 4^3)}{12 \cdot 3^2} 2 = 144000\ \text{MN} \qquad \{2.29\}$$

With the two components, the shear stiffness is

$$K^* = K_b^* r^* = K_b^* \frac{K_c}{K_b^* + K_c} = 7659 \frac{144000}{7659 + 144000} = 7272\ \text{MN} \qquad \{2.72\}$$

where the reduction factor r^* is

$$r^* = \frac{K_c}{K_b^* + K_c} = \frac{144000}{7659 + 144000} = 0.9495$$

From now on, the modified values K^* and r^* will be used whenever K and r are needed.

The global second moment of area is

$$I_g = \sum_{j=1}^{n} A_{c,j} t_j^2 = 0.3(2 \cdot 9^2 + 4 \cdot 3^2)2 = 118.8\ \text{m}^4 \qquad \{2.32\}$$

The local second moment of area of the coupled shear walls is

$$I = r\sum_{j=1}^{n} I_{c,j} = 0.9495\frac{0.3(2^3+4^3)}{12}2 = 3.418\,\mathrm{m}^4 \qquad \{2.31\}$$

Finally, the sum of the local and global second moments of area will also be needed for the calculations:

$$I_f = I + I_g = I_c r + I_g = 3.418 + 118.8 = 122.22\ \mathrm{m}^4 \qquad \{2.23\}$$

Maximum deflection

In addition to the above stiffness characteristics, parameters s, κ and κH are also needed for the calculation of the maximum deflection:

$$s = 1 + \frac{I_c r}{I_g} = 1 + \frac{3.418}{118.8} = 1.02877 \qquad \{2.14\}$$

$$\kappa = \sqrt{\frac{Ks}{EI}} = \sqrt{\frac{7272\cdot 1.02877}{3\cdot 10^4\cdot 3.418}} = 0.270 \qquad \text{and} \qquad \kappa H = 24.3$$

With the above auxiliary quantities, the maximum top deflection of the coupled shear walls can now be calculated. Assuming a uniformly distributed horizontal load of intensity $w = 10$ kN/m, the maximum deflection of the coupled shear walls is obtained as

$$y_{\max} = y(H) = \frac{wH^4}{8EI_f} + \frac{wH^2}{2Ks^2} - \frac{wEI}{K^2 s^3}\left(\frac{1+\kappa H\sinh(\kappa H)}{\cosh(\kappa H)} - 1\right) \qquad \{2.24\}$$

$$= \frac{10\cdot 90^4}{8\cdot 3\cdot 10^7\cdot 122.22} + \frac{10\cdot 90^2}{2\cdot 7272\cdot 10^3\cdot 1.02877^2} -$$

$$-\frac{10\cdot 3\cdot 10^7\cdot 3.418}{7272^2\cdot 10^6\cdot 1.02877^3}\left(\frac{1+24.3\sinh 24.3}{\cosh 24.3} - 1\right) = 0.0224 + 0.0052 - 0.0004 = 0.0272\ \mathrm{m}$$

The Finite Element solution, obtained using Axis VM X5 (2019), is $y_{\max} = 0.0287$ m.

As in most practical cases, the effect of the interaction between the bending and shear modes is small—third term in Equation {2.24}. Neglecting the third term leads to a very simple, truly back-of-the-envelope formula with the first two terms.

Fundamental frequency

Assuming uniformly distributed vertical floor load of intensity $p = 60$ kN/m, the mass density per unit length is

$$m = \frac{pL}{gh} = \frac{60 \cdot 18}{9.81 \cdot 3} = 36.7 \text{ kg/m} \tag{2.41}$$

With the above mass density, the square of the fundamental frequency of the coupled shear walls due to shear deformation can be calculated as

$$f_{s'}^2 = \frac{1}{(4H)^2} \frac{r_f^2 K}{m} = \frac{0.9673^2 \cdot 7272 \cdot 10^3}{(4 \cdot 90)^2 \cdot 36.7} = 1.4306 \text{ Hz}^2 \tag{2.42}$$

where mass distribution factor r_f was obtained using Table 4.1:

$$r_f = \sqrt{\frac{n}{n+2.06}} = \sqrt{\frac{30}{30+2.06}} = 0.9673 \qquad \{\text{Table 4.1}\}$$

The square of the fundamental frequency that belongs to the global bending stiffness is

$$f_g^2 = \frac{0.313 r_f^2 EI_g}{H^4 m} = \frac{0.313 \cdot 0.9673^2 \cdot 3 \cdot 10^7 \cdot 118.8}{90^4 \cdot 36.7} = 0.4335 \text{ Hz}^2 \tag{2.43}$$

The effectiveness factor shows the extent the global bending deformation erodes the shear stiffness:

$$s_f = \sqrt{\frac{f_g^2}{f_{s'}^2 + f_g^2}} = \sqrt{\frac{0.4335}{1.4306 + 0.4335}} = \sqrt{0.2326} = 0.4823 \tag{2.46}$$

With the effectiveness factor, the effective shear stiffness is

$$K_e = s_f^2 K = 0.2326 \cdot 7272 \cdot 10^3 = 1692 \cdot 10^3 \text{ kN} \tag{2.45}$$

The square of the fundamental frequency that belongs to the effective shear stiffness can now be calculated:

$$f_s^2 = \frac{1}{(4H)^2} \frac{r_f^2 K_e}{m} = \frac{1}{(4 \cdot 90)^2} \frac{0.9673^2 \cdot 1692 \cdot 10^3}{36.7} = 0.3329 \text{ Hz}^2 \tag{2.44}$$

The fundamental frequency which is associated with the local bending stiffness is given as

$$f_b^2 = \frac{0.313 r_f^2 EI}{H^4 m} = \frac{0.313 \cdot 0.9673^2 \cdot 3 \cdot 10^7 \cdot 3.418}{90^4 \cdot 36.7} = 0.01247 \text{ Hz}^2 \qquad \{2.47\}$$

As a function of the non-dimensional parameter

$$k = H\sqrt{\frac{K_e}{EI}} = 90\sqrt{\frac{1692 \cdot 10^3}{3 \cdot 10^7 \cdot 3.418}} = 11.56 \qquad \{2.51\}$$

the frequency parameter is obtained using Table 4.2 as

$$\eta = 3.172 + \frac{3.295 - 3.172}{12 - 11.5}(11.56 - 11.5) = 3.187 \qquad \{\text{Table 4.2}\}$$

Finally, the fundamental frequency is

$$f = \sqrt{f_b^2 + f_s^2 + \left(\frac{\eta^2}{0.313} - \frac{k^2}{5} - 1\right) s_f f_b^2} \qquad \{2.52\}$$

$$= \sqrt{0.01247 + 0.3329 + \left(\frac{3.187^2}{0.313} - \frac{11.56^2}{5} - 1\right) 0.4823 \cdot 0.01247} = 0.611 \text{ Hz}$$

The Finite Element based solution, obtained using Axis VM X5 (2019), is $f = 0.579$ Hz.

Critical load

It is assumed for the stability analysis that the structure is subjected to uniformly distributed floor load. Load distribution factor r_s is obtained from Table 5.1 as $r_s = 0.95$.

Before the calculation of the critical load is carried out, the shear stiffness that belongs to the wall sections has to be recalculated. Its value is

$$K_c = \sum_{j=1}^{n} \frac{\pi^2 EI_{c,j}}{h^2} = \frac{\pi^2 \cdot 3 \cdot 10^4 \cdot 0.3(2^3 + 4^3)}{12 \cdot 3^2} 2 = 118435 \text{ MN} \qquad \{2.54\}$$

Three more characteristics are affected by this change.

The shear stiffness is

$$K = K_b r = K_b \frac{K_c}{K_b + K_c} = 7659 \frac{118435}{7659 + 118435} = 7194 \text{ MN} \qquad \{2.55\}$$

where the reduction factor r is

$$r = \frac{K_c}{K_b + K_c} = \frac{118435}{7659 + 118435} = 0.9393 \qquad \{2.56\}$$

The local second moment of area of the coupled shear walls, amended by r, is

$$I = r\sum_{1}^{n} I_{c,i} = 0.9393\frac{0.3(2^3 + 4^3)}{12}2 = 3.381\,\text{m}^4 \qquad \{2.58\}$$

The characteristic ratio defined by Equation (2.62) is greater than 3 and condition (2.66) is fulfilled:

$$\frac{KH^2}{EI_f r_s} = \frac{7194 \cdot 90^2}{3 \cdot 10^4(3.381 + 118.8)0.95} = 16.73 > 3 \qquad \{2.62; 2.66\}$$

In such cases, the simple method introduced in Section 2.4.1 can be used for the determination of the critical load.

The global bending critical load is needed for the calculation:

$$N_g = \frac{7.837 r_s EI_g}{H^2} = \frac{7.837 \cdot 0.95 \cdot 3 \cdot 10^4 \cdot 118.8}{90^2} = 3276\ \text{MN} \qquad \{2.65\}$$

As a function of

$$\beta_s = \frac{K}{N_g} = \frac{7194}{3276} = 2.196 \qquad \{2.64\}$$

the critical load parameter is obtained from Table 2.2 as

$$\alpha_s = 0.3852 - \frac{0.3852 - 0.3711}{2.2 - 2.1}(2.196 - 2.1) = 0.3717 \qquad \{\text{Table 2.2}\}$$

Finally, the critical load of the coupled shear walls is

$$N_{cr} = \alpha_s K = 0.3717 \cdot 7194 = 2674\ \text{MN} \qquad \{2.63\}$$

The Finite Element based solution, obtained using Axis VM X5 (2019), is $N_{cr} = 2582$ MN.

2.6 SHEAR WALLS

The situation is very simple with shear walls. The formulae derived for the deflection, fundamental frequency and critical load of rigid sway-frames already contain the solutions for shear walls. Without connecting beams, shear walls act as the individual columns of frames. It follows that of the three characteristic stiffnesses only local bending stiffness EI is relevant and it corresponds to the bending stiffness of the shear wall. (The resistance of shear walls to global bending and shear—using frame terminology—can normally be considered infinitely great.) The well-known formulae for the deflection, maximum deflection, the fundamental frequency and the critical load are as follows.

Deflection:

$$y(z) = \frac{w}{EI}\left(\frac{H^3 z}{6} - \frac{z^4}{24}\right) \qquad \text{and} \qquad y_{\max} = y(H) = \frac{wH^4}{8EI} \tag{2.74}$$

where w [kN/m] is the intensity of the uniformly distributed horizontal load and z is measured from ground floor level.

Fundamental frequency:

$$f = \frac{0.56 r_f}{H^2}\sqrt{\frac{EI}{m}} \tag{2.75}$$

where m [kg/m] is the mass density per unit length of the shear wall and r_f is the mass distribution factor according to Figure 2.11 and Table 4.1.

Critical load:

$$N_{cr} = qH = \frac{7.837 EI r_s}{H^2} \tag{2.76}$$

where r_s is the load distribution factor given in Figure 2.15 and in Table 5.1.

It should be noted here that for low shear walls the effect of shear deformation may be significant and may need to be taken into account.

2.7 CORES

Shear walls are often built together to create three-dimensional units. A prime example is the U-shaped elevator core but many different shapes exist in building structures. The second moments of area of a reinforced concrete core are normally large and a small number of cores are often sufficient to provide the building with the necessary stiffness to resist lateral loading. In the two principal directions they act as shear walls and, knowing the second moments of area, Equations (2.74), (2.75) and (2.76) can be readily used to calculate the deflection, the maximum top deflection, the fundamental frequency and the critical load of a core.

As opposed to shear walls (and frames), however, cores are three-dimensional structures and they may also have considerable torsional resistance themselves which may constitute a significant part of the overall torsional resistance of the building. In the case of the five-storey building investigated in detail in Section 6.4, for example, the torsional resistance of the building is solely provided by a single U-core.

2.7.1 Torsional stiffness characteristics

As far as torsional behaviour is concerned, in view of their dimensions (height of core, thickness of the wall sections of the cross-section of the core), cores can be considered thin-walled columns and their torsional resistance originates from two sources: the Saint-Venant—or pure—torsional stiffness (GJ) and the warping torsional stiffness (EI_ω).

For cores of open cross-section (Figure 2.28/a), the Saint-Venant torsional constant is obtained from

$$J = \frac{1}{3}\sum_{i=1}^{m} h_i t_i^3 \tag{2.77}$$

where

- h_i is the length of the ith wall section
- t_i is the thickness of the ith wall section
- m is the number of wall sections

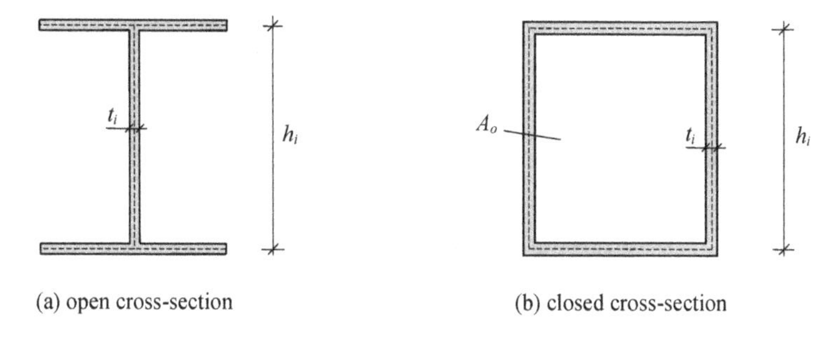

Figure 2.28 Cores of thin-walled sections.

On the rare occasion when the core has closed cross-section (or its openings are so small that they can be ignored), Bredt's formula should be used:

$$J = \frac{4A_o^2}{\sum_{i=1}^{m} \frac{h_i}{t_i}} \tag{2.78}$$

where A_o is the area enclosed by the mean centre lines of the wall sections

(Figure 2.28/b) and again h_i is the length of the ith wall section, t_i is the thickness of the ith wall section and m is the number of wall sections.

The relationship between the modulus of elasticity in shear and Young's modulus is

$$G = \frac{E}{2(1+\nu)} \tag{2.79}$$

where ν is the Poisson ratio.

Warping torsion is associated with the bending of the wall sections of the core—c.f. Section 3.3.1—but the determination of the stiffness associated with it is much more complicated than with Saint-Venant—or pure—torsion (Vlasov, 1961; Zbirohowski-Koscia; 1967, Kollbrunner and Basler, 1969). No simple procedure of general validity is available for the calculation of the warping constant (I_ω) but closed-form solutions exist for several cross-sections. Some of these formulae are collected in Table 2.6 where, in addition to the warping constant, formulae for the Saint-Venant constant and the location of the shear centre (e_o) are also presented for the most commonly used cores.

Tables 2.7 and 2.8 cover more complex (or asymmetric) cases.

Typical bracing cores are shown in Figure 2.29. The value of the warping stiffness of the first four cores (Figure 2.29/a) is so small that it can safely be ignored for practical calculations, while the other four cores (Figure 2.29/b) have sizeable warping stiffness.

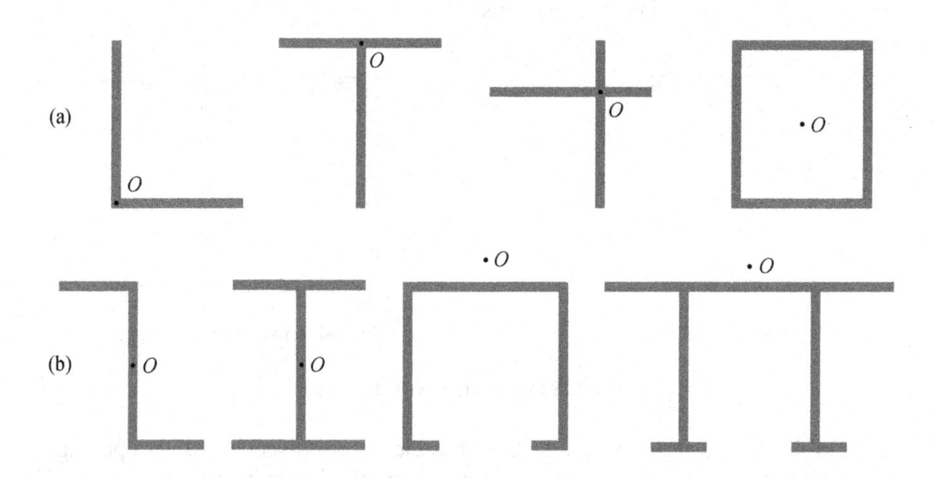

Figure 2.29 Cores. (a) with $I_\omega \approx 0$, (b) with considerable I_ω. O: shear centre.

U-cores are perhaps the most commonly used types (for elevator and service shafts) but they are in most cases partially closed at storey levels by floor slabs or beams (Figure 2.30). The effect of the connecting elements can always be safely ignored, but their contribution may be significant and the structural engineer may wish to take it into consideration for economic reasons.

Table 2.6 Torsional characteristics for common bracing cores of simple cross-section.

Type of core	Torsional characteristics
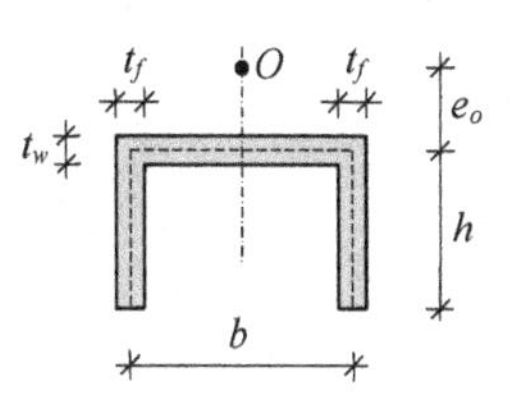	$e_o = \dfrac{3t_f h^2}{6t_f h + t_w b}$, $J = \dfrac{1}{3}(2ht_f^3 + bt_w^3)$ $I_\omega = \dfrac{t_f h^3 b^2}{12} \dfrac{3t_f h + 2t_w b}{6t_f h + t_w b}$
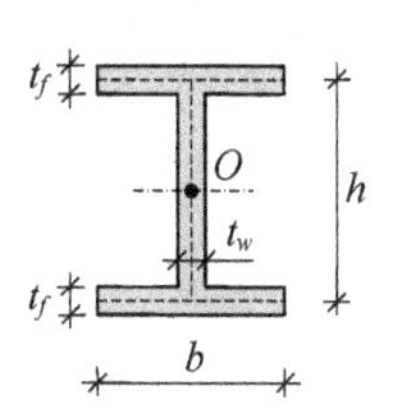	$J = \dfrac{1}{3}(2bt_f^3 + ht_w^3)$ $I_\omega = \dfrac{t_f b^3 h^2}{24}$
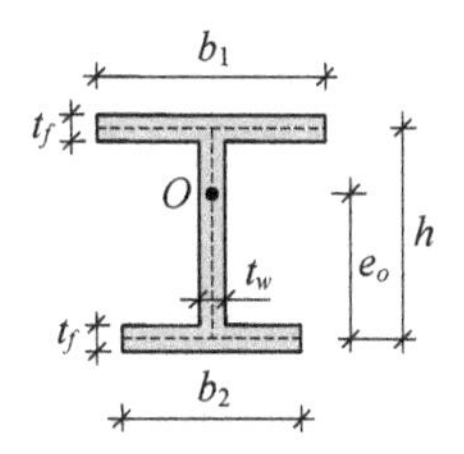	$e_o = h\dfrac{b_1^3}{b_1^3 + b_2^3}$, $J = \dfrac{1}{3}\left[(b_1 + b_2)t_f^3 + ht_w^3\right]$ $I_\omega = \dfrac{t_f h^2}{12} \dfrac{b_1^3 b_2^3}{b_1^3 + b_2^3}$
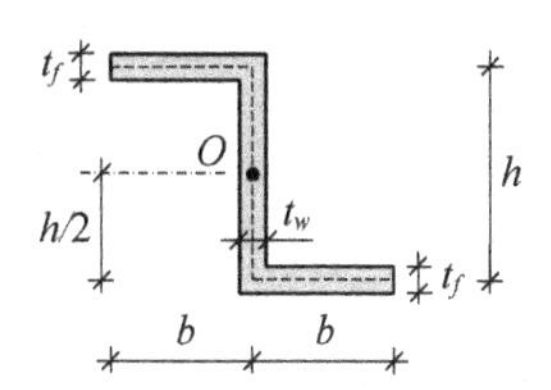	$J = \dfrac{1}{3}(2bt_f^3 + ht_w^3)$ $I_\omega = \dfrac{b^3 h^2}{12(2b + h)^2}\left[2t_f(b^2 + bh + h^2) + 3t_w bh\right]$
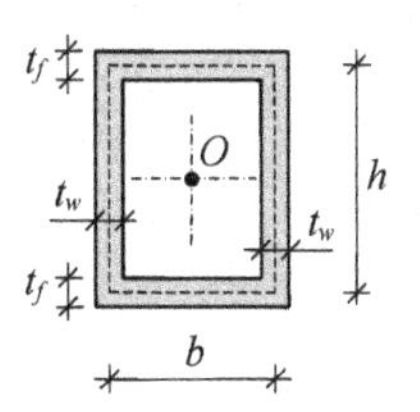	$J = \dfrac{2b^2 h^2 t_f t_w}{ht_f + bt_w}$ $I_\omega \cong 0$

Table 2.7 Cross-sectional characteristics for TT-sections.

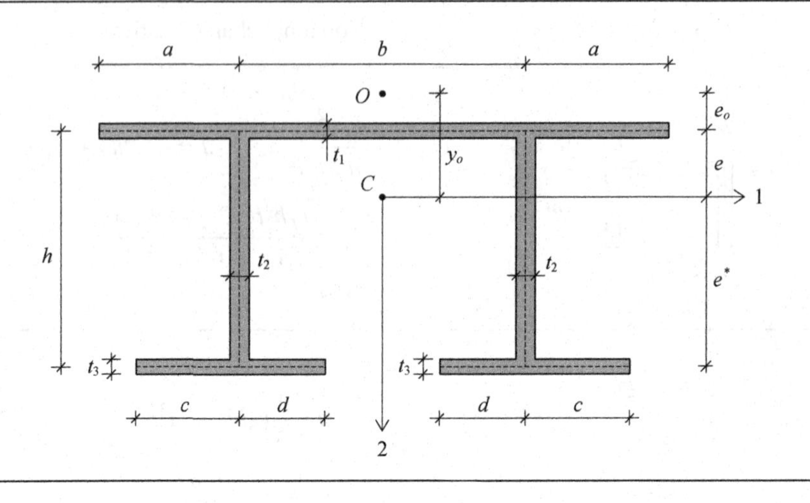

$$A = A_f + 2A_g + 2A_a \qquad \text{with}$$

$$A_f = t_1(2a+b), \quad A_g = t_2\left(h - \frac{t_1}{2} - \frac{t_3}{2}\right), \quad A_a = t_3(c+d)$$

$$e = \frac{1}{A}\left(2A_a h + 2A_g\left(\frac{h}{2} - \frac{t_3}{4} + \frac{t_1}{4}\right)\right), \qquad y_o = e + \frac{eAb^2}{4I_2} - \frac{2hI_{ag}}{I_2}$$

$$\text{where} \qquad I_{ag} = \frac{t_3}{3}(c^3 + d^3); \qquad e^* = h - e; \qquad e_o = y_o - e$$

$$I_1 = \frac{1}{12}\left(A_f t_1^2 + 2A_a t_3^2 + 2A_g\left(h - \frac{t_1}{2} - \frac{t_3}{2}\right)^2\right) + A_f e^2 + 2A_a e^{*2} + 2A_g\left(\frac{h}{2} - \frac{t_3}{4} + \frac{t_1}{4} - e\right)^2$$

$$I_2 = \frac{1}{12}\left(A_f(2a+b)^2 + t_3(b+2c)^3 - t_3(b-2d)^3 + 2A_g t_2^2\right) + \frac{b^2}{2}A_g$$

$$I_{1,2} = 0, \qquad J = \frac{1}{3}\left(A_f t_1^2 + 2A_g t_2^2 + 2A_a t_3^2\right)$$

$$I_\omega = \frac{b^2 I_1}{4} + \frac{b^2 e^2 A}{4}\left(1 - \frac{b^2 A}{4I_2}\right) + 2h^2 I_{ag} - 2bfh^2 A_a + b^2 ehA\frac{I_{ag}}{I_2} - 4h^2\frac{I_{ag}^2}{I_2}$$

$$\text{where } f = \frac{c-d}{2}$$

Table 2.8 Cross-sectional characteristics for □-sections.

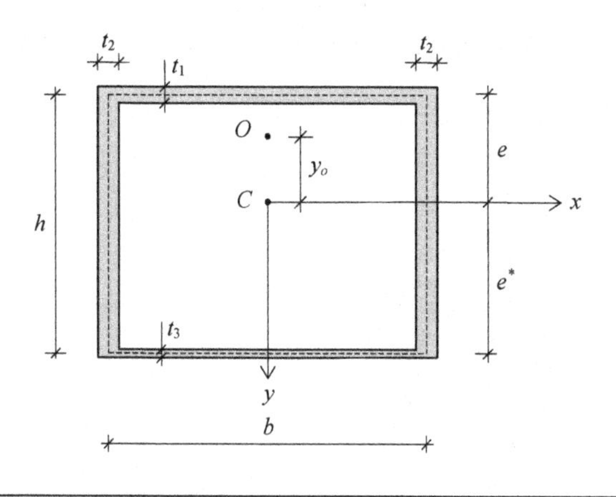

$$A = A_f + 2A_g + A_a \qquad \text{with}$$

$$A_f = t_1(b+t_2), \quad A_g = t_2(h - \frac{t_1}{2} - \frac{t_3}{2}), \quad A_a = t_3(b+t_2)$$

$$e = \frac{1}{A}\left(A_a h + A_g(h - \frac{t_3}{2} + \frac{t_1}{2})\right), \quad e^* = h - e$$

$$I_x = \frac{1}{12}\left(A_f t_1^2 + A_a t_3^2 + 2A_g(h - \frac{t_1}{2} - \frac{t_3}{2})^2\right) + A_f e^2 + A_a e^{*2} + 2A_g\left(\frac{h}{2} - \frac{t_3}{4} + \frac{t_1}{4} - e\right)^2$$

$$I_y = \frac{1}{12}\left((A_f + A_a)(b+t_2)^2 + 2A_g t_2^2\right) + \frac{b^2}{2}A_g, \qquad I_{xy} = 0$$

$$y_o = -e - \frac{I_{\omega x}}{I_y} \qquad \text{where} \qquad I_{\omega x} = \frac{b^2}{6}(\omega_1 t_1 + \omega_2 t_3) + \frac{bht_2}{2}(\omega_1 + \omega_2)$$

$$\omega_1 = -\Psi\frac{b}{2t_1}, \qquad \omega_2 = \frac{bh}{2} - \Psi\left(\frac{b}{2t_1} + \frac{h}{t_2}\right), \qquad \Psi = \frac{2A_o}{\int \frac{ds}{t}} = \frac{2hb}{\frac{b}{t_1} + \frac{b}{t_3} + \frac{2h}{t_2}}$$

$$J = 2A_o\Psi = \frac{4h^2b^2}{\frac{b}{t_1} + \frac{b}{t_3} + \frac{2h}{t_2}}, \qquad I_\omega = \frac{2}{3}\left(ht_2(\Omega_1^2 + \Omega_1\Omega_2 + \Omega_2^2) + \frac{1}{2}b(\Omega_1^2 t_1 + \Omega_2^2 t_3)\right)$$

$$\text{where} \quad \Omega_1 = -\frac{bI_{\omega x}}{2I_y} + \omega_1, \qquad \Omega_2 = -\frac{bI_{\omega x}}{2I_y} + \omega_2 \qquad \text{and} \qquad A_o = bh$$

The connecting elements restrain the section of the U-core from warping and increase its torsional resistance. Vlasov's (1961) investigations show that the phenomenon can be taken into account by amending the governing differential equation of torsion [Equation (2.86), to be discussed later on] in the form of

$$EI_{\omega}\varphi''''(z) - G(J + \bar{J})\varphi''(z) = m_z \tag{2.80}$$

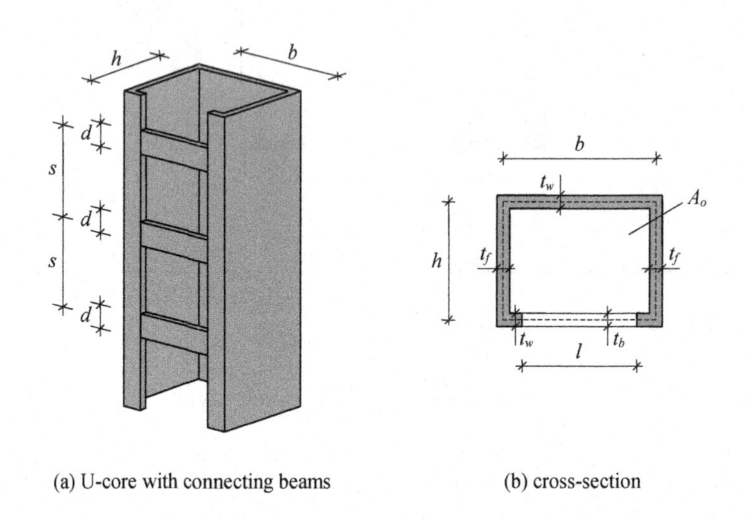

(a) U-core with connecting beams (b) cross-section

Figure 2.30 Partially closed U-core.

In Equation (2.80) moment m_z [kNm/m] is the torsional moment around the shear centre axis, J is the Saint-Venant constant defined by Equation (2.77) and

$$\bar{J} = \frac{4A_o^2}{\dfrac{l^3 sG}{12EI_b} + \dfrac{1.2ls}{A_b}} \tag{2.81}$$

represents the effect of the connecting beams.

In the above equation

$A_o = bh$	is the area enclosed by the mean centre lines of the wall sections (Figure 2.30/b)
b and h	are the lengths of the wall sections of the U-core
l	is the span of the connecting beams
s	is the vertical distance of the connecting beams (storey height in most cases, Figure 2.30/a)
E	is the modulus of elasticity of connecting beam
G	is the modulus of elasticity in shear of connecting beam

and

$$A_b = t_b d \qquad \text{and} \qquad I_b = \frac{t_b d^3}{12}$$

are the area and the second moment of area of the cross-section of the connecting beams with t_b and d being the thickness and depth of the connecting beams.

If the amended torsional stiffness ($J + \bar{J}$) is used, then all the formulae originating from the governing differential equation (2.86), and given in this section later on, can be used for the determination of the rotation, fundamental frequency and critical load of the partially closed core.

Numerical investigations indicate that Equation (2.81) tends to overestimate the effect of the connecting beams in certain cases. When the depth of the connecting beam (d) is relatively great, Equation (2.81) may even result in a value for the torsional stiffness that is greater than that of an entirely closed section—which is clearly impossible. In such cases, the approximation

$$\bar{J} = \frac{4A_o^2}{\dfrac{2b-l}{t_w} + \dfrac{l}{t_w^*} + \dfrac{2h}{t_f}} \tag{2.82}$$

may be used where t_w^* is an equivalent thickness whose value is determined using

$$t_w^* = \frac{d}{s} t_b \tag{2.83}$$

where, again, d is the depth of the connecting beams and s is the vertical distance of the connecting beams (Figure 2.30).

2.7.2 Deflection and rotation under uniformly distributed horizontal load

When the resultant of the horizontal load passes through the shear centre, Equations (2.74) can be applied, using the relevant second moment of area, for the determination of the deflection and maximum deflection of the core:

$$u(z) = \frac{w_x}{EI_y}\left(\frac{H^3 z}{6} - \frac{z^4}{24}\right) \qquad \text{and} \qquad u_{\max} = u(H) = \frac{w_x H^4}{8EI_y} \tag{2.84}$$

$$v(z) = \frac{w_y}{EI_x}\left(\frac{H^3 z}{6} - \frac{z^4}{24}\right) \qquad \text{and} \qquad v_{\max} = v(H) = \frac{w_y H^4}{8EI_x} \tag{2.85}$$

When the torsional behaviour of a core subjected to m_z [kNm/m] uniformly distributed torsional moment around its shear centre axis is investigated (Figure 2.31), the governing differential equation assumes the form

$$EI_\omega \varphi''''(z) - GJ\varphi''(z) = m_z \tag{2.86}$$

The boundary conditions of the problem are

$$\varphi(0) = 0\,, \qquad \varphi'(0) = 0$$

$$\varphi''(H) = 0\,, \qquad EI_\omega \varphi'''(H) - GJ\varphi'(H) = 0$$

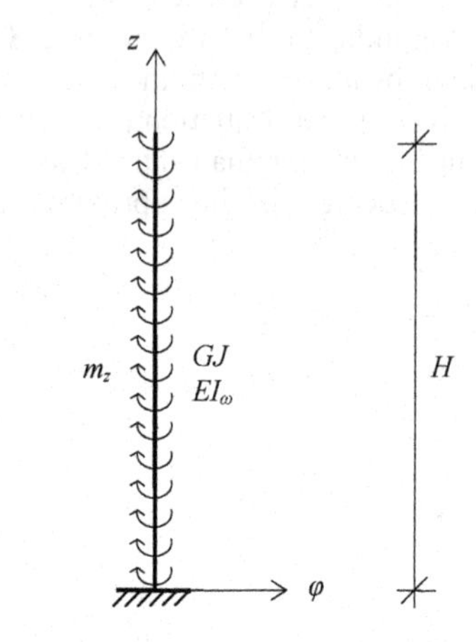

Figure 2.31 Bracing core for the torsional analysis, subjected to uniformly distributed torsional moment.

The solution can be expressed in closed form:

$$\varphi(z) = \frac{m_z}{GJ\cosh k}\left[\frac{H^2}{k^2}\left(\cosh\frac{kz}{H} + k\sinh(k - \frac{kz}{H}) - k\sinh k - 1\right) + \right.$$

$$\left. + z(H - \frac{z}{2})\cosh k\right] \tag{2.87}$$

where

$$k = H\sqrt{\frac{GJ}{EI_\omega}} \tag{2.88}$$

is the torsion parameter.

Maximum rotation develops at the top:

$$\varphi_{\max} = \varphi(H) = \frac{m_z H^2}{GJ}\left(\frac{\cosh k - 1}{k^2 \cosh k} - \frac{\tanh k}{k} + \frac{1}{2}\right) \tag{2.89}$$

Two special cases will now be considered.

In most practical cases, the effect of the Saint-Venant stiffness (GJ) is small compared to the effect of the warping stiffness (EI_ω) and

$$k << 1$$

holds. In such cases, keeping in mind that $GJ/EI_\omega \approx 0$ holds, the governing differential equation of the torsional problem simplifies to

$$\varphi''''(z) = \frac{m_z}{EI_\omega}$$

and the formula for the maximum rotation is obtained as

$$\varphi_{\max} = \varphi(H) = \frac{m_z H^4}{8EI_\omega} \tag{2.90}$$

When the core has no warping stiffness (Figure 2.29/a), the above solutions cannot be used as the denominator in Equations (2.88) and (2.90) vanishes. In such cases the original governing differential equation simplifies to

$$\varphi''(z) = -\frac{m_z}{GJ}$$

whose solution results in

$$\varphi_{\max} = \varphi(H) = \frac{m_z H^2}{2GJ} \tag{2.91}$$

for the maximum rotation.

2.7.3 Fundamental frequency

When the vibration of the core is investigated, the frequencies for the lateral vibration can again be readily obtained using the solution given for shear walls:

$$f_x = \frac{0.56 r_f}{H^2}\sqrt{\frac{EI_y}{m}} \qquad \text{and} \qquad f_y = \frac{0.56 r_f}{H^2}\sqrt{\frac{EI_x}{m}} \tag{2.92}$$

where m [kg/m] is the mass density per unit length of the material of the core and r_f is the mass distribution factor by Figure 2.11 or Table 4.1.

The analysis of pure torsional vibration is carried out by investigating the equilibrium of an elementary section of the core (Figure 2.32). Its governing differential equation emerges as

$$r_f^2 EI_\omega \varphi''''(z,t) - r_f^2 GJ\varphi''(z,t) + mi_p^2 \ddot{\varphi}(z,t) = 0$$

where primes and dots mark differentiation by z and t (time).

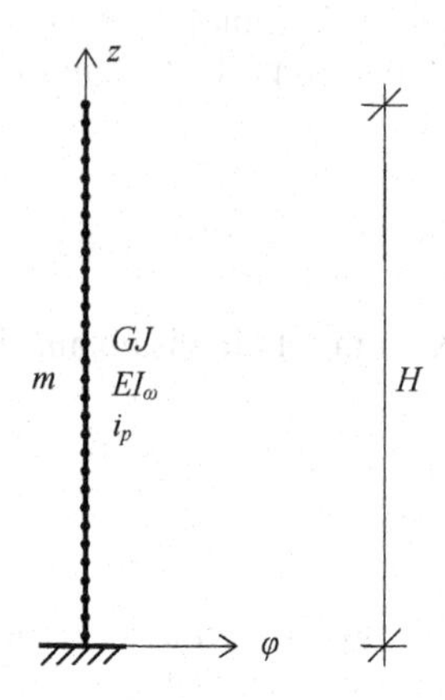

Figure 2.32 Bracing core for the analysis of pure torsional vibration.

After seeking the solution in a product form, separating the variables and eliminating the time dependent functions, the above governing differential equation leads to the boundary value problem

$$r_f^2 EI_\omega \varphi_1''''(z) - r_f^2 GJ\varphi_1''(z) - \omega^2 mi_p^2 \varphi_1(z) = 0$$

where

$$i_p = \sqrt{\frac{I_x + I_y}{A}} \tag{2.93}$$

is the radius of gyration of the cross-section of the core.

This differential equation is identical to Equation (2.49) in structure, has the same boundary conditions, and so the solution to Equation (2.49) can be used if the stiffness characteristics in Equation (2.49) are replaced with those in the above equation. In doing so, the formula for the pure torsional frequency is obtained as

$$f_\varphi = \frac{\eta r_f}{i_p H^2}\sqrt{\frac{EI_\omega}{m}} \tag{2.94}$$

where values for frequency parameter η are given in Figure 2.12 and in Table 4.2 as a function of k [Equation (2.88)]:

$$k = H\sqrt{\frac{GJ}{EI_\omega}}$$

Equation (2.94) for the fundamental frequency of pure torsional vibration cannot be used for cores with zero warping stiffness. For such cores, the fundamental frequency for pure torsion is calculated from

$$f_\varphi = \frac{r_f}{4Hi_p}\sqrt{\frac{GJ}{m}} \tag{2.95}$$

It is important to point out that when Equations (2.94) and (2.95) are used for the frequency analysis of a *building* braced by a single core, then the radius of gyration refers to the layout of the building (rather than to the cross-section of the core).

It should be noted here that cores normally behave in a true three-dimensional fashion and the above three fundamental frequencies (f_x, f_y and f_φ) can only be considered *basic* fundamental frequencies and they normally combine during vibration resulting in the fundamental frequency of the core. This combination is very important as the fundamental frequency of the core is always smaller than (or equal to) the smallest one of the three basic frequencies. For the coupling of the basic modes, see Chapter 4 that deals with three-dimensional behaviour.

2.7.4 Critical load

When the stability of a core under uniformly distributed vertical load of intensity q is investigated, three things have to be considered: lateral buckling in the two principal directions and pure torsional buckling. For lateral buckling, Equation (2.76) given for shear walls can be used and the sway critical loads in the principal directions can be calculated from

$$N_{cr,x} = \frac{7.837 EI_y r_s}{H^2} \quad \text{and} \quad N_{cr,y} = \frac{7.837 EI_x r_s}{H^2} \tag{2.96}$$

The situation with pure torsional buckling is more complicated. It is advantageous for the origin of the coordinate system to be placed at and attached to the upper free end of the core (Figure 2.33). The governing differential equation of the problem assumes the form

$$r_s EI_\omega \varphi''''(z) + \left[\left(N(z)i_p^2 - GJ\right)\varphi'(z)\right]' = 0$$

In the above equation i_p is the radius of gyration of the cross-section of the core, r_s is the load distribution factor (Figure 2.15 and Table 5.1) and $N(z) = qz$

with q being the intensity of the uniformly distributed vertical load.

The boundary conditions for the governing differential equation are

$$\varphi(0) = 0\,, \qquad \varphi''(0) = 0$$

$$\varphi'(H) = 0\,, \qquad EI_\omega \varphi'''(H) - GJ\varphi'(H) = 0$$

in the coordinate system shown in Figure 2.33.

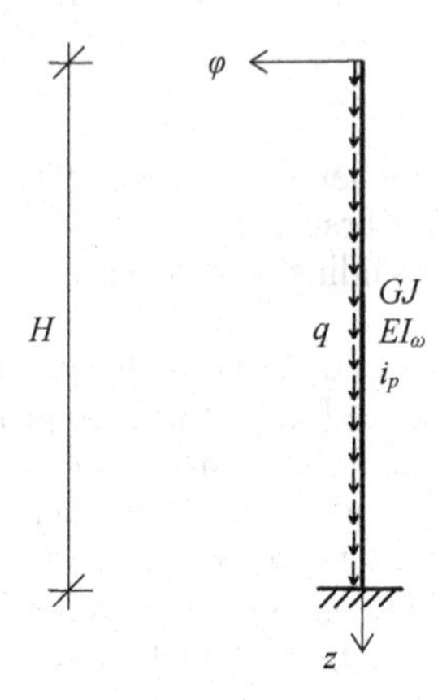

Figure 2.33 Bracing core for the analysis of pure torsional buckling.

The solution of the above governing differential equation gives the critical load for pure torsional buckling as

$$N_{cr,\varphi} = \frac{\alpha r_s EI_\omega}{i_p^2 H^2} \tag{2.97}$$

where α is the critical load factor. Its values are given in Table 2.9 and in Figure 2.34 as a function of

$$k_s = \frac{k}{\sqrt{r_s}} = H\sqrt{\frac{GJ}{r_s EI_\omega}} \tag{2.98}$$

When the warping stiffness of the core is zero, Equations (2.97) and (2.98) cannot be used. For such cases the critical load is obtained from

$$N_{cr,\varphi} = \frac{GJ}{i_p^2} \tag{2.99}$$

It is interesting to note that in this case the value of the critical load does not depend on the height of the structure.

Table 2.9 Critical load parameter α as a function of parameter k_s.

k_s	α	k_s	α	k_s	α	k_s	α
0.00	7.837	1.3	12.72	2.8	28.03	50	2984.7
0.01	7.838	1.4	13.47	2.9	29.30	60	4209.3
0.05	7.845	1.5	14.27	3.0	30.59	70	5640.9
0.10	7.867	1.6	15.11	4.0	44.69	80	7278.1
0.20	7.957	1.7	15.99	5.0	60.75	90	9120.7
0.30	8.107	1.8	16.91	6.0	78.80	100	11168
0.40	8.316	1.9	17.87	7.0	98.94	200	42864
0.50	8.583	2.0	18.87	8.0	121.2	300	94863
0.60	8.909	2.1	19.91	9.0	145.7	400	167093
0.70	9.291	2.2	20.98	10	172.4	500	259498
0.80	9.730	2.3	22.08	15	338.6	1000	1023750
0.90	10.22	2.4	23.21	20	558.6	2000	4059499
1.00	10.77	2.5	24.38	25	831.8	3000	9101926
1.10	11.37	2.6	25.57	30	1157.8	4000	16149383
1.20	12.02	2.7	26.79	40	1967.1	>4000	k_s^2

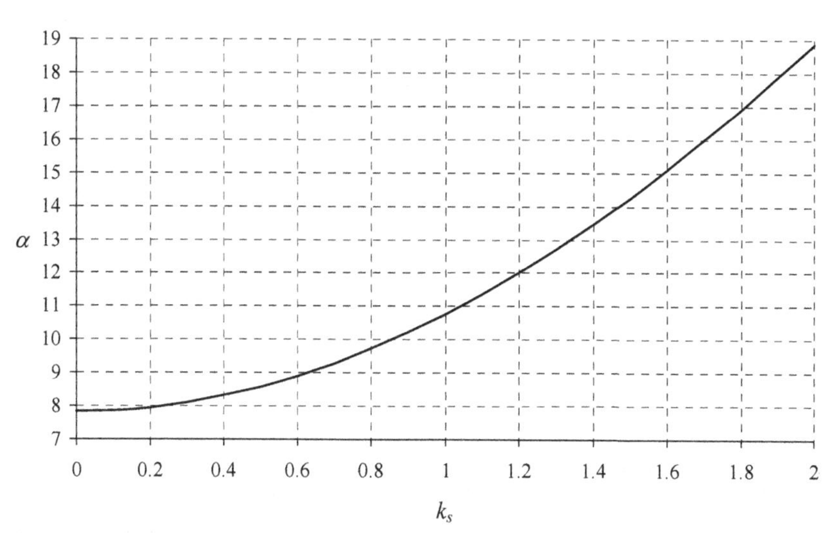

Figure 2.34 Critical load parameter α as a function of parameter k_s.

It is important to point out that when Equations (2.97) and (2.99) are used for the stability analysis of a *building* braced by a single core, then the radius of gyration refers to the layout of the building (rather than to the cross-section of the core). See also Section 6.4 where the global behaviour of a five-storey building is investigated in detail. The torsional resistance of that building is solely provided by a single U-core.

As cores normally behave in a three-dimensional manner, the above critical loads characterising torsional behaviour can only be considered *basic* critical loads. Together with the other two basic critical loads, the three basic critical loads ($N_{cr,x}$, $N_{cr,y}$ and $N_{cr,\varphi}$) may, and normally will, combine during buckling resulting in the global critical load of the core. This combination is very important as the actual critical load of the core is always smaller than (or equal to) the smallest one of the three basic critical loads. For the coupling of the basic critical loads, see Chapter 5 that deals with three-dimensional behaviour.

3

Deflection and rotation analysis of buildings under horizontal load

In addition to the general assumptions listed in Chapter 1, it is also assumed in this chapter that the multi-storey buildings are subjected to uniformly distributed horizontal load. Based on the deflection analysis of a single frame, the characteristic unit of a bracing system, whole structures braced by frames, coupled shear walls, shear walls and cores (Figure 3.1) will now be investigated.

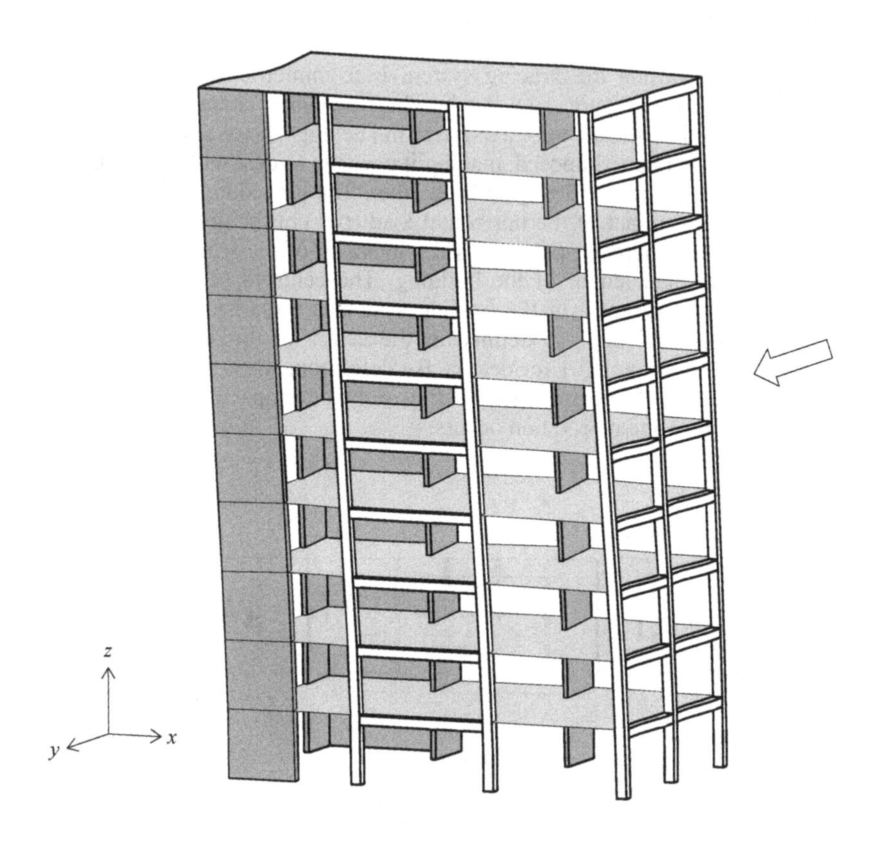

Figure 3.1 Three-dimensional behaviour of ten-storey building under horizontal load. (See Section 6.5 for its comprehensive global structural analysis.)

The task is made more complicated than with a simple unit as, in addition to the interaction among the elements of a frame, interaction also occurs among the bracing units themselves, often in a three-dimensional manner. Approximate methods have been developed for the investigation of bracing systems under horizontal load (Pearce and Matthews, 1971; Dowrick, 1976; Schueller, 1977 and 1990; Irwin, 1984; Stafford Smith and Coull, 1991; Coull and Wahab, 1993) but they often have restrictive assumptions and their accuracy and reliability have not been comprehensively investigated. Sporadic checks show that in certain cases they may lead to very poor estimates, including unconservative estimates of unacceptable magnitude (up to 70%).

The solutions to be presented in this chapter are not only relatively simple but they are also expressive; their structure is such that they show how the geometrical arrangement and the different stiffness characteristics influence the deflection and rotation of the building. According to the results of a comprehensive accuracy analysis (in Chapter 7), they are reliable to use for practical structural engineering purposes with good accuracy well within practical requirements.

3.1 THREE-DIMENSIONAL BEHAVIOUR

In the general case when the bracing system is asymmetric, due to the three-dimensional behaviour (Figure 3.1), the building develops lateral displacements *and* rotation (around its shear centre axis) over the height. The two phenomena can be separated, and later superimposed, making it possible to deal with the deflection and rotation problems independently of each other. The procedure is demonstrated in Figure 3.2. The resultant of the horizontal load (per unit height) is represented by $w = w^*L$ [kN/m], where w^* [kN/m^2] is the intensity of the wind load (per unit area) and L is the plan length of the building. The centroid of the plan of the building and the shear centre of the bracing system are marked with C and O, respectively. The shear centre is defined as the centre of the stiffnesses of the bracing units—see Section 3.3.1 for details. By definition, when the external force passes through the shear centre, the building only develops translation (in the direction of the force) and no rotation occurs.

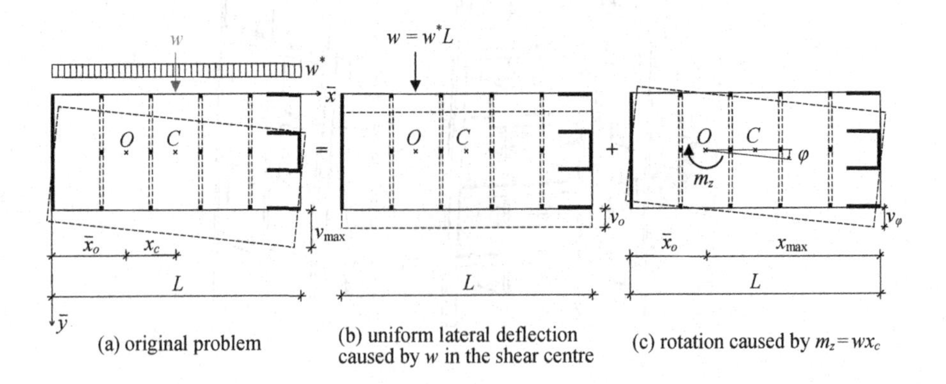

Figure 3.2 Deflection and rotation of asymmetric multi-storey buildings under horizontal load.

The resultant of the horizontal load (w) passes through the centroid of the plan of the building (C) (Figure 3.2/a). This situation represents our original problem. Let's now transfer force w to the shear centre (O) (Figure 3.2/b). This force in the shear centre must be accompanied by torque $m_z = wx_c$ (Figure 3.2/c), where x_c is the distance between the centroid and the shear centre. Force w in the shear centre only causes $v_o(z)$ uniform lateral displacements while torque m_z only develops rotations around the shear centre axis causing additional $v_\varphi(z)$ displacements. The original problem has thus been separated into two parts and, accordingly, the deflection of the building at any location can be expressed by

$$v(z) = v_o(z) + v_\varphi(z)$$

Maximum deflection develops at the top of the building at one of the corner points of the plan of the building:

$$v_{\max} = v(H) = v_o(H) + x_{\max}\varphi(H) = v_o(H) + (L - \bar{x}_o)\varphi(H)$$

where H is the height of the building, $x_{\max}$ is the distance of the corner point (where the maximum deflection occurs) from the shear centre, $\varphi(H)$ is the angle of rotation and $\bar{x}_o$ defines the location of the shear centre. See Section 3.3.1 for details regarding the shear centre.

From a practical point of view, these equations also offer clear guidance as to the calculation of the maximum deflection of a multi-storey building. According to the equations, two separate problems have to be solved in the general (asymmetrical) case and then their combination results in the deflection of the building. The case when the lateral load passes through the shear centre leads to a torsion-free situation and it can be called the planar problem whose solution leads to the uniform deflection $v_o(z)$. The second problem can be called the torsional problem with the angle of rotation $\varphi(z)$ as its solution. These two problems will be dealt with separately in the next two sections where the characteristic deformations and stiffnesses introduced in Section 2.1 will be used.

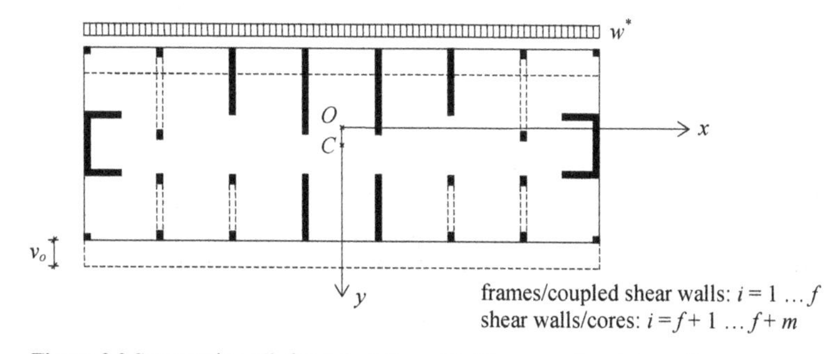

Figure 3.3 Symmetric wall–frame building with f frames and m shear walls.

It is worth noting that in the special case when the bracing system is symmetric (with respect to the line of the resultant of the lateral load), the resultant of the horizontal load *ab initio* passes through the shear centre of the bracing system and no rotation develops (Figure 3.3). It follows that in such cases the solution of the planar problem [$v_o(z)$] directly results in the deflection of the building. In all other cases when the bracing system is asymmetric, the maximum deflection is only obtained after taking into account the rotation of the building.

3.2 THE PLANAR PROBLEM: LATERAL DEFLECTION ANALYSIS OF TORSION-FREE BUILDINGS

Consider for the planar analysis a bracing system that consists of frames, coupled shear walls, shear walls and cores. From now on, for the sake of simplicity, frames and coupled shear walls will be grouped together under the term "frames" and shear walls and cores will be referred to as "shear walls".

The resultant of the external load passes through the shear centre of the bracing system [either because the system is symmetric (Figure 3.3) or because the resultant was transferred there (Figure 3.2/b)]. The horizontal load on the façade of the building is first transmitted to the floor slabs and then to the bracing units. The floor slabs, being stiff in their plane, make the bracing units work together, which develop uniform lateral displacements. The building can be modelled by a planar system of bracing units that are linked by incompressible pinned bars representing the floor slabs.

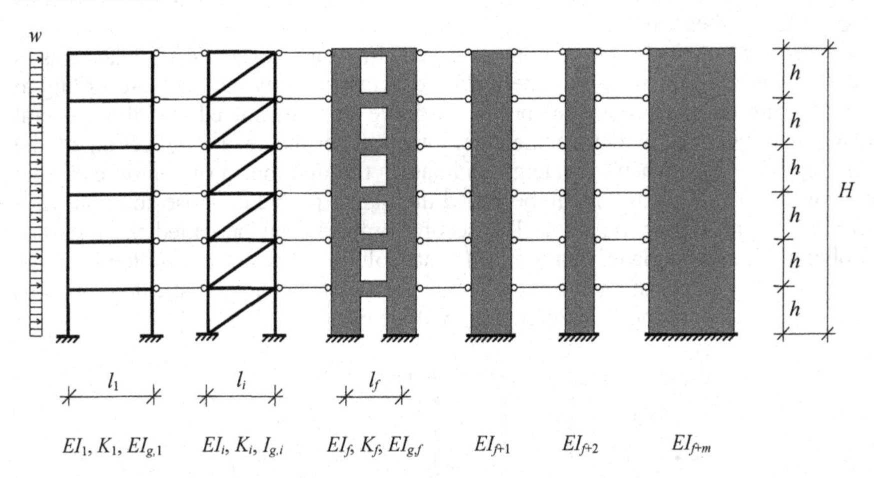

Figure 3.4 Model for the planar analysis with f frames/coupled shear walls and m shear walls/cores.

Figure 3.4 shows such a typical model where the first f bracing units may represent frames and coupled shear walls and the remaining m bracing units may be shear walls and cores.

3.2.1 The governing differential equations of the problem

When the lateral load of a multi-storey building is resisted by a system of f frames and m shear walls, the behaviour of the system is complex. As a rule, the frames develop a deflection shape which is a combination of bending and shear deformation. The nature of the resulting deflection shape depends on several factors. The deflection shape of the shear walls and cores is of "pure" bending. The floor slabs of the building, being stiff in their horizontal plane, make the bracing units assume the same deflection shape. As the two types would have different shapes on their own, they interact and this interaction results in the "compromise" deflection of the building.

The characteristics of the interaction can be best investigated by using the governing differential equations of the system and analysing the different roles that the two different bracing types play. The system of governing differential equations of f frames and m shear walls consists of two sets of equations. The first set represents f compatibility conditions for the f frames expressing continuity at the vertical lines of contraflexure of the beams of the frames. (The frames at this stage are single-bay structures but, as it is demonstrated in Section 2.1.3, the final results can be extended to cover multi-bay frames as well.) The second set relates to the bending of the vertical elements, i.e. to the full-height columns of the frames and the shear walls and cores.

Based on the derivation presented in Section 2.1.2 and making use of Equation (2.9), the first set of f compatibility equations is given below, of which, for the sake of clarity, only the first, the ith and the last one are listed:

$$y_1''(z) - \frac{l_1}{K_1} N_1''(z) + \frac{l_1}{EI_{g,1}} N_1(z) = 0 \tag{3.1}$$

$$y_i''(z) - \frac{l_i}{K_i} N_i''(z) + \frac{l_i}{EI_{g,i}} N_i(z) = 0 \tag{3.2}$$

$$y_f''(z) - \frac{l_f}{K_f} N_f''(z) + \frac{l_f}{EI_{g,f}} N_f(z) = 0 \tag{3.3}$$

In the above equations $EI_{g,i}$ is the global bending stiffness, K_i is the shear stiffness and l_i is the size of the bay of the ith frame (with $i = 1 \ldots f$). Force $N_i(z)$ stands for the normal force on the columns of the ith frame that originates from the bending of the beams.

A second set of equations is needed as, in addition to deflections $y_i(z)$, normal forces $N_i(z)$ are also unknown quantities in the above equations. This second set (of $f+m$ equations) is concerned with the bending of the vertical elements. The first f equations represent the bending of the full-height columns of the f frames:

$$y_1''(z) EI_1 = -M_1(z) + l_1 N_1(z) \tag{3.4}$$

$$y_i''(z) EI_i = -M_i(z) + l_i N_i(z) \tag{3.5}$$

$$y_f''(z) EI_f = -M_f(z) + l_f N_f(z) \tag{3.6}$$

The second group of m equations represent the bending of the shear walls:

$$y''_{f+1}(z)EI_{f+1} = -M_{f+1}(z) \tag{3.7}$$

$$y''_{f+2}(z)EI_{f+2} = -M_{f+2}(z) \tag{3.8}$$

$$y''_{f+m}(z)EI_{f+m} = -M_{f+m}(z) \tag{3.9}$$

where the first shear wall is marked by subscript $f+1$. (Of the m shear walls only the first, the second and the last one are listed.)

Local bending stiffness EI_i for the frames $(1 \le i \le f)$ is determined using the sum of the second moments of area of the columns ($I_{c,i}$), adjusted by parameter r_i—see Equations (2.5), (2.10) and (2.13) for one-bay frames. The bending stiffness of the shear walls ($f+1 \le i \le m$) is determined in the usual manner.

Bending moment $M_i(z)$ in the above equations is the moment share on the ith bracing unit, according to

$$M_i(z) = q_i M(z) \tag{3.10}$$

where $1 \le i \le f+m$ holds. In this equation

$$M(z) = \frac{wz^2}{2} \tag{3.11}$$

is the total external bending moment on the system and q_i is the apportioner of the external load. [Although Equation (3.10) refers to moment share, because of the linear relationship between forces and moments, the apportioner can be used with moments as well as with forces.] The value of the apportioner is determined by

$$q_i = \frac{S_i}{\sum_{i=1}^{f+m} S_i} \tag{3.12}$$

where S_i is the "overall" lateral stiffness of the ith bracing unit. The "overall" lateral stiffness of a bracing unit is defined by

$$S_i = \frac{1}{y_i(H)} \tag{3.13}$$

where $y_i(H)$ is the maximum top deflection of the ith bracing unit.

When the value of the ith proportioner is known, the load share on the ith bracing unit can also be determined:

$$w_i = q_i w \tag{3.14}$$

The above two sets of differential equations represent the complete system of governing differential equations of the bracing system consisting of frames and shear walls. The first set consists of f equations and these equations are responsible for fulfilling the compatibility conditions. The second set consists of $f+m$ equations in two parts. The first part (with f equations) represents the bending of the full-height columns of the frames and the second part (with m equations) stands for the bending of the shear walls.

There are two possibilities to proceed from here. One approach leads to a very simple solution and the other approach results in a more accurate solution. Both solutions will be given in the following as both are important. Although the more accurate solution will be recommended for use regarding this planar problem, the simple solution will play an important role when the torsional behaviour of asymmetric buildings is investigated (in Section 3.3). It is also the simple solution that is used in the special case when the bracing system only contains frames/coupled shear walls (but not shear walls/cores).

3.2.2 Method "A": The simple method

In order to create a simple solution, the original bracing system will be regrouped and the governing differential equations of the bracing system will be rearranged. When the planar behaviour of the system is considered, rearranging the bracing units does not change the behaviour of the system as the units are linked by incompressible pinned bars and the order of the bracing units is irrelevant.

Depending on the type of bracing unit, two groups will be created: the "frame" group and the "wall" group. The frame group also contains the coupled shear walls and the wall group includes the cores. Naturally, both sub-systems have their own load share. The load that belongs to the frames is defined by apportioners q_1, q_i, ... q_f and the load on the shear walls/cores is determined by q_{f+1}, q_{f+2}, ... q_{f+m}.

Consider first the first sub-system of f frames. Equations (3.1), (3.2) and (3.3) represent the compatibility conditions of the f frames and they contain $y_i(z)$ and $N_i(z)$ as unknown quantities. Equations (3.4), (3.5) and (3.6), representing the bending of the full-height vertical elements of the frames, contain the same unknown quantities. The $2f$ equations necessary for the solution are therefore available. It follows that the normal forces from the compatibility equations can be eliminated. After substituting for $N(z)$ and $N''(z)$, the governing equation of the ith frame of the first sub-system (with $1 \le i \le f$) is obtained as

$$y_i''(z) - \frac{1}{K_i}\left[y_i''(z)EI_i + M_i(z)\right]'' + \frac{1}{EI_{g,i}}\left[y_i''(z)EI_i + M_i(z)\right] = 0 \tag{3.15}$$

In substituting for $M_i(z)$ [using Equations (3.10) and (3.11)] and after some rearrangement, Equation (3.15) can be written as

$$y_i''''(z) - y_i''(z)\left(\frac{K_i}{EI_i} + \frac{K_i}{EI_{g,i}}\right) = \frac{q_i w}{EI_i}\left(\frac{z^2}{2}\frac{K_i}{EI_{g,i}} - 1\right) \tag{3.16}$$

When bracing systems are investigated by analytical methods, the equivalent column approach is often used and the bracing units are "pushed together" to create an equivalent column. The structure of Equation (3.16) clearly shows that, as a rule, it is now not possible to create an equivalent column in such a way that the corresponding stiffnesses of the frames (EI_i, $EI_{g,i}$ and K_i) are simply added up. This is a significant observation as the situation with the stability and frequency analyses is different: the solution of the frequency and stability problems (in Chapters 4 and 5) is based on an equivalent column whose characteristic stiffnesses are obtained by adding up the relevant stiffnesses of the individual bracing units. That—very convenient—approach works there, but it cannot be applied here to the deflection analysis of the bracing system under horizontal load.

The governing differential equations of the second sub-system of m shear walls [Equations (3.7), (3.8) and (3.9)] can be expressed in a similar (but much simpler) form. The ith equation assumes the form:

$$y_i''(z)EI_i = -\frac{q_i w z^2}{2} \tag{3.17}$$

where $f+1 \le i \le m$ holds.

Equations (3.16) and (3.17), representing the ith unit of the two sub-systems with altogether $f+m$ bracing units, define the *complete* system of governing differential equations of the bracing system. Solving this system of $f+m$ differential equations would lead to a rather complicated procedure. However, there is no need for this procedure. Two important observations can be made that make it possible to simplify the deflection problem. These observations are as follows.

(a) According to the assumption regarding the floor slabs, all the bracing units assume the same deflection shape, i.e. $y_1(z) = y_i(z) = y_{f+m}(z) = y(z)$.

(b) Knowing the load share on the ith bracing unit [Equations (3.12), (3.13) and (3.14)], it is possible to concentrate on one bracing unit only.

If a frame is to be used for the determination of the deflection of the building, then the solution of Equation (3.16) is needed. The short form of Equation (3.16) is

$$y_i''''(z) - \kappa_i^2 y_i''(z) = \frac{q_i w}{EI_i}\left(\frac{a_i z^2}{2} - 1\right) \tag{3.18}$$

where

$$a_i = \frac{K_i}{EI_{g,i}}, \qquad b_i = \frac{K_i}{EI_i}, \qquad \kappa_i = \sqrt{a_i + b_i} = \sqrt{\frac{K_i(I_i + I_{g,i})}{EI_i I_{g,i}}} = \sqrt{\frac{K_i s_i}{EI_i}} \tag{3.19}$$

and

$$s_i = 1 + \frac{a_i}{b_i} = \frac{I_{g,i} + I_i}{I_{g,i}} = 1 + \frac{I_i}{I_{g,i}} \tag{3.20}$$

are auxiliary stiffness parameters.

The structure of the above differential equation is identical to that of a single, *independent* frame and therefore its solution—Equation (2.18)—can be directly applied (Zalka, 2013). Bearing in mind that the deflection of the ith frame is identical to the deflection of the whole system, the formula for the deflection of the system is obtained as

$$y(z) = y_i(z) = \frac{q_i w}{EI_{f,i}}\left(\frac{H^3 z}{6} - \frac{z^4}{24}\right) + \frac{q_i w z^2}{2K_i s_i^2} -$$

$$-\frac{q_i w EI_i}{2K_i^2 s_i^3}\left(\frac{\cosh[\kappa_i(H-z)] + \kappa_i H \sinh(\kappa_i z)}{\cosh(\kappa_i H)} - 1\right) \qquad (3.21)$$

where

$$I_{f,i} = I_i + I_{g,i} \qquad (3.22)$$

is the sum of the local and global second moments of area of the ith bracing unit.

Maximum deflection develops at $z = H$:

$$y_{\max} = y_{i,\max} = \frac{q_i w H^4}{8EI_{f,i}} + \frac{q_i w H^2}{2K_i s_i^2} - \frac{q_i w EI_i}{K_i^2 s_i^3}\left(\frac{1 + \kappa_i H \sinh(\kappa_i H)}{\cosh(\kappa_i H)} - 1\right) \qquad (3.23)$$

When the above equations are used for multi-bay frames, stiffnesses I_i, $I_{g,i}$ and K_i are to be calculated according to Section 2.1.3.

If a shear wall is to be used for the determination of the deflection of the building, then the solution of Equation (3.17) is needed:

$$y(z) = y_i(z) = \frac{q_i w}{EI_i}\left(\frac{H^3 z}{6} - \frac{z^4}{24}\right) \qquad (3.24)$$

The maximum top-level deflection at $z = H$ is

$$y_{\max} = y_{i,\max} = \frac{q_i w H^4}{8EI_i} \qquad (3.25)$$

Whichever bracing unit is chosen as the ith bracing unit, the result is the same. It looks therefore advisable to choose a shear wall as the ith bracing unit, as Equation (3.25) is simpler than Equation (3.23). It should be noted, however, that the determination of the load share on the bracing unit that is used for the calculation of the deflection of the building requires the determination of the maximum deflection of *every* bracing unit of the bracing system—see Equations

(3.12) and (3.13). Equations (3.23) and (3.25) can be used for this purpose. An arbitrary apportioner, say $q_i = 1$, can be used for these calculations as the intensity of the load drops out of the formulae.

The beauty of this solution is in its simplicity. The drawback of this procedure lies in the fact that in the process of separating the two sub-systems—the frames and the shear walls—the direct interaction between the shear walls and the frames is tacitly ignored. (The shear walls develop pure bending deformation and the frames develop a combination of bending and shear deformation.) According to a comprehensive accuracy analysis (in Section 7.3.1) involving 198 test structures, the accuracy of this simple solution is still acceptable in most practical cases. However, the method can be improved as the direct interaction between the shear walls and frames can be taken into account. This procedure is presented in the following section.

3.2.3 Method "B": The more accurate method

By comparing the governing differential equations of bracing systems of different types and examining the effect of the interaction of the deformations of different types, i.e. frames-frames, shear walls-shear walls and frames-shear walls, it is possible to draw certain conclusions enabling the creation of a procedure that is more sophisticated (and consequently more accurate) than the simple method introduced in the previous section.

When the bracing system consists of frames of different geometrical characteristics, the frames—due to their different stiffnesses—*would* develop different deflection shapes. But they do not, because the floor slabs force them to take the same "compromise" deflection. It follows that interaction occurs among the frames. When the bracing system only contains shear walls, there is no interaction because each wall develops pure bending deformation, each wall has the same deflection shape and the "size" of the deflection is proportional to the load share on the wall (which is proportional to the bending stiffness of the wall). When the bracing system is a mixture of frames and shear walls, there is *always* interaction as the original type of their deflection is always different.

It should be noted that the equations in the second part of the second set [i.e. Equations (3.7), (3.8) and (3.9)] are of the same structure. They only contain the local bending stiffness and, more importantly, they do not contain normal forces $N_i(z)$ that are needed for the overall solution. The mathematical consequence of this is that these equations are not needed technically for the solution of the deflection problem from the point of view of the frames. (This fact was utilized in the previous section when the two groups—frames and shear walls—were effectively separated.) The practical consequence of this is that any number of shear walls can be "put together" into a single shear wall (by adding up their bending stiffnesses) for the deflection analysis. This also follows from the fact that there is no interaction among the shear walls themselves, whose deflection *shapes* (in pure bending) are of the same nature and their load is proportional to their stiffnesses.

The problem of f frames and m shear walls can thus be reduced to a system of f frames and one shear wall, accompanied by differential equations (3.1), (3.2), (3.3), (3.4), (3.5), (3.6) and (3.7). From now on, for practical reasons, subscript

$f+1$ is replaced with the more meaningful w as it refers to the shear wall (Figure 3.5).

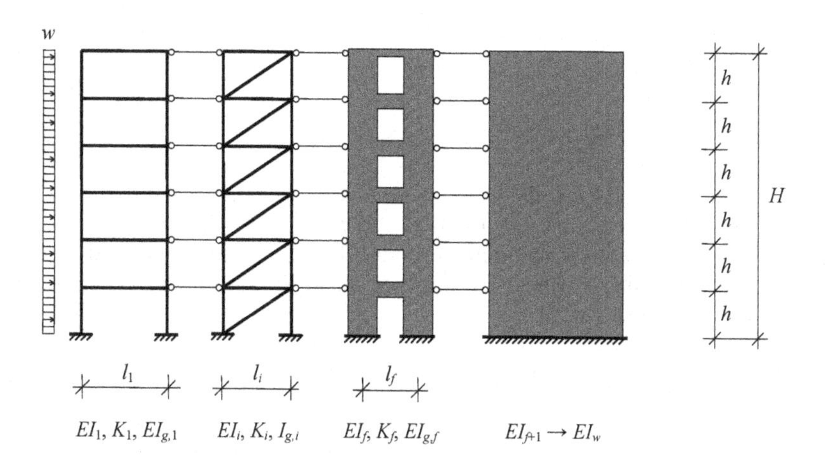

Figure 3.5 Bracing system with f frames/coupled shear walls and a single shear wall.

Instead of separating the different types of bracing units (and losing the effect of direct interaction), the shear wall will now be incorporated into the system of frames. The investigation of a single frame and one shear wall (Figure 3.6) shows how this can be achieved.

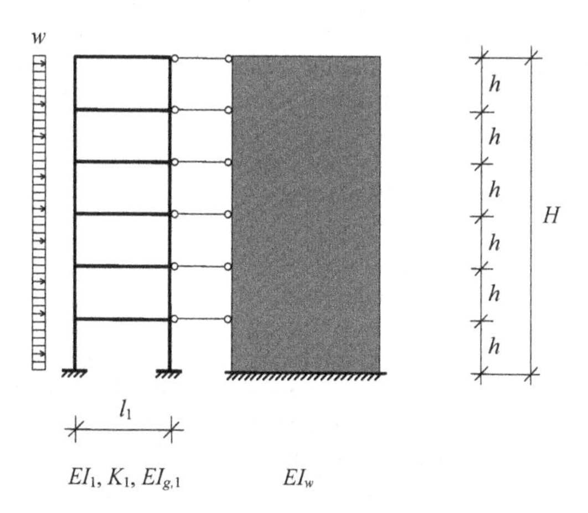

Figure 3.6 A system of one frame and one shear wall.

The behaviour of a system consisting of one frame and one shear wall is defined by Equations (3.1), (3.4) and (3.7). With $y_1(z) = y_{f+1}(z) = y(z)$, and using subscript w referring to the shear wall, they assume the form

$$y''(z)-\frac{l_1}{K_1}N_1''(z)+\frac{l_1}{EI_{g,1}}N_1(z)=0 \qquad (3.26)$$

$$y''(z)EI_1=-M_1(z)+l_1N_1(z) \qquad (3.27)$$

$$y''(z)EI_w=-M_w(z) \qquad (3.28)$$

where $M_1(z)$ and $M_w(z)$ are the moment shares on the frame and the shear wall, respectively. It is clear that Equations (3.27) and (3.28) can be combined: they represent the same type of bending (i.e. pure bending), their left-hand side only contains local bending stiffness EI and they stand for the same deflection shape $y(z)$. In doing so, we arrive at

$$y''(z)E(I_1+I_w)=-M(z)+l_1N_1(z) \qquad (3.29)$$

where $M(z) = M_1(z) + M_w(z)$ is the total external bending moment on the system.

Altogether, two equations are needed for the final solution—$y(z)$ and $N_1(z)$ being the two unknowns—and Equations (3.26) and (3.29) furnish these two equations. In practical terms, it can be said that the shear wall has been "pushed" into the frame, increasing its local bending stiffness. There is another important aspect of this procedure. By incorporating the shear wall into the frame, it is automatically ensured that the interaction between the frame and the shear wall is taken into account. It occurs when Equations (3.26) and (3.29) are solved as I_w is now part of the system. This is what we have referred to earlier as "direct interaction". (In the previous section when we presented "the simple solution", the second moments of area of the walls were not part of the system to be solved as the shear walls were separated into another sub-system.)

The above equations also demonstrate the precise meaning of the popular term "wall-frame interaction". The term is normally interpreted as the interaction between the two bracing units, i.e., the shear wall and the frame. It may be more to the point to refer to this phenomenon as the interaction between the bending and shear deformations.

The situation is similar, although slightly more complicated when the system consists of f frames and one shear wall (that, as we saw above, may represent several shear walls). The number of equations needed for the solution is $2f$. The choice for one set of f equations is obvious: the compatibility equations represented by Equations (3.1), (3.2) and (3.3). The question arises: how to obtain the second set of f equations. Proceeding as in the case of the single frame–single wall system above, the differential equation of the shear wall [Equation (3.28)] should be combined with those representing the bending of the vertical elements of the frames [Equations (3.4), (3.5) and (3.6)]. This task seems to be difficult—if not impossible—as there is only one shear wall and there are f frames, and f equations are needed. However, understanding the behaviour of the system during interaction points at the solution. Due to the floor slabs, the shear wall interacts with all the frames during deflection as, as a rule, their individual deflection shapes are different. It follows that all the frames participate when—as with the single frame–

single wall case—the bending stiffness of the shear wall is added to the frame system. The "intensity" of the interaction depends on the stiffnesses of the participants. It follows that the bending stiffness of the shear wall should be apportioned among the frames according to their relative stiffnesses. This is achieved if such apportioners are used that are only related to the first f bracing units, i.e. to the original frames. These apportioners are defined as

$$\bar{q}_i = \frac{S_i}{\sum_{i=1}^{f} S_i} \tag{3.30}$$

where S_i is the "overall" lateral stiffnesses of the ith frame.

The system of f frames and m shear walls has now been reduced to a system of f frames. However, these are not the original frames as the local bending stiffness of each frame is now amended by its share of the bending stiffness of the shear wall(s). Accordingly, the second set of equations assume the form

$$y''(z)E(I_1 + \bar{q}_1 I_w) = -M_1^*(z) + l_1 N_1(z) \tag{3.31}$$

$$y''(z)E(I_i + \bar{q}_i I_w) = -M_i^*(z) + l_i N_i(z) \tag{3.32}$$

$$y''(z)E(I_f + \bar{q}_f I_w) = -M_f^*(z) + l_f N_f(z) \tag{3.33}$$

Including the stiffness of the shear wall(s) in the above equations also means that the interaction between the shear wall(s) and the frames is directly taken into account.

It should be noted that $M_1^*(z)$, $M_i^*(z)$ and $M_f^*(z)$ in the above equations are different from their equivalents in Equations (3.4), (3.5) and (3.6), as the frames themselves are now different from the original frames. Their value

$$M_i^*(z) = q_i^* M(z) \tag{3.34}$$

is determined using the new apportioner

$$q_i^* = \frac{S_i^*}{\sum_{i=1}^{f} S_i^*} \tag{3.35}$$

whose values are determined using the "new" frames. The "overall" lateral stiffness of the ith "new" frame is defined as

$$S_i^* = \frac{1}{y_i^*(H)} \tag{3.36}$$

where $y_i^*(H)$ is the maximum deflection of the *i*th (new) frame. The load share on this frame is now

$$w_i^* = q_i^* w \tag{3.37}$$

The star in the above equations indicates that the frames in question differ from the original ones in that they also contain a portion of the bending stiffness of the shear wall(s).

It is now feasible to combine the two sets of differential equations: Equations (3.1), (3.2) and (3.3) representing the compatibility conditions of the *f* frames, and Equations (3.31), (3.32) and (3.33) representing the bending of the vertical elements of the bracing system *including* the shear walls that are now incorporated into the frames. In doing so, the governing equation of the *i*th frame of the system is obtained as

$$y''(z) - \frac{1}{K_i}\left[y''(z)EI_i^* + M_i^*(z)\right]'' + \frac{1}{EI_{g,i}}\left[y''(z)EI_i^* + M_i^*(z)\right] = 0 \tag{3.38}$$

where

$$I_i^* = I_i + \overline{q}_i I_w \tag{3.39}$$

and $i = 1 \ldots f$. In the above equations K_i, EI_i and $EI_{g,i}$ are the stiffnesses of the *i*th (original) frame and EI_w is the bending stiffness of the shear wall(s).

Equation (3.38) is clearly analogous with Equation (3.15) and therefore the procedure presented in Section 3.2.2 can be repeated. This leads to the governing differential equation

$$y''''(z) - \kappa_i^{*2} y''(z) = \frac{q_i^* w}{EI_i^*}\left(\frac{a_i z^2}{2} - 1\right) \tag{3.40}$$

where

$$a_i = \frac{K_i}{EI_{g,i}}, \quad b_i^* = \frac{K_i}{EI_i^*}, \quad \kappa_i^* = \sqrt{a_i + b_i^*} = \sqrt{\frac{K_i(I_i^* + I_{g,i})}{EI_i^* I_{g,i}}} = \sqrt{\frac{K_i s_i^*}{EI_i^*}} \tag{3.41}$$

and

$$s_i^* = 1 + \frac{a_i}{b_i^*} = \frac{I_{g,i} + I_i^*}{I_{g,i}} = 1 + \frac{I_i^*}{I_{g,i}} \tag{3.42}$$

The solution of Equation (3.15)—after amending the relevant bending stiffnesses—can also be used. The formulae for the deflection of the system is obtained as

$$y(z) = \frac{q_i^* w}{EI_{f,i}^*}\left(\frac{H^3 z}{6} - \frac{z^4}{24}\right) + \frac{q_i^* w z^2}{2K_i s_i^{*2}} - \frac{q_i^* w EI_i^*}{K_i^2 s_i^{*3}}\left(\frac{\cosh[\kappa_i^*(H-z)] + \kappa_i^* H \sinh(\kappa_i^* z)}{\cosh(\kappa_i^* H)} - 1\right) \tag{3.43}$$

where

$$I_{f,i}^* = I_i^* + I_{g,i} \tag{3.44}$$

is the sum of the local and global second moments of area.

Maximum deflection develops at $z = H$:

$$y_{\max} = y(H) = \frac{q_i^* w H^4}{8EI_{f,i}^*} + \frac{q_i^* w H^2}{2K_i s_i^{*2}} - \frac{q_i^* w EI_i^*}{K_i^2 s_i^{*3}}\left(\frac{1 + \kappa_i^* H \sinh(\kappa_i^* H)}{\cosh(\kappa_i^* H)} - 1\right) \tag{3.45}$$

The situation is similar to that with Method "A" in Section 3.2.2 in that the determination of the load share on the frame $(q_i^* w)$ that is used for the calculation of the deflection of the building requires the determination of the maximum deflection of *each* frame [cf. Equations (3.35) and (3.36)]. These values are calculated using Equation (3.45) with an arbitrary apportioner, say, $q_i^* = 1$, as the intensity of the load eventually drops out of the formulae.

Again, the above equations spectacularly demonstrate that, as a rule, it is not possible to carry out the lateral deflection analysis of a building by adding up the corresponding stiffnesses of the bracing units in order to create an equivalent column, as is circulated in the literature. The method based on adding up the corresponding stiffnesses does work for the frequency and stability analyses—as it will be seen in Chapters 4 and 5—but not for the deflection analysis. There is only one exception: a system of shear walls and a single frame.

This—more accurate—method requires slightly more calculation than that of the simple method (in Section 3.2.2) but its accuracy is much better and therefore Method "B" is recommended for the analysis of the planar (rotation-free) problem. A comprehensive accuracy analysis is presented in Section 7.3.

3.3 THE TORSIONAL PROBLEM

When a multi-storey building under horizontal load has a symmetric bracing system, the resultant of the external load passes through the shear centre of the bracing system and the building develops uniform deflection (and no torsion occurs). In such cases one of the methods presented in Section 3.2 can be used for the determination of the top deflection of the building. As the name of the methods indicates, Method "A" (The simple method) offers a simple procedure and Method "B" (The more accurate method) leads to a slightly more complicated but more accurate procedure.

However, in most practical cases the bracing system has an asymmetrical arrangement (Figure 3.1) and, as a consequence, in addition to the uniform deflection, the building also undergoes torsion around the shear centre axis (Figure 3.2). In such cases the angle of rotation should also be determined. Knowing the angle of rotation, the maximum deflection of the building is obtained in a simple manner as described in Section 3.1 and summarized again in Section 3.4.

Because of the complexity of the torsional behaviour, not many authors deal with the problem. There are some excellent publications that offer relatively simple solutions for the global torsional problem (Council..., 1978; Irwin, 1984; Schueller, 1990; Coull and Wahab, 1993; Hoenderkamp, 1995; Nadjai and Johnson, 1998; Howson and Rafezy, 2002), but they are either still too complicated or of limited applicability and neither of them is backed up with a comprehensive accuracy analysis.

The aim of this section is therefore to develop a simple and reliable procedure for the determination of the rotation of the bracing system subjected to a system of uniformly distributed torsional moments. In addition to the general assumptions listed in Chapter 1, it is also assumed that the torsional stiffnesses of the individual bracing cores are negligible compared to the global torsional stiffnesses of the bracing system. This is a conservative assumption that is fulfilled in most practical cases. A rare exception is investigated in Section 6.4.

Two totally different approaches can be considered for handling the torsional problem. Following the "traditional" avenue, starting from scratch by examining the equilibrium of the system, a lengthy and complicated procedure leads to the solution. A similar—but simpler—procedure demonstrates this option in (Zalka and Armer, 1992) where the bracing system only consists of shear walls and cores. The other possibility is the application of the well-known analogy in the stress analysis of thin-walled structures in bending and torsion (Vlasov, 1961; Kollbrunner and Basler, 1969). This avenue is followed here. According to the analogy, translations, bending moments and shear forces correspond to rotations, warping moments and torsional moments. This analogy leads to a relatively simple solution when the corresponding characteristics can be established (Zalka, 2014).

3.3.1 Torsional behaviour and basic characteristics

As with thin-walled bars, multi-storey buildings react to torsional effects by utilising their torsional resistance. As with thin-walled bars, the torsional resistance of multi-storey buildings may originate from two main sources: in the vast majority of practical cases, both the warping stiffness and the Saint-Venant stiffness contribute to the torsional resistance of the building. The warping stiffness again may come from two sources, leading to three characteristic types of torsional stiffness altogether: local warping, global warping and Saint-Venant stiffnesses.

The "thin-walled bar"–"multi-storey building" analogy can be used to demonstrate spectacularly how the three characteristic stiffnesses operate when the structures undergo torsion. By way of illustration, the torsional behaviour of a single I-column is investigated below. The column has a fixed base and a free upper end. It is subjected to torsional moment M_t around axis z passing through

re O (Figure 3.7).
n the I-column has solid flanges (Figure 3.7/a), its torsional resistance iginates from its two flanges. The web of the column makes the two ork together. The two flanges respond to the external torsional moment g in their plane and their in-plane second moment of area is "activated" oment arm. The resulting warping stiffness of the column is calculated lying the second moment of area and the square of the perpendicular f the flanges from the shear centre of the column:

$$EI_{flange}\left(\frac{h}{2}\right)^2 2 = E\frac{tb^3}{12}\frac{h^2}{4}2 = E\frac{tb^3h^2}{24}$$

e above equation E is the modulus of elasticity of the material of the d I_ω is the local warping constant.

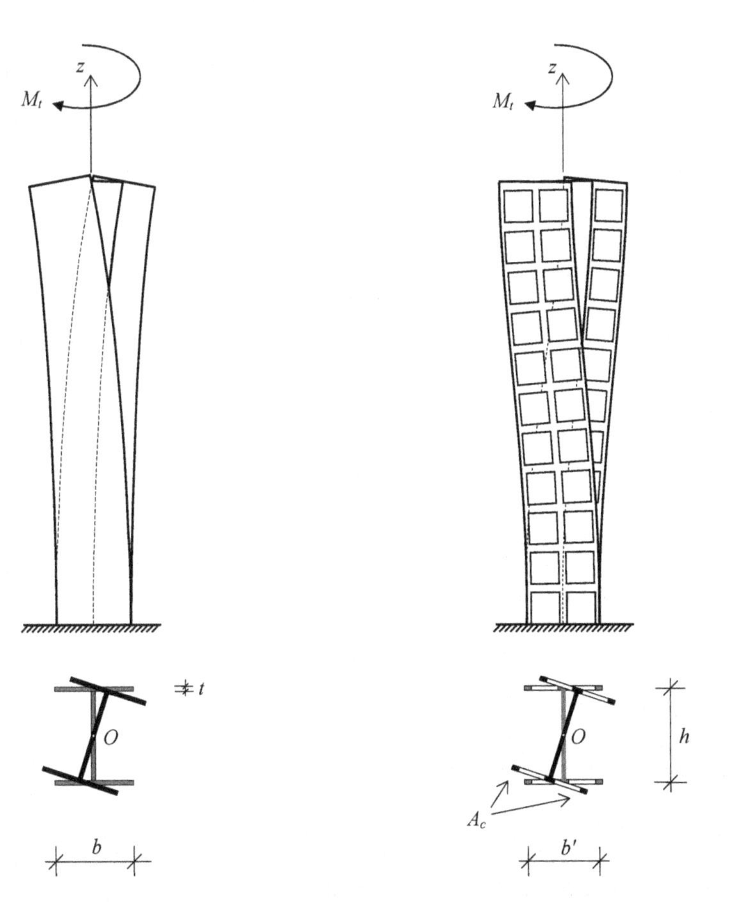

I-column with solid flanges (b) I-column with flanges with openings

The behaviour of the shear walls and cores in a multi-storey building sponds to this phenomenon. It is easy to see that in multi-storey buildings the slabs of the building (with their great in-plane stiffness) play the role of the of the I-column in making the shear walls/cores—the flanges—work together. investigating the torsional behaviour of multi-storey buildings, the above of warping stiffness EI_ω is defined as the *local warping stiffness.*

Consider again the same I-column but this time the flanges are pierced with penings of rectangular shape—the flanges are in fact frames (Figure 3.7/b). frames have two types of second moment of area: local and global. As far as cal second moment of area $I = \Sigma I_{c,i}$ is concerned, the three individual columns frames act exactly in the same manner as the two solid flanges above—ugh, because of their much smaller size, with much less efficiency.

However, these frames also have global second moments of area I_g. entrating on these second moments of area, and taking into account that they gain activated through their moment arms, the warping stiffness of the column v calculated as

$$I_{g\omega} = EI_g\left(\frac{h}{2}\right)^2 2 = EA_c\left(\frac{b'}{2}\right)^2 2\left(\frac{h}{2}\right)^2 2 = E\frac{A_c b'^2 h^2}{4}$$

A_c is the cross-sectional area of the "columns" of the flanges (Figure 3.7/b).

With multi-storey buildings, where the role of the web is taken over by the slabs, the above type of warping stiffness $EI_{g\omega}$ is defined as the *global ng stiffness* with $I_{g\omega}$ being the global warping constant.

Finally, the two flanges in Figure 3.7/b also have shear stiffness K. When the mn undergoes torsion, the flanges are made to activate their shear stiffness web. The torsional resistance is calculated by multiplying the shear stiffness flanges (i.e. the frames) and the square of the perpendicular distance of the es from the shear centre of the I-column as

$$GJ) = K\left(\frac{h}{2}\right)^2 2 = \frac{Kh^2}{2}$$

With buildings, the floor slabs again act as the web with the I-column, and pe of torsional stiffness is called the *Saint-Venant stiffness.*

The three formulae above show that in addition to the stiffnesses of the ng units, their distance from the shear centre of the bracing system is also d. The location of the shear centre is originally defined as the centre of esses of the bracing units. As the frames have three stiffnesses, the stiffness of bracing unit is (re)defined as the reciprocal of the top (in-plane) deflection of it.

With the stiffnesses of the units available—see Equation (3.13)—the ation of the location of the shear centre is conveniently carried out in the inate system $\bar{x} - \bar{y}$ whose origin lies in the upper left corner of the plan of ilding and whose axes are aligned with the sides of the building (Figure 3.8).

$$\frac{\sum_{i=1}^{f+m} S_{y,i}\bar{x}_i}{\sum_{i=1}^{f+m} S_{y,i}} \quad \text{and} \quad \bar{y}_o = \frac{\sum_{i=1}^{f+m} S_{x,i}\bar{y}_i}{\sum_{i=1}^{f+m} S_{x,i}} \tag{3.46}$$

are the perpendicular distances of the *i*th bracing unit from $\bar{y}$ and $\bar{x}$
is the number of frames and coupled shear walls
is the number of shear walls and cores
$S_{y,i}$ are the "overall" stiffnesses by Equation (3.13) in directions x and y

the calculation of the location of the shear centre, only the in-plane f the frames and shear walls needs to be taken into account. Once the f the shear centre is determined, coordinate system $\bar{x}-\bar{y}$ has fulfilled its new coordinate system $x-y$ is introduced with its origin in the shear gure 3.8).

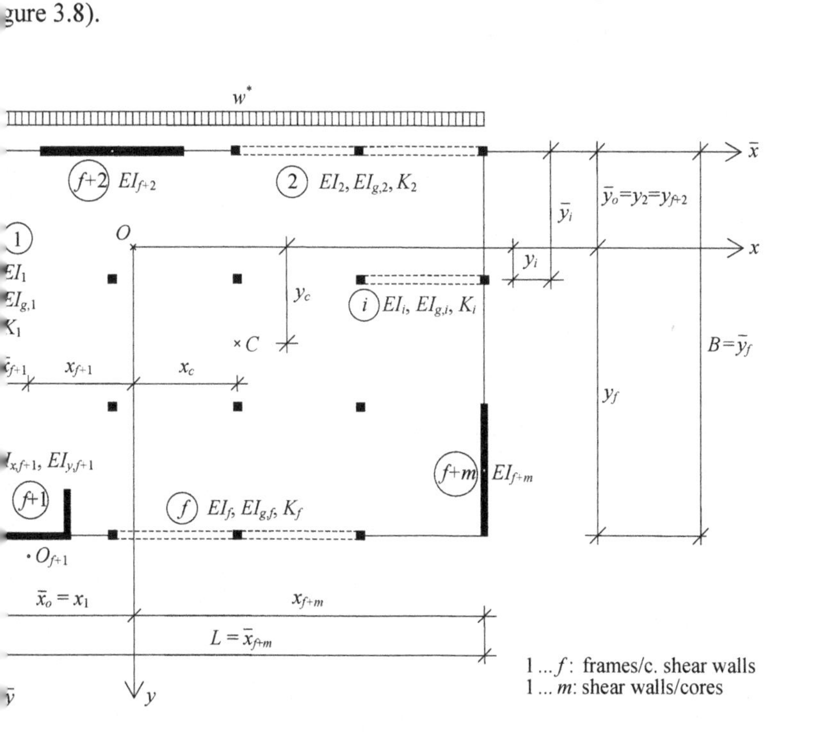

Figure 3.8 Plan arrangement of the bracing system for the torsional analysis.

easily seen that when the units of the bracing system only have bending and no or negligible shear stiffness) as is the case with shear walls/cores

$$_o = \frac{\sum_{i=1}^{f+m} I_{x,i}\bar{x}_i}{\sum_{i=1}^{f+m} I_{x,i}} \qquad \text{and} \qquad \bar{y}_o = \frac{\sum_{i=1}^{f+m} I_{y,i}\bar{y}_i}{\sum_{i=1}^{f+m} I_{y,i}} \tag{3.47}$$

In such situations the task of establishing the location of the shear centre ifies considerably.

Torsional analysis

e solution of the lateral deflection problem is available (in Section 3.2), v's analogy can be used to present the solution for the torsional problem. rding to Vlasov's analogy, the replacement of the stiffnesses in the lateral tion solution by the corresponding torsional stiffnesses directly leads to the on of the bracing system.

The three characteristic stiffnesses with the lateral deflection analysis are the and global bending stiffnesses and the shear stiffness. The corresponding nal stiffnesses are established as follows.

Stiffness EI is the local bending stiffness with the deflection analysis. The sponding stiffness with the torsional analysis is the local warping stiffness:

$$I_\omega = EIt^2 \tag{3.48}$$

t is the perpendicular distance of the bracing unit in question from the shear and $I_\omega = It^2$ is the local warping constant. For (coupled) shear walls and s, distance t represents distance x *or* distance y, and when the bracing unit in ion is a core, t stands for distance x *and* distance y (as the cores normally have ng stiffnesses in both directions x and y—see Figure 3.8).

Stiffness EI_g is the global bending stiffness with the deflection analysis. The sponding stiffness with the torsional analysis is the global warping stiffness:

$$I_{g\omega} = EI_g t^2 \tag{3.49}$$

$I_{g\omega} = I_g t^2$ is the global warping constant.

The total warping stiffness is the sum of the local and global warping esses:

$$I_{f\omega} = EI_\omega + EI_{g\omega} = E(I + I_g)t^2 = EI_f t^2 \tag{3.50}$$

$I_{f\omega} = (I + I_g)t^2$ is the total warping constant.

Stiffness K is the shear stiffness with the deflection analysis. The sponding stiffness with the torsional analysis is the Saint-Venant stiffness:

Saint-Venant stiffness is also called pure torsional stiffness.

ddition to stiffnesses (3.48), (3.49) and (3.51), the individual bracing pecially the cores—may have their "own" warping and Saint-Venant but they are normally small and, in accordance with the basic n made in Section 3.3, they are neglected here.

he stiffnesses and other geometrical characteristics are now available for ation of the analogy. Looking at the derivations of the lateral (sway) n Section 3.2, two possibilities may be considered. With the sway two procedures were developed: "The simple method" and "The more nethod". As the terms suggest, "The simple method" offers a simple vhile "The more accurate method" results in a more accurate albeit more ed solution. Careful investigation of the two procedures automatically e question "Which procedure to adopt?". The more accurate method was by incorporating the stiffnesses of the shear walls and cores into the his approach made it possible to take into consideration the effect of more accurately than with the other method, leading to a more accurate With the torsional problem, however, another phenomenon enters the e location of the shear centre plays a very important role and becomes geometrical characteristics. When the stiffnesses of the shear walls and incorporated into the frames, the procedure indeed leads to a more andling of the effect of interaction between the shear and bending modes same time, somewhat distorts the behaviour, as far as the location of the re of the bracing system is concerned. This follows from the fact that by some of the bracing units from their original places and creating new, rames (by incorporating the shear walls and cores into the frames), the f the shear centre of the bracing system is altered. This would be ble with the torsional problem. (As the location of the shear centre was with the planar problem, it was possible to make use of the advantage of accurate method" without its detrimental effect.)

llows that "The simple method" should be applied and the analysis must the original system of f frames and m shear walls when the analogy is e solution of the torsional problem. In doing so, and using the torsional introduced by Equations (3.48), (3.49), (3.50) and (3.51), and the differential equation (3.18) of the planar problem, the governing l equation of the torsional problem is obtained as

$$) - \kappa_i^2 \varphi_i''(z) = \frac{q_{\omega,i} m_z}{EI_{\omega,i}} \left(\frac{a_i z^2}{2} - 1 \right) \qquad (3.52)$$

script i refers to the ith bracing unit and $i = 1, 2, \ldots f + m$. Auxiliary κ_i and a_i are identical to those used in the planar problem, i.e., they are Equation (3.19). This follows from the fact that when the step-by-step nts of the analogy are met and the corresponding stiffnesses are matched, ms t_i drop out of the formulae.

e general case when the horizontal load has an angle with coordinate e torsional moment (per unit height) on the bracing system is

, the value of the torsional moment is

$$_z = wx_c = w\left(\frac{L}{2} - \bar{x}_o\right) \tag{3.53}$$

Torsional apportioner $q_{\omega,i}$ in Equation (3.52) plays the same role as q_i in the r case. Its value is obtained using the global torsional stiffnesses of the ng units. The global torsional stiffness of the *i*th bracing unit is defined as

$$_{\omega,i} = S_i t_i^2 = \frac{t_i^2}{d_i(H)} \tag{3.54}$$

e t_i is the perpendicular distance of the *i*th bracing unit from the shear centre $_i(H)$ is its maximum deflection. The total torsional stiffness of the system is

$$_\omega = \sum_{i=1}^{f+m} S_{\omega,i} \tag{3.55}$$

It should be noted that although the summation goes from 1 to $f+m$, nation (3.55) may contain more than $f+m$ items. This follows from the fact cores normally have bending stiffnesses in both principal directions and also may have two valid coordinates in the coordinate system whose origin is the centre and therefore one core generally appears twice in the summation as and $S_{\omega y,i}$. As a consequence, the number of torsional apportioners given v normally exceeds the number of bracing units $f+m$ when the bracing system ores.

The torsional apportioner can now be defined as

$$_{\omega,i} = \frac{S_{\omega,i}}{S_\omega} \tag{3.56}$$

The torsional moment share on the *i*th bracing unit is

$$_{z,i} = q_{\omega,i} m_z \tag{3.57}$$

Finally, in completing the application of the analogy, the formula for the on emerges as

$$_i(z) = \frac{m_{z,i}}{EI_{f\omega,i}}\left(\frac{H^3 z}{6} - \frac{z^4}{24}\right) + \frac{m_{z,i} z^2}{2(GJ)_i s_i^2} -$$

$$\frac{m_{z,i} EI_{\omega,i}}{} \left(\cosh[\kappa_i (H - z)] + \kappa_i H \sinh(\kappa_i z) \right. \quad 1 \Big) \qquad (3.58)$$

where auxiliary parameters κ_i and s_i are defined by Equations (3.19) and (3.20).

Bearing in mind that the above rotation calculated using the characteristics of the ith bracing unit is identical to the rotation of the building and that maximum rotation develops at $z = H$, the formula for the maximum rotation emerges as

$$\varphi_{\max} = \varphi(H) = \frac{m_{z,i} H^4}{8EI_{f\omega,i}} + \frac{m_{z,i} H^2}{2(GJ)_i s_i^2} - \frac{m_{z,i} EI_{\omega,i}}{(GJ)_i^2 s_i^3} \left(\frac{1 + \kappa_i H \sinh(\kappa_i H)}{\cosh(\kappa_i H)} - 1 \right) \quad (3.59)$$

Naturally, Equation (3.59) is identical to Equation (3.23) in structure.

The torsional mode is characterized by the warping and Saint-Venant torsional modes [first and second terms in Equations (3.58) and (3.59)]. The interaction between the two modes—third term—always reduces the rotation of the building. The interaction is therefore always beneficial.

Equation (3.59) [together with Equations (3.48), (3.49) and (3.51)] offers guidance about creating effective bracing system arrangements by reducing the rotation of the building. The rotation of the building can be reduced by increasing the bending and shear stiffnesses of the bracing units and, perhaps more importantly, by increasing the effective (perpendicular) distances of the bracing units from the shear centre. However, the most efficient way of reducing rotations (preferably to zero) is to minimise the external torque. It can be done by reducing the perpendicular distance of the line of action of the external load (preferably to zero) from the shear centre. Optimum is achieved by eliminating the arm of the horizontal load altogether, in other words, by creating a bracing system where the resultant of the horizontal load passes through the shear centre.

3.3.3 Discussion and special cases

The evaluation of Equations (3.58) and (3.59) and the results of a comprehensive accuracy analysis (in Section 7.3.2) lead to the following useful observations regarding the torsional behaviour of multi-storey buildings under horizontal load.

(a) The torsional behaviour and the rotation shape of the building can be characterized by three distinctive parts. The first part, warping torsion (Figure 3.9/a), is defined by the first term in Equations (3.58) and (3.59). It is associated with the bending stiffness of the bracing units that is activated through the torsion of the system around the shear centre of the bracing system. The second part, identified as Saint-Venant torsion (Figure 3.9/b), is given by the second term in Equations (3.58) and (3.59). The Saint-Venant torsion is linked to the shear stiffness of the bracing units that is again activated through the torsion of the system around the shear centre of the bracing system. The two rotation shapes are different in nature but the floor slabs force the bracing units to develop the same shape (Figure 3.9/d). It follows that interaction occurs between the two phenomena (Figure 3.9/c). The third term in Equations (3.58) and (3.59) reflects this interaction. (Figure 3.9 shows the rotation shape of a forty-storey building with a mixture of different types of bracing units. Note that z is measured from ground floor level upwards.)

(b) The interaction between the warping torsion and Saint-Venant torsion is always beneficial, as it reduces the rotation of the building.

(c) The effect of interaction is becoming smaller as the height of the building increases.

(d) With a building of given height, the effect of interaction can be considered constant over the height (Figure 3.9/c).

Finally, some special cases are investigated in the following. In these cases, the formula for the maximum rotation [Equation (3.59)] is further simplified.

Case A: Bracing systems whose horizontal elements (including the connecting beams in the frames and the floor slabs) have negligibly small bending stiffness.

This case is characterized by $K_b \to 0$ (for the frames). Consequently, the shear stiffness of the system tends to zero ($K \to 0$), which leads to $a \to 0$, $b \to 0$ and $\kappa \to 0$. Governing differential equation (3.52) simplifies to

$$\varphi_i''''(z) = -\frac{m_z}{EI_{\omega,i}}$$

As the system only has local warping stiffness, the stiffnesses belonging to the individual bracing units can be added up, further simplifying the situation. The solutions for the rotation and top rotation assume the form

$$\varphi(z) = \frac{m_z}{EI_\omega}\left(\frac{H^3 z}{6} - \frac{z^4}{24}\right) \tag{3.60}$$

and

$$\varphi_{\max} = \varphi(H) = \frac{m_z H^4}{8EI_\omega} \tag{3.61}$$

where EI_ω is the local warping stiffness. The rotation shape that is associated with this case is shown in Figure 3.9/a.

The use of Equations (3.60) and (3.61) should be considered when the shear stiffness of the bracing units is very small and/or when the bracing system is dominated by shear walls/cores. It should be noted that in this case m_z is the *total* torsional moment and I_ω represents the *sum* of the warping stiffnesses linked to the shear walls and cores with some contribution from the columns of the frames (which tends to be negligible).

Case B: Bracing systems comprising multi-bay, low-rise frames.
These structures tend to develop predominantly Saint-Venant-type rotation and the effect of the warping stiffness becomes insignificant.

This case is characterised by $a \to 0$ and $b \to \infty$ and governing differential equation (3.52) cannot be used directly. However, after some rearrangement, the original derivation leads to

$$\varphi_i''(z) = \frac{m_{z,i}}{(GJ)_i}$$

This differential equation, together with the boundary conditions $\varphi(0) = 0$ and $\varphi'(0) = 0$, lead to the rotation and the top rotation as

$$\varphi(z) = \frac{m_{z,i} z^2}{2(GJ)_i} \tag{3.62}$$

and

$$\varphi_{\max} = \varphi(H) = \frac{m_{z,i} H^2}{2(GJ)_i} \tag{3.63}$$

The characteristic rotation shape is shown in Figure 3.9/b. It is certainly worth considering the use of Equations (3.62) and (3.63) when the building is relatively low and the bracing system only consists of (mainly multi-bay) frames.

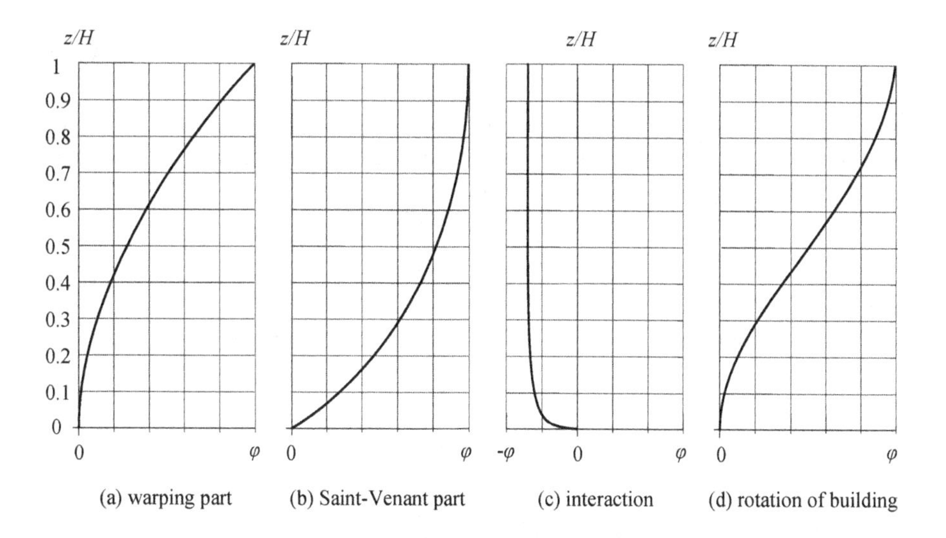

Figure 3.9 Typical rotation shapes of a forty-storey building with mixed bracing units.

Case C: The structure is relatively slender (with great height/width ratio).
The structure develops predominantly warping-type rotation. The second and third terms in Equations (3.58) and (3.59) tend to be by orders of magnitude smaller than the first term and the solutions for the rotation and the top rotation effectively become

$$\varphi(z)=\frac{m_{z,i}}{EI_{f\omega,i}}\left(\frac{H^3z}{6}-\frac{z^4}{24}\right) \tag{3.64}$$

and

$$\varphi_{\max}=\varphi(H)=\frac{m_{z,i}H^4}{8EI_{f\omega,i}} \tag{3.65}$$

This case is illustrated in Figure 3.9/a. It is interesting to note that both Case A and Case C are characterised by warping-type rotation.

3.4 MAXIMUM DEFLECTION

Everything is now available for the determination of the deflection of a multi-storey building. It is normally the maximum deflection that is of interest. Maximum deflection develops at the top and at a corner point of the plan of the building. As was shown in Section 3.1, in the general case the maximum deflection consists of two parts. Assuming horizontal load in direction y, the maximum deflection is calculated from

$$v_{\max}=v_o(H)+x_{\max}\varphi(H)=v_o(H)+(L-\bar{x}_o)\varphi(H) \tag{3.66}$$

where H is the height of the building, $v_o(H)$ is the top deflection of the shear centre axis (i.e. the solution of the planar problem), $x_{\max}$ marks the location of the corner point, $\varphi(H)$ is the maximum rotation, L is the size of the building perpendicular to the wind direction and $\bar{x}_o$ locates the shear centre (Figure 3.10/b).

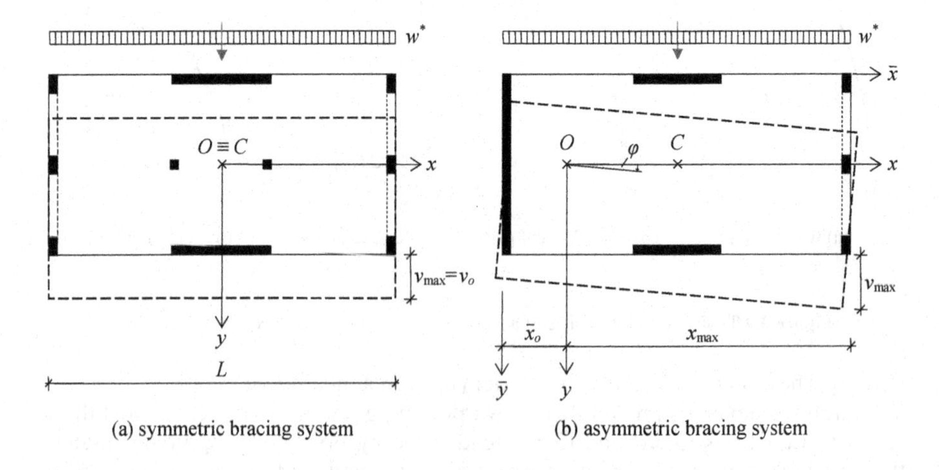

Figure 3.10 Typical bracing system arrangements.

In the special case when the bracing system is symmetric (with respect to the direction of the horizontal load) and the resultant of the horizontal load passes through the shear centre of the bracing system (Figure 3.10/a), the situation is relatively simple. The building develops a uniform deflection and Equation (3.66) simplifies to its first term, i.e. the maximum deflection is $v_{max} = v_o(H)$.

In the general case, however, the resultant of the horizontal load does not pass through the shear centre of the bracing system. In practice, this is the "asymmetric case" when the bracing system is asymmetric with respect to the line of the resultant of the horizontal load (Figure 3.10/b). In addition to the uniform deflection, rotation around the shear centre also occurs resulting in additional deflection. To determine the maximum deflection in this case, Equation (3.66) is used where the value of distance x_{max} is also needed. The value of the maximum rotation is obtained using Equation (3.59).

Maximum deflections are restricted in multi-storey buildings in order to ensure the comfort of the occupants and to avoid damage to structural and non-structural elements and mechanical systems. In different regions, different regulations and recommendations are available. The commonly accepted range for maximum deflection is from $0.0016H$ to $0.0035H$, depending on the height of the building and the magnitude of the wind pressure. According to the recommendation of the Committee on Wind Bracing of the American Society of Civil Engineers, the bracing system of the building is considered adequate if the condition

$$d_{max} \le d_{ASCE} = \frac{H}{500} \tag{3.67}$$

is fulfilled, where H is the height of the building (Schueller, 1977).

3.5 LOAD DISTRIBUTION AMONG THE BRACING UNITS

Multi-storey buildings subjected to horizontal load respond to the load by utilizing their bracing system. The external load is first transmitted from the façade to the floor slabs and through the floor slabs to the units of the bracing system before it is passed on to the foundation. Once the arrangement of the bracing system and the size of the bracing units are established, the main task of the structural engineer is to determine the distribution of the external load among the bracing units. In other words, the structural engineer has to establish the load share on each bracing unit in order to be able to finalize details of the bracing system.

The amount of load a bracing unit takes of the total load depends on several factors: the stiffness of the bracing unit, the stiffness of the whole bracing system, the location of the bracing unit and the plan arrangement of the whole bracing system. The situation is made fairly complicated as the term "stiffness" may now cover different types of stiffness such as local bending stiffness, global bending stiffness, shear stiffness, Saint-Venant stiffness and warping stiffness. This complexity was first faced in Section 3.2 (when the planar behaviour of the bracing system was analysed). Regarding lateral stiffness, the definition used in Section 3.2.1 will be used here. According to this definition, the in-plane stiffness of the *i*th bracing unit equals the reciprocal of the top deflection of the unit in the plane in question, i.e.:

$$S_{x,i} = \frac{1}{u_i(H)} \qquad \text{and} \qquad S_{y,i} = \frac{1}{v_i(H)} \tag{3.68}$$

where $u_i(H)$ and $v_i(H)$ are the top deflections of the bracing unit in directions x and y.

Stiffnesses $S_{x,i}$ and $S_{y,i}$ now automatically take into consideration the combined effect of the local bending, global bending and shear stiffnesses of the bracing unit.

The torsional stiffness of the ith bracing unit is introduced in Section 3.3.2 (where the torsional behaviour of the bracing system is investigated). In line with Equations (3.13) and (3.54), the torsional stiffness of the ith bracing unit is expressed as

$$S_{\omega,i} = S_i t_i^2 = \frac{x_i^2}{v_i(H)} + \frac{y_i^2}{u_i(H)} \tag{3.69}$$

where x_i and y_i are the coordinates of the bracing unit in the coordinate system whose origin lies in the shear centre and $v_i(H)$ and $u_i(H)$ are the top deflections of the bracing unit in directions y and x.

Knowing the above stiffnesses, the load shares on the ith bracing unit are calculated from

$$w_{x,i} = \frac{S_{x,i}}{S_x} w_x - \frac{S_{x,i} y_i}{S_\omega} m_z \tag{3.70}$$

$$w_{y,i} = \frac{S_{y,i}}{S_y} w_y + \frac{S_{y,i} x_i}{S_\omega} m_z \tag{3.71}$$

In the above equations, the denominators refer to the sum of the stiffnesses:

$$S_x = \sum_{i=1}^{f+m} S_{x,i}, \qquad S_y = \sum_{i=1}^{f+m} S_{y,i} \qquad \text{and} \qquad S_\omega = \sum_{i=1}^{f+m} S_{\omega,i} \tag{3.72}$$

and m_z is the torsional moment around the shear centre axis:

$$m_z = w_x y_c + w_y x_c \tag{3.73}$$

Distances x_c and y_c mark the position of the centroid (Figure 3.8). It should be emphasized for practical application that w_x, w_y, x_i, y_i and m_z are signed quantities in Equations (3.70), (3.71) and (3.73). Clockwise torsional moment m_z is positive.

Knowing $w_{x,i}$ and $w_{y,i}$, the maximum shear force and bending moment on the ith bracing unit at base level can be determined:

$$T_{x,i}^{\max} = w_{x,i} H \quad \text{and} \quad T_{y,i}^{\max} = w_{y,i} H \tag{3.74}$$

and

$$M_{x,i}^{\max} = \frac{w_{x,i} H^2}{2} \quad \text{and} \quad M_{y,i}^{\max} = \frac{w_{y,i} H^2}{2} \tag{3.75}$$

Those bracing units that have torsional stiffness of their own—the cores—also take torsional moments from the external load. However, in most practical cases when the building has an efficient and well-balanced bracing system, this torsional moment share on a core is small and can normally be ignored.

3.6 THE BEHAVIOUR OF BUILDINGS UNDER HORIZONTAL LOAD

The derivations and formulae presented in this chapter make it possible to identify the most important characteristics of the bracing system and to give recommendations for the creation of efficient bracing systems.

The most important characteristics are:

- the location and orientation of the bracing units
- the eccentricity of the external load: x_c, y_c
- local bending stiffness: EI_x, EI_y
- global bending stiffness: EI_g {with frames/coupled shear walls}
- shear stiffness: K {with frames/coupled shear walls}
- torsional stiffness: $EI_{f\omega}$ {of great importance}
  (GJ) {normally of limited importance}

The deflection of the building can be reduced and the efficiency of the bracing system can be increased in several ways, e.g., by

- reducing the eccentricity of the horizontal load (thus reducing torsion)
- increasing warping stiffness $EI_{f\omega}$ by placing the bracing units as far from the shear centre as possible in such a way that their *effective* (i.e. perpendicular) distance is maximum (Figure 3.11) (thus increasing torsional resistance)
- increasing local bending stiffness (EI_x, EI_y) in accordance with the magnitude of the horizontal load (w_y, w_x)
- increasing global bending stiffness (EI_g) {with frames/coupled shear walls}
- increasing shear stiffness (K) {with frames/coupled shear walls}
- making use of the beneficial effect of interaction between bending and shear deformation, in other words, using a mixture of shear walls and frames. It should be noted, however, that the effect of interaction tends to be negligible for tall buildings

Finally, some observations can be made regarding the load shares on the bracing units:

- the bracing units primarily take their load share according to their stiffness {first term of Equations (3.70) and (3.71)}
- the bracing units may take additional load share due to the rotation of the

bracing system {second term of Equations (3.70) and (3.71)}

- the apportionment of the lateral load among the bracing units does not depend on the distribution of the horizontal load over the height of the building

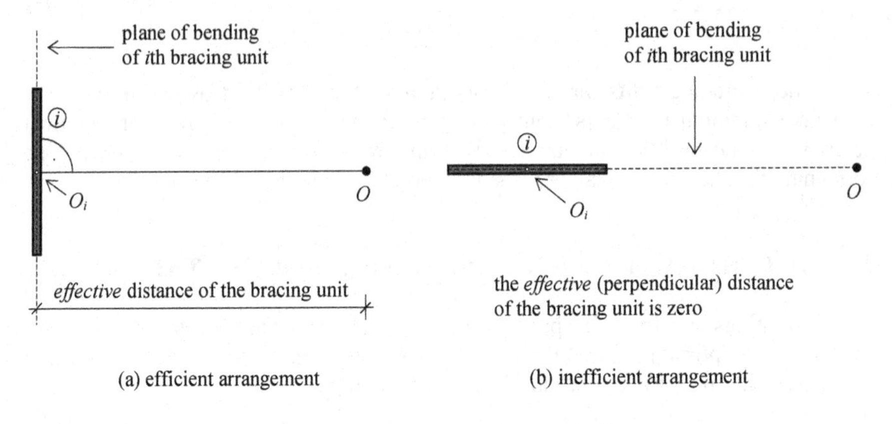

Figure 3.11 The placement of the bracing unit in relation to shear centre *O* of the bracing system.

The single most important observation regarding the behaviour of the bracing system is the importance of torsion or, more precisely, the importance of reducing torsion as much as possible in order to create a bracing system with the least deflection and the most balanced load distribution, i.e. the most economic bracing system. The ideal situation is a doubly symmetric bracing system when the centroid of the plan of the building and the shear centre of the bracing system coincide, i.e. when $x_c = 0$ and $y_c = 0$. As it will be seen in Chapters 4 and 5, the doubly symmetric arrangement also represents the optimum situation regarding the stability and the fundamental frequency of the building. It is not always possible—or feasible—to achieve a doubly symmetric arrangement but any reduction in the eccentricity of the horizontal load may go a long way in creating a more efficient bracing system.

3.7 WORKED EXAMPLES

When the formulae for the maximum deflection were derived in the preceding sections, the presentation followed an order that was most suitable for, and in line with, the theoretical considerations. For practical applications, however, it is advisable to follow a different order to simplify and minimize the amount of calculation. This is shown here using two worked examples. The calculations are based on the material presented in Chapters 2 and 3, and the numbers of the equations used will be given on the right-hand side in curly brackets. The first example presents a symmetric building, while the second example deals with a building with an asymmetric bracing system arrangement.

3.7.1 Maximum deflection of twenty-eight storey symmetric building

The maximum deflection of the twenty-eight storey building whose layout is shown in Figure 3.12 is calculated in this section. Both Method "A" (Section 3.2.2) and Method "B" (Section 3.2.3) are used. The building is subjected to a uniformly distributed horizontal load of intensity $w^* = 1$ kN/m^2 in direction y.

The building has a symmetric bracing system that consists of four frames and two cores. The storey-height is $h = 3$ m and the total height of the building is $H = 28 \cdot 3 = 84$ m. The modulus of elasticity is $E = 25 \cdot 10^6$ kN/m^2. Because of symmetry, it is possible to deal with half of the system (with half of the external load: $w = w^* L/2 = 15$ kN/m).

Table 3.1 Cross-sectional characteristics for frames F5 and F7.

Bracing unit	Cross-section of columns [m]	Cross-section of beams [m]	$I_{c,i}$ [m^4]	$I_{b,i}$ [m^4]	$I_{g,i}$ [m^4]
1: F5	0.4×0.7	0.4×0.4	0.0343	$0.0042\dot{6}$	20.16
2: F7	0.4×0.4	0.4×0.4	0.0064	$0.0042\dot{6}$	11.52

The cross-sectional characteristics of the frames are given in Table 3.1. The relevant second moment of area of the U-core (with respect to axis x) is

$$I_w = \frac{1}{12}\left(4.15 \cdot 4.3^3 - 3.85 \cdot 3.7^3\right) = 11.245 \text{ m}^4$$

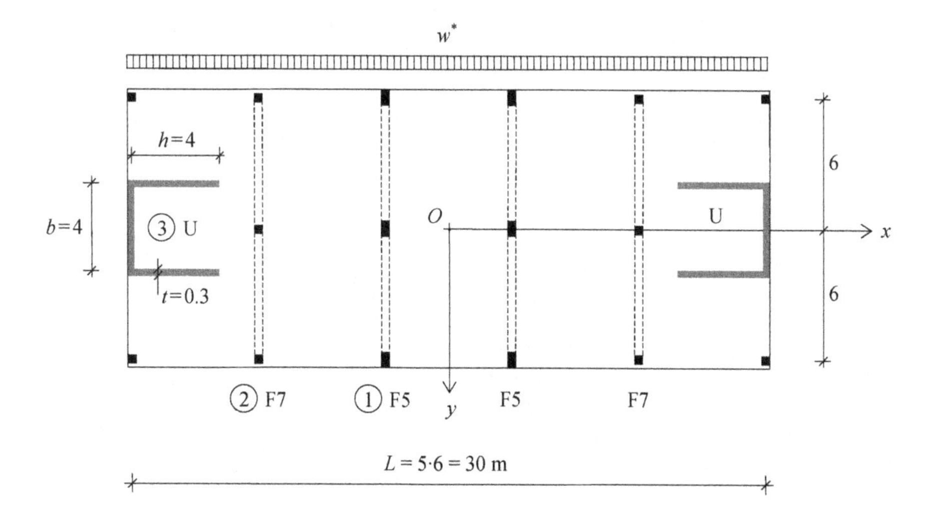

Figure 3.12 Layout of twenty-eight storey symmetric building.

Method "A": The simple method

The calculation is best carried out in two steps:
(1) The basic stiffness characteristics, the maximum deflection and the "overall" stiffness for each bracing unit are calculated (EI, EI_g, K, y_{max}, S)
(2) The maximum deflection of the building is determined using any of the bracing units [Equation (3.23) or Equation (3.25)] with its apportioner

(1) The basic characteristics for each bracing unit.
Bracing Unit 1: frame F5

The part of the shear stiffness which is associated with the beams is

$$K_{b,1} = \sum_{j=1}^{n-1} \frac{12EI_{b,j}}{l_j h} = 2\frac{12 \cdot 25 \cdot 10^6 \cdot 0.4^4}{12 \cdot 6 \cdot 3} = 71111\,\text{kN} \qquad \{2.28\}$$

The part of the shear stiffness which is associated with the columns is

$$K_{c,1} = \sum_{j=1}^{n} \frac{12EI_{c,j}}{h^2} = 3\frac{12 \cdot 25 \cdot 10^6 \cdot 0.4 \cdot 0.7^3}{12 \cdot 3^2} = 1143333\,\text{kN} \qquad \{2.29\}$$

The above two parts define reduction factor r as

$$r_1 = \frac{K_{c,1}}{K_{b,1} + K_{c,1}} = \frac{1143333}{71111 + 1143333} = 0.9414 \qquad \{2.30\}$$

The shear stiffness of the frame can now be determined:

$$K_1 = K_{b,1}\frac{K_{c,1}}{K_{b,1} + K_{c,1}} = 71111\frac{1143333}{71111 + 1143333} = 66947\,\text{kN} \qquad \{2.27\}$$

For the local bending stiffness ($EI = EI_c r$), the sum of the second moments of area of the columns should be produced (and multiplied by reduction factor r). As the columns of the frame are identical, the second moment of area of one column is simply multiplied by n and r:

$$I_1 = r_1 \sum_{j=1}^{n} I_{c,j} = 0.9414 \cdot 3 \cdot \frac{0.4 \cdot 0.7^3}{12} = 0.03229\,\text{m}^4 \qquad \{2.31\}$$

The global second moment of area is

$$I_{g,1} = \sum_{j=1}^{n} A_{c,j} t_j^2 = 0.4 \cdot 0.7 \cdot 6^2 \cdot 2 = 20.16\,\text{m}^4 \qquad \{2.32\}$$

The total second moment of area for the bending stiffness is

$$I_{f,1} = I_1 + I_{g,1} = I_{c,1}r_1 + I_{g,1} = 0.03229 + 20.16 = 20.1923\ \mathrm{m}^4 \qquad \{2.23\}$$

Parameters s, κ and κH are also needed for the calculation of the maximum deflection:

$$s_1 = 1 + \frac{I_{c,1}r_1}{I_{g,1}} = 1 + \frac{0.03229}{20.16} = 1.0016 \qquad \{2.14\}$$

$$\kappa_1 = \sqrt{\frac{K_1 s_1}{EI_1}} = \sqrt{\frac{66947 \cdot 1.0016}{25 \cdot 10^6 \cdot 0.03229}} = 0.2882 \frac{1}{\mathrm{m}} \quad \text{and} \quad \kappa_1 H = 24.2 \qquad \{2.14\}$$

With the above auxiliary quantities, the maximum top deflection of the frame can now be calculated (with $q_1 = 1$):

$$y_1 = \frac{15 \cdot 84^4}{8 \cdot 25 \cdot 10^6 \cdot 20.1923} + \frac{15 \cdot 84^2}{2 \cdot 66947 \cdot 1.0016^2} - \qquad \{3.23\}$$

$$-\frac{15 \cdot 25 \cdot 10^6 \cdot 0.03229}{66947^2 \cdot 1.0016^3}\left(\frac{1 + 24.2 \sinh 24.2}{\cosh 24.2} - 1\right) = 0.185 + 0.788 - 0.063 = 0.910\ \mathrm{m}$$

The overall stiffness of Bracing Unit 1 is

$$S_1 = \frac{1}{y_1} = \frac{1}{0.91} = 1.10 \frac{1}{\mathrm{m}} \qquad \{3.13\}$$

Bracing Unit 2: frame F7

A similar calculation leads to the overall stiffness of the frame.

The part of the shear stiffness which is associated with the beams is

$$K_{b,2} = \sum_{j=1}^{n-1} \frac{12EI_{b,j}}{l_j h} = 2\frac{12 \cdot 25 \cdot 10^6 \cdot 0.4 \cdot 0.4^3}{12 \cdot 6 \cdot 3} = 71111\ \mathrm{kN} \qquad \{2.28\}$$

The part of the shear stiffness which is associated with the columns is

$$K_{c,2} = \sum_{j=1}^{n} \frac{12EI_{c,j}}{h^2} = 3\frac{12 \cdot 25 \cdot 10^6 \cdot 0.4^4}{12 \cdot 3^2} = 213333\ \mathrm{kN} \qquad \{2.29\}$$

The above two part stiffnesses define reduction factor r as

$$r_2 = \frac{K_{c,2}}{K_{b,2} + K_{c,2}} = \frac{213333}{71111 + 213333} = 0.75 \qquad \{2.30\}$$

The shear stiffness of the frame can now be determined:

$$K_2 = K_{b,2} r_2 = K_{b,2} \frac{K_{c,2}}{K_{b,2} + K_{c,2}} = 71111 \cdot 0.75 = 53333 \text{ kN} \qquad \{2.27\}$$

For the local bending stiffness ($EI = EI_c r$), the sum of the second moments of area of the columns should be produced (and multiplied by reduction factor r). As the columns of the frame are identical, the second moment of area of one column is simply multiplied by n and r:

$$I_2 = r_2 \sum_{j=1}^{n} I_{c,j} = 0.75 \cdot 3 \cdot \frac{0.4^4}{12} = 0.0048 \text{ m}^4 \qquad \{2.31\}$$

The global second moment of area is

$$I_{g,2} = \sum_{j=1}^{n} A_{c,j} t_j^2 = 0.4 \cdot 0.4 (6^2 + 6^2) = 11.52 \text{ m}^4 \qquad \{2.32\}$$

The total second moment of area for the bending stiffness is

$$I_{f,2} = I_2 + I_{g,2} = I_{c,2} r_2 + I_{g,2} = 11.5248 \text{ m}^4 \qquad \{2.23\}$$

Parameters s, κ and κH are also needed for the calculation of the maximum deflection:

$$s_2 = 1 + \frac{I_{c,2} r_2}{I_{g,2}} = 1 + \frac{0.0048}{11.52} = 1.000417 \qquad \{2.14\}$$

$$\kappa_2 = \sqrt{\frac{K_2 s_2}{EI_2}} = \sqrt{\frac{53333 \cdot 1.000417}{25 \cdot 10^6 \cdot 0.0048}} = 0.6668 \frac{1}{\text{m}} \quad \text{and} \quad \kappa_2 H = 56 \qquad \{2.14\}$$

With the above auxiliary quantities, the maximum top deflection of the frame can now be calculated (with $q_2 = 1$):

$$y_2 = \frac{15 \cdot 84^4}{8 \cdot 25 \cdot 10^6 \cdot 11.5248} + \frac{15 \cdot 84^2}{2 \cdot 53333 \cdot 1.000417^2} - \qquad \{3.23\}$$

$$- \frac{15 \cdot 25 \cdot 10^6 \cdot 0.0048}{53333^2 \cdot 1.000417^3} \left(\frac{1 + 56 \sinh 56}{\cosh 56} - 1 \right) = 0.324 + 0.991 - 0.035 = 1.28 \text{ m}$$

The overall stiffness of Bracing Unit 2 is

$$S_2 = \frac{1}{y_2} = \frac{1}{1.28} = 0.78 \frac{1}{\text{m}} \qquad \{3.13\}$$

Bracing Unit 3: U-core
The maximum deflection of the core (with $q_3 = 1$) is

$$y_3 = \frac{q_3 w H^4}{8EI_3} = \frac{15 \cdot 84^4}{8 \cdot 25 \cdot 10^6 \cdot 11.245} = 0.332 \text{ m} \qquad \{3.25\}$$

and the stiffness of the core is

$$S_3 = \frac{1}{y_3} = \frac{1}{0.332} = 3.01 \frac{1}{\text{m}} \qquad \{3.13\}$$

(2) The maximum deflection of the building.
The core is used for the calculation of the maximum deflection. Its apportioner is

$$q_3 = \frac{S_3}{\sum_{i=1}^{f+m} S_i} = \frac{3.01}{1.1 + 0.78 + 3.01} = 0.615 \qquad \{3.12\}$$

The maximum deflection of the building is

$$y_{\max} = y_3 = \frac{q_3 w H^4}{8EI_3} = \frac{0.615 \cdot 15 \cdot 84^4}{8 \cdot 25 \cdot 10^6 \cdot 11.245} = 0.204 \text{ m} \qquad \{3.25\}$$

Naturally, the same value is obtained using the other two bracing units (F5 or F7). The Finite Element based solution, obtained using Axis VM X5 (2019), is $y = 0.1844$ m.

Method "B": The more accurate method

The procedure for the more accurate solution can be organized into three steps:
(1) The basic stiffness characteristics, the maximum deflection, the "overall" stiffness and the apportioner for each bracing unit are calculated (EI, EI_g, K, $y_{\max}$, S, $\overline{q}$)
(2) Using apportioners $\overline{q}$, the bending stiffness of each frame is amended ($EI \rightarrow EI^*$). All characteristics that are affected are re-calculated for each frame
(3) Using one of the new apportioners (q^*), the maximum deflection of the building is determined using any of the frames [Equation (3.45)]

(1) The basic characteristics for each frame.
This task has already been completed above with *Method "A"*, and the results will

be used below.

(2) New bending stiffness and new characteristics for the frames.

*Frame F5**

According to Equation (3.39), a portion of the second moment of area of the shear wall that is proportional to the overall stiffness of frame F5 is added to the original second moment of area of the frame. The apportioner is given by Equation (3.30). The amended local second moment of area is

$$I_1^* = I_1 + \bar{q}_1 I_w = 0.0343 \cdot 0.9414 + \frac{1.1}{1.1+0.78} 11.245 = 6.612\,\text{m}^4 \qquad \{3.39\}$$

Because of this change, three other parameters have to be amended, according to Equations (3.41) and (3.42):

$$s_1^* = 1 + \frac{I_1^*}{I_{g,1}} = 1 + \frac{6.612}{20.16} = 1.328 \qquad \{3.42\}$$

$$\kappa_1^* = \sqrt{\frac{K_1 s_1^*}{EI_1^*}} = \sqrt{\frac{66947 \cdot 1.328}{25 \cdot 10^6 \cdot 6.612}} = 0.02319 \qquad \kappa_1^* H = 1.948 \qquad \{3.41\}$$

The sum of the local and global second moments of area is

$$I_{f,1}^* = I_1^* + I_{g,1} = 6.61 + 20.16 = 26.77\,\text{m}^4 \qquad \{3.44\}$$

Using the maximum deflection (with $q_1^* = 1$)

$$y_1^*(H) = \frac{15 \cdot 84^4}{8 \cdot 25 \cdot 10^6 \cdot 26.77} + \frac{15 \cdot 84^2}{2 \cdot 66947 \cdot 1.328^2} - \qquad \{3.45\}$$

$$- \frac{15 \cdot 25 \cdot 10^6 \cdot 6.612}{66947^2 \cdot 1.328^3} \left(\frac{1 + 1.948 \sinh 1.948}{\cosh 1.948} - 1 \right) = 0.316\,\text{m}$$

the overall stiffness of F5* is

$$S_1^* = \frac{1}{y_1^*(H)} = \frac{1}{0.316} = 3.164 \,\frac{1}{\text{m}} \qquad \{3.36\}$$

*Frame F7**

The procedure for the other frame is the same. Its amended local second moment of area is

$$I_2^* = I_2 + \bar{q}_2 I_w = 0.0048 + \frac{0.78}{1.1+0.78} 11.245 = 4.67\,\text{m}^4 \qquad \{3.39\}$$

Because of this change, three other parameters have to be amended, according to Equations (3.41) and (3.42):

$$s_2^* = 1 + \frac{I_2^*}{I_{g,2}} = 1 + \frac{4.67}{11.52} = 1.405 \qquad \{3.42\}$$

$$\kappa_2^* = \sqrt{\frac{K_2 s_2^*}{EI_2^*}} = \sqrt{\frac{53333 \cdot 1.405}{25 \cdot 10^6 \cdot 4.67}} = 0.0253\,, \qquad \kappa_2^* H = 2.128 \qquad \{3.41\}$$

The sum of the local and global second moments of area is

$$I_{f,2}^* = I_2^* + I_{g,2} = 4.67 + 11.52 = 16.19\,\text{m}^4 \qquad \{3.44\}$$

Using the maximum deflection (with $q_2^* = 1$)

$$y_2^*(H) = \frac{15 \cdot 84^4}{8 \cdot 25 \cdot 10^6 \cdot 16.19} + \frac{15 \cdot 84^2}{2 \cdot 53333 \cdot 1.405^2} - \qquad \{3.45\}$$

$$- \frac{15 \cdot 25 \cdot 10^6 \cdot 4.67}{53333^2 \cdot 1.405^3} \left(\frac{1 + 2.128 \sinh 2.128}{\cosh 2.128} - 1 \right) = 0.444\,\text{m}$$

the overall stiffness of frame F7* is

$$S_2^* = \frac{1}{y_2^*(H)} = \frac{1}{0.444} = 2.252\,\frac{1}{\text{m}} \qquad \{3.36\}$$

(3) The maximum deflection of the building.
Using one of the new apportioners

$$q_1^* = \frac{S_1^*}{\sum_{i=1}^{f} S_i^*} = \frac{3.164}{3.164 + 2.252} = 0.584\,, \qquad q_2^* = \frac{2.252}{3.164 + 2.252} = 0.416 \qquad \{3.35\}$$

the maximum deflection of the building can now be calculated. The maximum deflection of the two frames has already been calculated under a horizontal load of $w = 15$ kN/m. According to Equation (3.45), the same calculation—but with the real load share of the frame—gives the maximum deflection of the building. Using frame F5*, this is

$$y_{\max} = q_1^* y_1^*(H) = 0.584 \cdot 0.316 = 0.1845\ \text{m} \qquad \{3.45\}$$

This value is practically identical with the "exact" (computer-based) FE solution (0.1844 m). Naturally, using the other frame (F7*) with its load share (q_2^*) leads to the same value.

3.7.2 Maximum deflection of twenty-eight storey asymmetric building

For one reason or another, the bracing systems of multi-storey buildings are often not symmetrical and the rotation of the building also has to be taken into account when the maximum deflection of the building is calculated.

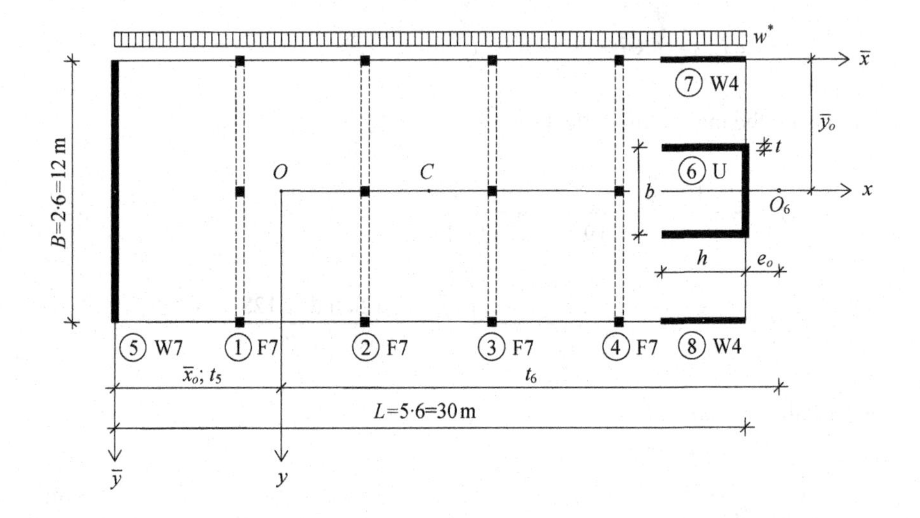

Figure 3.13 Layout of twenty-eight storey asymmetric building.

The deflection analysis of such asymmetric structures can be carried out as described in Sections 3.3 and 3.4, but a more practical order should be followed. To simplify and minimize the amount of calculation, it is advised that the procedure is carried out in the following four steps.

(1) The basic stiffness characteristics, the maximum deflection, the "overall" stiffness (and, if needed, the apportioner) for each bracing unit are calculated (EI, EI_g, K, $v_{\max}$, S, q).
(2) The maximum deflection of the shear centre axis is determined after incorporating the stiffnesses of the shear walls and cores into the frames in the relevant direction {Equation (3.45)}.
(3) Having determined the location of the shear centre and then the torsional

stiffnesses of the bracing system, the maximum rotation of the system is determined {Equation (3.59)}.
(4) The maximum deflection of the building is obtained by adding up the two components of the deflection {Equation (3.66)}.

This procedure is demonstrated below using a twenty-eight storey building whose layout is shown in Figure 3.13. The building is subjected to a uniformly distributed horizontal load of intensity $w^* = 1$ kN/m^2 in direction y. The bracing system consists of four frames, three shear walls and a U-core. The storey-height is $h = 3$ m and the total height of the building is $H = 28 \cdot 3 = 84$ m. The modulus of elasticity is $E = 25 \cdot 10^6$ kN/m^2. The cross-sectional characteristics of the bracing units are given in Table 3.2. The stiffness of the shear walls perpendicular to their plane is ignored.

Table 3.2 Cross-sectional characteristics of the bracing units.

Bracing unit	Cross-section of columns [m]	Cross-section of beams [m]	$I_{c,i}$ [m^4]	$I_{b,i}$ [m^4]	$I_{g,i}$ [m^4]
F7	0.4×0.4	0.4×0.4	0.0064	$0.0042\dot{6}$	11.52
W4	0.2×4.0	–	$1.06\dot{6}$	–	–
W7	0.3×12.0	–	43.2	–	–
U	$h = b = 4.0,\ t = 0.3,\ e_o = 1.714$		$I_x = 11.245$	–	–

PART 1: The basic characteristics of the bracing units.

Frame F7 (Bracing Units 1, 2, 3 and 4)
The part of the shear stiffness which is associated with the beams is

$$K_{b,1} = \sum_{j=1}^{n-1} \frac{12EI_{b,j}}{l_j h} = 2\frac{12 \cdot 25 \cdot 10^6 \cdot 0.4 \cdot 0.4^3}{12 \cdot 6 \cdot 3} = 71111 \text{ kN} \qquad \{2.28\}$$

The part of the shear stiffness which is associated with the columns is

$$K_{c,1} = \sum_{j=1}^{n} \frac{12EI_{c,j}}{h^2} = 3\frac{12 \cdot 25 \cdot 10^6 \cdot 0.4^4}{12 \cdot 3^2} = 213333 \text{ kN} \qquad \{2.29\}$$

The above two part stiffnesses define reduction factor r as

$$r_1 = \frac{K_{c,1}}{K_{b,1} + K_{c,1}} = \frac{213333}{71111 + 213333} = 0.75 \qquad \{2.30\}$$

The shear stiffness of the frame can now be determined:

$$K_1 = K_{b,1}r_1 = K_{b,1}\frac{K_{c,1}}{K_{b,1}+K_{c,1}} = 71111\cdot 0.75 = 53333 \text{ kN} \qquad \{2.27\}$$

For the local bending stiffness ($EI = EI_c r$), the sum of the second moments of area of the columns should be produced (and multiplied by reduction factor r). As the columns of the frame are identical, the second moment of area of one column is simply multiplied by n and r:

$$I_1 = r_1\sum_{j=1}^{n} I_{c,j} = 0.75\cdot 3\cdot\frac{0.4^4}{12} = 0.0048 \text{ m}^4 \qquad \{2.31\}$$

The global second moment of area is

$$I_{g,1} = \sum_{j=1}^{n} A_{c,j}t_j^2 = 0.4\cdot 0.4(6^2+6^2) = 11.52 \text{ m}^4 \qquad \{2.32\}$$

The total second moment of area for the bending stiffness is

$$I_{f,1} = I_1 + I_{g,1} = I_{c,1}r_1 + I_{g,1} = 11.5248 \text{ m}^4 \qquad \{2.23\}$$

Parameters s, κ and κH are also needed for the calculation of the maximum deflection:

$$s_1 = 1+\frac{I_{c,1}r_1}{I_{g,1}} = 1+\frac{0.0048}{11.52} = 1.000417 \qquad \{2.14\}$$

$$\kappa_1 = \sqrt{\frac{K_1 s_1}{EI_1}} = \sqrt{\frac{53333\cdot 1.000417}{25\cdot 10^6\cdot 0.0048}} = 0.6668\frac{1}{\text{m}} \qquad \text{and} \qquad \kappa_1 H = 56$$

With the above auxiliary quantities, the maximum top deflection of the frame can now be calculated (with $q_1 = 1$):

$$v_1 = \frac{30\cdot 84^4}{8\cdot 25\cdot 10^6\cdot 11.5248} + \frac{30\cdot 84^2}{2\cdot 53333\cdot 1.000417^2} - \qquad \{3.23\}$$

$$-\frac{30\cdot 25\cdot 10^6\cdot 0.0048}{53333^2\cdot 1.000417^3}\left(\frac{1+56\sinh 56}{\cosh 56}-1\right) = 0.648+1.983-0.070 = 2.561 \text{ m}$$

The overall stiffness of the frame is:

$$S_1 = \frac{1}{v_1} = \frac{1}{2.561} = 0.39 \frac{1}{\text{m}} \qquad \{3.13\}$$

As $v_2 = v_3 = v_4 = v_1 = 2.561$ m holds,

$$S_2 = S_3 = S_4 = 0.39 \frac{1}{\text{m}}$$

Shear wall W7 (Bracing Unit 5)
The maximum (in-plane) deflection of shear wall W7 (with $q_5 = 1$) is

$$v_5 = \frac{q_5 w H^4}{8EI_5} = \frac{30 \cdot 84^4}{8 \cdot 25 \cdot 10^6 \cdot 43.2} = 0.1729 \text{ m} \qquad \{3.25\}$$

and its stiffness is

$$S_5 = \frac{1}{v_5} = \frac{1}{0.1729} = 5.784 \frac{1}{\text{m}} \qquad \{3.13\}$$

U-core (Bracing Unit 6) {Only I_{6x} and deflection in plane zy are relevant}
Using the maximum deflection (with $q_6 = 1$)

$$v_6 = \frac{q_6 w H^4}{8EI_{6x}} = \frac{30 \cdot 84^4}{8 \cdot 25 \cdot 10^6 \cdot 11.245} = 0.664 \text{ m} \qquad \{3.25\}$$

the stiffness of the core is

$$S_6 = \frac{1}{v_6} = \frac{1}{0.664} = 1.506 \frac{1}{\text{m}} \qquad \{3.13\}$$

Shear wall W4 (Bracing Units 7 and 8)
The maximum (in-plane) deflection of shear wall W4 (with $q_7 = 1$) is

$$v_7 = v_8 = \frac{q_7 w H^4}{8EI_7} = \frac{30 \cdot 84^4}{8 \cdot 25 \cdot 10^6 \cdot 1.0\dot{6}} = 7.0 \text{ m} \qquad \{3.25\}$$

and its stiffness is

$$S_7 = S_8 = \frac{1}{v_7} = \frac{1}{7.0} = 0.143 \frac{1}{\text{m}} \qquad \{3.13\}$$

PART 2: The maximum deflection of the shear centre axis.
The participating bracing units are the four frames (F7: 1, 2, 3, 4), shear wall W7 (5) and the U-core (6). There is no need for the calculation of load shares as the four frames are identical. It is sufficient to consider one frame only which takes one fourth of the external load. It is also sufficient to consider one frame (when the shear wall and core are incorporated into the frames) which takes one fourth of the bending stiffnesses of the shear wall and the core. The local bending stiffness of this new frame, marked $F7^*$, is

$$I_1^* = I_1 + \frac{1}{4}(I_5 + I_{6x}) = 0.0048 + \frac{1}{4}(43.2 + 11.245) = 13.62\,\mathrm{m}^4 \qquad \{3.39\}$$

The global second moment of area and the shear stiffness are unchanged at

$$I_{g,1} = 11.52\,\mathrm{m}^4 \qquad \text{and} \qquad K_1 = 53333\ \mathrm{kN}$$

and the total second moment of area is

$$I_{f,1}^* = I_1^* + I_{g,1} = 13.62 + 11.52 = 25.14\ \mathrm{m}^4 \qquad \{3.44\}$$

With the new auxiliary quantities

$$s_1^* = 1 + \frac{I_1^*}{I_{f,1}} = 1 + \frac{13.62}{11.52} = 2.18 \qquad \{3.42\}$$

$$\kappa_1^* = \sqrt{\frac{K_1 s_1^*}{EI_1^*}} = \sqrt{\frac{53333 \cdot 2.18}{25 \cdot 10^6 \cdot 13.62}} = 0.0185 \qquad \text{and} \qquad \kappa_1^* H = 1.552 \qquad \{3.41\}$$

the maximum deflection of the shear centre axis (with $q_1^* = 1/4$) is

$$v_o = \frac{0.25 \cdot 30 \cdot 84^4}{8 \cdot 25 \cdot 10^6 \cdot 25.14} + \frac{0.25 \cdot 30 \cdot 84^2}{2 \cdot 53333 \cdot 2.18^2} - \qquad \{3.45\}$$

$$- \frac{0.25 \cdot 30 \cdot 25 \cdot 10^6 \cdot 13.62}{53333^2 \cdot 2.18^3}\left(\frac{1 + 1.552 \sinh 1.552}{\cosh 1.552} - 1\right) = 0.107\ \mathrm{m}$$

PART 3: Maximum rotation around the shear centre.
The participating bracing units are the four frames (F7), the three shear walls (W7, W4 and W4) and the U-core. The U-core is only "active" in plane yz (with I_{6x}) as the length of its other (perpendicular) moment arm is zero ($t_{6y} = 0$).

The location of the shear centre and then the torsional stiffnesses of the bracing units are needed first. Because of symmetry, only one of the two coordinates needs calculation. The shear centre coordinates are

$$\bar{x}_o = \frac{\sum_{i=1}^{f+m} S_{y,i}\bar{x}_i}{\sum_{i=1}^{f+m} S_{y,i}} = \frac{S_1(6+12+18+24)+S_6(L+e_o)}{S_5+4S_1+S_6} \qquad \{3.46\}$$

$$= \frac{0.39\cdot 60+1.506\cdot 31.714}{5.784+4\cdot 0.39+1.506} = 8.04\text{ m} \qquad \text{and} \qquad \bar{y}_o = 6\text{ m}$$

The torsional moment causing rotation around the shear centre axis is

$$m_z = wx_c = w\left(\frac{L}{2}-\bar{x}_o\right) = 30(15-8.04) = 208.8\text{ kNm/m} \qquad \{3.53\}$$

A new x-y coordinate system is now established whose O origin is in the shear centre (Figure 3.13). The locations of the bracing units in this coordinate system are:

$$t_1 = \bar{x}_o - 6 = 2.04\text{ m},\; t_2 = 12-\bar{x}_o = 3.96\text{ m},\; t_3 = 18-\bar{x}_o = 9.96\text{ m},\; t_5 = \bar{x}_o = 8.04\text{ m}$$

$$t_4 = 24-\bar{x}_o = 15.96\text{ m}, \qquad t_6 = L+e_o-\bar{x}_o = 23.674\text{ m}, \qquad t_7 = t_8 = \bar{y}_o = 6.0\text{ m}$$

The "governing" torsional stiffnesses of the bracing units are obtained using their perpendicular distance from the shear centre and their maximum deflection:

$$S_{\omega,1} = \frac{t_1^2}{v_1} = \frac{2.04^2}{2.561} = 1.625\text{ m}, \qquad S_{\omega,2} = \frac{t_2^2}{v_2} = \frac{3.96^2}{2.561} = 6.123\text{ m}$$

$$S_{\omega,3} = \frac{t_3^2}{v_3} = \frac{9.96^2}{2.561} = 38.73\text{ m}, \qquad S_{\omega,4} = \frac{t_4^2}{v_4} = \frac{15.96^2}{2.561} = 99.46\text{ m}$$

$$S_{\omega,5} = \frac{t_5^2}{v_5} = \frac{8.04^2}{0.1729} = 373.87\text{ m}, \qquad S_{\omega,6} = \frac{t_6^2}{v_6} = \frac{23.674^2}{0.6641} = 843.94\text{ m}$$

$$S_{\omega,7} = S_{\omega,8} = \frac{t_7^2}{v_7} = \frac{6.0^2}{7.0} = 5.143\text{ m} \qquad \{3.54\}$$

With the sum of the torsional stiffnesses

$$S_{\omega} = \sum_{i=1}^{f+m} S_{\omega,i} = 1374.0\,\text{m} \qquad \{3.55\}$$

the torsional apportioners can now be determined. Choosing, say, Bracing Unit 5, its torsional apportioner is

$$q_{\omega,5} = \frac{S_{\omega,5}}{S_{\omega}} = \frac{373.87}{1374.0} = 0.2721 \qquad \{3.56\}$$

The torsional moment share on Bracing Unit 5 is

$$m_{z,5} = q_{\omega,5} m_z = 0.2721 \cdot 208.8 = 56.8\,\text{kNm/m} \qquad \{3.57\}$$

Using Bracing Unit 5, the maximum rotation of the building can now be determined from Equation (3.59). As Bracing Unit 5 is a shear wall, the equation is reduced to its first term (and $I_{f\omega}$ is reduced to I_{ω}):

$$\varphi_{\max} = \varphi_5(H) = \frac{m_{z,5} H^4}{8EI_{f\omega,5}} = \frac{56.8 \cdot 84^4}{8 \cdot 25 \cdot 10^6 \cdot 43.2 \cdot 8.04^2} = 0.005063\,\text{rad} \qquad \{3.59\}$$

PART 4: Maximum deflection of the building.
Figure 3.13 indicates that maximum deflection develops at the right-hand side of the building where the two components of the deflection add up:

$$v_{\max} = v_o(H) + (L - \bar{x}_o)\varphi(H) = 0.107 + (30 - 8.04)0.005063 = 0.218\,\text{m} \qquad \{3.66\}$$

The Finite Element based computer program Axis VM X5 (2019) gives $v_{\max} = 0.205$ m as the maximum deflection of the building.

4

Frequency analysis of buildings

A great number of methods have been developed for the dynamic analysis of individual frames, coupled shear walls and shear walls. Fewer methods are available to deal with a *system* of these bracing units. The reason for having a much limited choice for systems is simple: interaction among the elements (beams/lintels and columns/walls) of a single frame or coupled shear walls is complex enough but then the bracing units interact with one another not only in planar behaviour but normally also in a three-dimensional fashion. This is why the available analytical methods make one or more simplifying assumptions regarding the characteristic stiffnesses of the bracing units or the geometry of the building.

Based on drift calculations and assuming a doubly symmetric structural arrangement, Goldberg (1973) presented several simple methods for the calculation of the fundamental frequency of (uncoupled) lateral vibration and pure torsional vibration. The effect of the axial deformation of the vertical elements was taken into account by a correction factor in his method. The continuous connection method enabled more rigorous analyses (Coull, 1975; Rosman, 1973 and 1981; Kollár, 1992). Using a single-storey torsional analogy, Glück *et al.* (1979) developed a matrix-based solution for buildings having uncoupled stiffness matrixes. A simple procedure with design tables was made available for asymmetrical buildings developing predominantly bending deformation (Zalka, 2000). Ng and Kuang (2000) presented a simple method for the triply coupled vibration of asymmetric wall-frame structures. However, their method is only applicable to buildings whose vertical bracing elements develop no or negligible axial deformation.

In taking into consideration all the characteristic stiffnesses of the bracing frames, shear walls and cores, as well as the interaction among the elements of the bracing structures and among the bracing units themselves, the aim of this chapter is to introduce a simple analytical method for the calculation of the natural frequencies of regular multi-storey buildings braced by a system of frames, (coupled) shear walls and cores (Zalka, 2001).

In addition to the general assumptions listed in Chapter 1, it is also assumed for the analysis that the mass of the building is uniformly distributed over the floors of the building.

The equivalent column approach shall be used for the analysis. The equivalent mass and stiffnesses shall be established first, taking into account deformations due to bending, shear, the lengthening and shortening of the vertical elements, and torsion. Closed-form solutions shall then be given for lateral, pure torsional and coupled vibration. Accuracy shall be investigated in Section 7.3.3.

The method is simple and accurate enough to be used both at the concept design stage and for final analysis. It can also be useful to verify the results of the computer-based FE method, where the time consuming procedure of handling all the data can always be a source of error.

A multi-storey building may develop lateral vibrations in the two principal directions and torsional vibration around its vertical shear centre axis. All the three corresponding frequencies have to be calculated before their coupling can be taken into account. The investigation here starts with the lateral vibration of the bracing system, which can be based on the vibration analysis of a single frame presented in Section 2.2.

4.1 LATERAL VIBRATION OF A SYSTEM OF FRAMES, (COUPLED) SHEAR WALLS AND CORES

Consider a *system* of frames and coupled shear walls ($i = 1...f$) and shear walls and cores ($k = 1...m$) shown in Figure 4.1/a. Based on the analysis of a single frame in Section 2.2.1, the whole bracing system can be characterised by the shear stiffness of the frames and coupled shear walls, the global bending stiffness of the frames and coupled shear walls and the local bending stiffness of the individual full-height columns/wall sections, shear walls and cores.

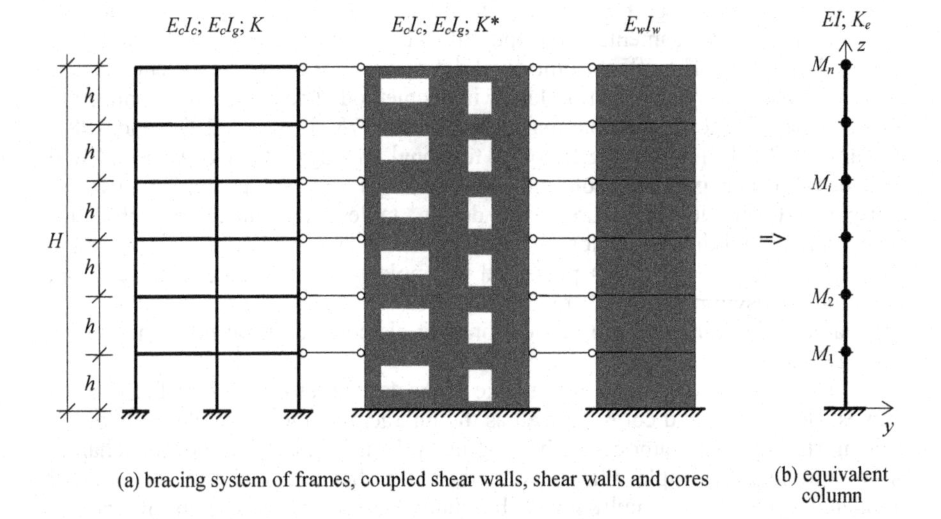

(a) bracing system of frames, coupled shear walls, shear walls and cores

(b) equivalent column

Figure 4.1 Model for the lateral vibration analysis.

By combining the individual bracing units, linked by the floor slabs, to form a single cantilever, an equivalent system can be established with shear stiffness K, global bending stiffness EI_g and local bending stiffness EI. The shear stiffness and the global bending stiffness are not independent of each other and can be incorporated into an effective shear stiffness, leading to a single equivalent column with effective shear stiffness K_e and bending stiffness EI (Figure 4.1/b). These stiffness characteristics shall be established below but first the non-uniform nature of the load of the system will be addressed. The procedure is similar to the one followed in the case of a single frame in Section 2.2.1.

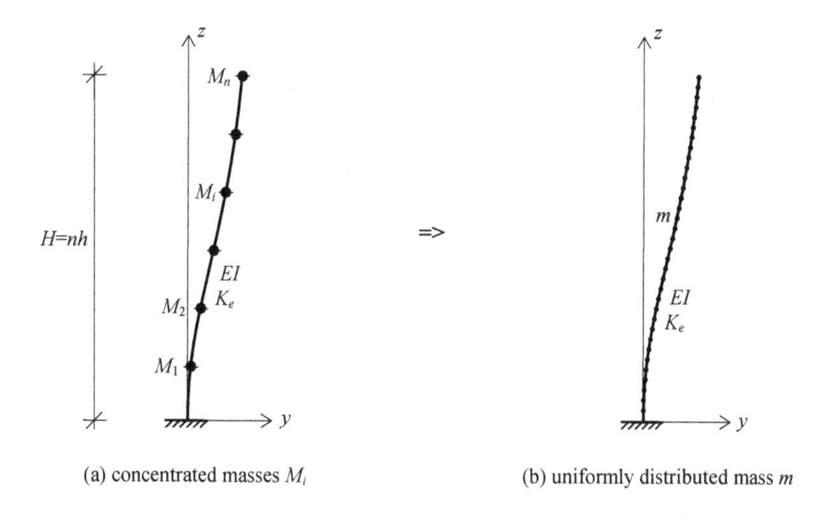

Figure 4.2 Load on the equivalent column.

The original load of the system—and of the equivalent column—is characterized by concentrated masses M_i [kg] as

$$M_i = \frac{QA}{g}$$

at floor levels (Figures 4.1/b and 4.2/a), where Q [kN/m^2] is the intensity of the uniformly distributed floor load, A is the plan area of the building and $g = 9.81$ m/s^2 is the gravity acceleration. To make the load of the equivalent column continuous, these concentrated masses are distributed over the height (Figure 4.2/b) leading to mass density per unit length of the column m [kg/m] as

$$m = \frac{M_i}{h} = \frac{QA}{gh} = \rho A \tag{4.1}$$

In the above equation, h is the storey height and ρ [kg/m^3] is the mass density per unit volume as

$$\rho = \frac{\gamma}{g}$$

with γ [kN/m^3] being the weight per unit volume of the building.

The shear stiffness of the system originates from the shear stiffnesses of the frames and coupled shear walls. Based on Equation (2.27), the "original" shear stiffness of the ith frame is

$$K_i = K_{b,i} r_i = K_{b,i} \frac{K_{c,i}}{K_{b,i} + K_{c,i}} \tag{4.2}$$

where the two contributors to the shear stiffness are

$$K_{b,i} = \sum_{j=1}^{n-1} \frac{12 E_b I_{b,j}}{l_j h} \tag{4.3}$$

and

$$K_{c,i} = \sum_{j=1}^{n} \frac{12 E_c I_{c,j}}{h^2} \tag{4.4}$$

In the above equations

E_c is the modulus of elasticity of the columns of the frames
E_b is the modulus of elasticity of the beams of the frames
$I_{c,j}$ is the second moment of area of the jth column of the ith frame
$I_{b,j}$ is the second moment of area of the jth beam of the ith frame
l_j is the jth bay of the ith frame
n is the number of columns of the ith frame

Factor r_i is introduced as a reduction factor:

$$r_i = \frac{K_{c,i}}{K_{b,i} + K_{c,i}} \tag{4.5}$$

The total "original" shear stiffness of f frames is

$$K = \sum_{i=1}^{f} K_i \tag{4.6}$$

[If coupled shear walls are also included in the system, their shear stiffness is determined using Equation (2.73).]

The square of the frequency of shear vibration associated with the "original" shear stiffness of the ith unit is

$$f_{s',i}^2 = \frac{1}{(4H)^2} \frac{r_f^2 K_i}{m} \tag{4.7}$$

Factor r_f is included in Equation (4.7). It is responsible for taking into account the fact that the mass of the original structures is concentrated at floor levels and is not uniformly distributed over the height (Figure 4.2) as was assumed for the model used for the original derivation. The situation is similar to that with individual frames. See detailed explanation in Section 2.2.1. Values for r_f are given in Figure 2.11 in Section 2.2.1 and in Table 4.1 below.

Table 4.1 Mass distribution factor r_f as a function of n (the number of storeys).

n	1	2	3	4	5	6	7	8	9	10	11
r_f	0.493	0.653	0.770	0.812	0.842	0.863	0.879	0.892	0.902	0.911	0.918
n	12	13	14	15	16	18	20	25	30	50	>50
r_f	0.924	0.929	0.934	0.938	0.941	0.947	0.952	0.961	0.967	0.980	$\sqrt{n/(n+2.06)}$

The full-height global bending vibration of the *i*th frame (or coupled shear walls) as a whole unit represents pure bending type vibration. The square of the fundamental frequency that is associated with this vibration is

$$f_{g,i}^2 = \frac{0.313 r_f^2 E_c I_{g,i}}{H^4 m} \tag{4.8}$$

where I_g is the global second moment of area of the cross-sections of the columns:

$$I_{g,i} = \sum_{j=1}^{n} A_{c,j} t_j^2 \tag{4.9}$$

with

$A_{c,j}$ the cross-sectional area of the *j*th column
t_j the distance of the *j*th column from the centroid of the cross-sections

According to the frequency analysis carried out in Section 2.2.1, there is an interaction between the shear and global bending modes that reduces the effectiveness of the shear stiffness. The factor of effectiveness for the *i*th unit can be calculated using the two relevant frequencies as

$$s_{f,i}^2 = \frac{f_{g,i}^2}{f_{g,i}^2 + f_{s',i}^2} \tag{4.10}$$

With this effectiveness factor, the effective shear stiffness for the whole system is obtained as

$$K_e = \sum_{i=1}^{f} s_{f,i}^2 K_i \tag{4.11}$$

Using the "original" and the effective shear stiffnesses, the effectiveness for the whole system is obtained as

$$s_f = \sqrt{\frac{K_e}{K}} \tag{4.12}$$

The actual lateral frequency of the system which is associated with shear deformation can now be determined using the effective shear stiffness:

$$f_s^2 = \frac{1}{(4H)^2}\frac{r_f^2 K_e}{m} \tag{4.13}$$

If higher frequencies are needed, factor 4 in Equation (4.13) should be replaced by 4/3 and 4/5, respectively, for the calculation of the second and third frequencies.

When the lateral frequency that is associated with the local bending deformation is to be determined, the bending stiffnesses of the columns of the frames and coupled shear walls, the shear walls and the cores have to be taken into consideration:

$$EI = E_c I_c + E_w I_w = E_c \sum_{i=1}^{f} I_{c,i} r_i + E_w \sum_{k=1}^{m} I_{w,k} \tag{4.14}$$

In Equation (4.14)

E_w is the modulus of elasticity of the shear walls/cores
$I_{c,i}$ is the sum of the second moments of area of the columns of the ith frame
r_i is the reduction factor for the ith frame [Equation (4.5)]
$I_{w,k}$ is the second moment of area of the kth shear wall/core

When the system has mixed bracing units—both frames and shear walls/cores—the contribution of the columns of the frames [first term on the right-hand side in Equation (4.14)] is normally very small compared to that of the shear walls/cores. In such cases this contribution can safely be ignored.

With the above bending stiffness, the lateral frequency of the system in bending is obtained from

$$f_b^2 = f_f^2 + f_w^2 = \frac{0.313 r_f^2}{H^4 m}\left(E_c \sum_{i=1}^{f} I_{c,i} r_i + E_w \sum_{k=1}^{m} I_{w,k}\right) = \frac{0.313 r_f^2 EI}{H^4 m} \tag{4.15}$$

with

f_f fundamental frequency of the frames/coupled shear walls
f_w fundamental frequency of the shear walls/cores

If higher frequencies are needed, factor 0.313 in Equation (4.15) should be replaced by 12.3 and 96.4, respectively, for the calculation of the second and third frequencies.

In Equations (4.14) and (4.15), the bending stiffness of the columns of the frames and coupled shear walls is adjusted by combination factor r_i. Theoretical investigations (Hegedűs and Kollár, 1999) demonstrate that this adjustment is necessary to prevent the over-representation of the second moments of area of the columns in the equivalent column where they are also represented in K_e [through K_c—c.f. Equations (4.4) and (4.11)].

The whole system can now be modelled by a single equivalent column with bending stiffness EI and effective shear stiffness K_e, subjected to uniformly distributed mass (Figure 4.2/b). The governing differential equation of this column is obtained by examining the equilibrium of its elementary section. This leads to

$$r_f^2 EIv''''(z,t) - r_f^2 K_e v''(z,t) + m\ddot{v}(z,t) = 0$$

where $v(z,t)$ denotes horizontal motion in direction y. Primes and dots mark differentiation by z and t (time). After seeking the solution in a product form, separating the variables and eliminating the time dependent functions, this equation results in the boundary value problem

$$r_f^2 EIv_1''''(z) - r_f^2 K_e v_1''(z) - \omega^2 m v_1(z) = 0 \tag{4.16}$$

where $v_1(z)$ only depends on place. In this equation

$$\omega = \frac{2\pi\eta}{H^2}\sqrt{\frac{EIr_f^2}{m}}$$

is the circular frequency and η is the frequency parameter, the eigenvalue of the problem.

The origin of the coordinate system is placed at the lower built-in end of the equivalent column (Figure 4.2/b) and the boundary conditions are as follows:

$$v_1(0) = 0, \qquad v_1'(0) = 0$$

and

$$v_1''(H) = 0, \qquad EIv_1'''(H) - K_e v_1'(H) = 0$$

Governing differential equation (4.16) is identical in form to Equation (2.49). As the corresponding boundary conditions are also identical, the solution to Equation (2.49) in Section 2.2.1 can be used, bearing in mind that the stiffnesses now refer to the whole system of bracing units (not to an individual frame).

Proceeding as in Section 2.2.1, and introducing the non-dimensional parameter

$$k = H\sqrt{\frac{K_e}{EI}} \tag{4.17}$$

the formula for the fundamental frequency in plane *y-z* (Figure 4.1) is obtained as

$$f_y = \sqrt{f_b^2 + f_s^2 + \left(\frac{\eta^2}{0.313} - \frac{k^2}{5} - 1\right) s_f f_b^2} \tag{4.18}$$

where f_b, f_s, k and s_f are calculated by taking into account the bracing units in the relevant direction, i.e. in direction *y*.

Values for frequency parameter η are given in Table 4.2 as a function of parameter k. Values of parameter η for the second and third frequencies are tabulated in (Zalka, 2000).

In replacing subscript *y* with *x* and considering the bracing units in direction *x*, Equation (4.18) can also be used for the calculation of the lateral frequency in direction *x*.

Table 4.2 Frequency parameters η and η_φ as a function of k and k_φ.

k or k_φ	η or η_φ	k or k_φ	η or η_φ	k or k_φ	η or η_φ	k or k_φ	η or η_φ	k or k_φ	η or η_φ
0.0	0.5596	1.4	0.7302	6.0	1.827	13.0	3.542	20	5.278
0.1	0.5606	1.5	0.7511	6.5	1.949	13.5	3.665	25	6.522
0.2	0.5638	1.6	0.7726	7.0	2.070	14.0	3.789	30	7.769
0.3	0.5690	1.7	0.7946	7.5	2.192	14.5	3.913	35	9.016
0.4	0.5761	1.8	0.8170	8.0	2.313	15.0	4.036	40	10.26
0.5	0.5851	1.9	0.8397	8.5	2.435	15.5	4.160	50	12.76
0.6	0.5959	2.0	0.8628	9.0	2.558	16.0	4.284	60	15.26
0.7	0.6084	2.5	0.9809	9.5	2.680	16.5	4.408	70	17.76
0.8	0.6223	3.0	1.1014	10.0	2.803	17.0	4.532	80	20.26
0.9	0.6376	3.5	1.2226	10.5	2.926	17.5	4.656	90	22.76
1.0	0.6542	4.0	1.3437	11.0	3.049	18.0	4.781	100	25.26
1.1	0.6718	4.5	1.4647	11.5	3.172	18.5	4.905	200	50.25
1.2	0.6905	5.0	1.5856	12.0	3.295	19.0	5.029	400	100.25
1.3	0.7100	5.5	1.7065	12.5	3.418	19.5	5.153	>400	$\frac{k}{4}$ or $\frac{k_\varphi}{4}$

The equivalent column for the lateral vibration analysis was created by simply adding up the effective shear stiffnesses and the local bending stiffnesses, respectively, of the individual bracing units (Figure 4.1). It is interesting to note that this simple approach only works with the effective shear stiffnesses and the local bending stiffnesses. When the system is characterized by the three original stiffnesses—local bending stiffness, global bending stiffness and shear stiffness—this simple procedure cannot be used. That is why a totally different procedure was applied for the rotation and deflection analysis in Chapter 2.

4.2 PURE TORSIONAL VIBRATION

Although the torsional vibration problem is more complex than that of lateral vibration, a solution may be obtained in a relatively simple way, due to an analogy between the three-dimensional torsional problem and the two-dimensional lateral vibration problem (discussed in the previous section). This analogy is well known in the stress analysis of thin-walled structures in bending and torsion (Vlasov, 1961; Kollbrunner and Basler, 1969). According to the analogy, translations, bending moments and shear forces correspond to rotations, warping moments and torsional moments, respectively. It will be demonstrated in the following that the analogy can be extended to the lateral vibration of an elastically supported cantilever (discussed in the previous section) and the pure torsional vibration of a cantilever of thin-walled cross-section (to be investigated in this section).

The model which is used for the pure torsional vibration analysis of the building is an equivalent cantilever of thin-walled, open cross-section which has effective Saint-Venant stiffness $(GJ)_e$ and warping stiffness EI_ω. This equivalent column replaces the bracing system of the building. The equivalent column is situated in the shear centre of the bracing system (Figure 4.3) and torsional vibration develops around the shear centre. The characteristic torsional stiffnesses can only be determined if the location of the shear centre is known so the first step is to determine the coordinates of the shear centre.

For bracing systems developing predominantly bending deformation, the location of the shear centre is calculated using the bending stiffness of the bracing elements [Equations (3.47)]. However, with bracing systems having frames and coupled shear walls as well, the shear deformation of some of the bracing units may be of considerable magnitude (in addition to their bending deformation). The behaviour of such systems is complex (and the location of the shear centre may even vary over the height). No exact solution is available for this case but, as a good approximation, the formulae given below can be used to determine the location of the shear centre.

As the lateral frequency of a bracing unit reflects both its bending and shear stiffnesses, the location of the shear centre is calculated using the lateral frequencies (f_x and f_y) of the bracing units:

$$\bar{x}_o = \frac{\sum_{i=1}^{f+m} f_{y,i}^2 \bar{x}_i}{\sum_{i=1}^{f+m} f_{y,i}^2}, \qquad \bar{y}_o = \frac{\sum_{i=1}^{f+m} f_{x,i}^2 \bar{y}_i}{\sum_{i=1}^{f+m} f_{x,i}^2} \tag{4.19}$$

where $\bar{x}_i$ and $\bar{y}_i$ are the perpendicular distances of the bracing units from axes $\bar{y}$ and $\bar{x}$ and f and m are the number of frames/coupled shear walls and shear walls/cores, respectively (Figure 4.3). Any suitable method can be used for the calculation of the lateral frequencies in Equation (4.19), including Equation (4.18) given in Section 4.1. The repeated application of the Southwell formula (Southwell, 1922) and the Föppl-Papkovich formula (Tarnai, 1999) offers another very simple alternative for the calculation of the fundamental lateral frequencies. According to this approach, a lower bound to the lateral frequency of a bracing unit is obtained from the summation formula

$$f = \sqrt{f_b^2 + \frac{f_g^2 f_{s'}^2}{f_g^2 + f_{s'}^2}} \tag{4.20}$$

where $f_{s'}$, f_g and f_b relate to the bracing unit. Their values are given by Equations (2.42), (2.43) and (2.47) in Section 2.2.1.

Knowing the location of the shear centre, the Saint-Venant and warping stiffnesses can be calculated in the coordinate system whose origin is in the shear centre (Figure 4.3).

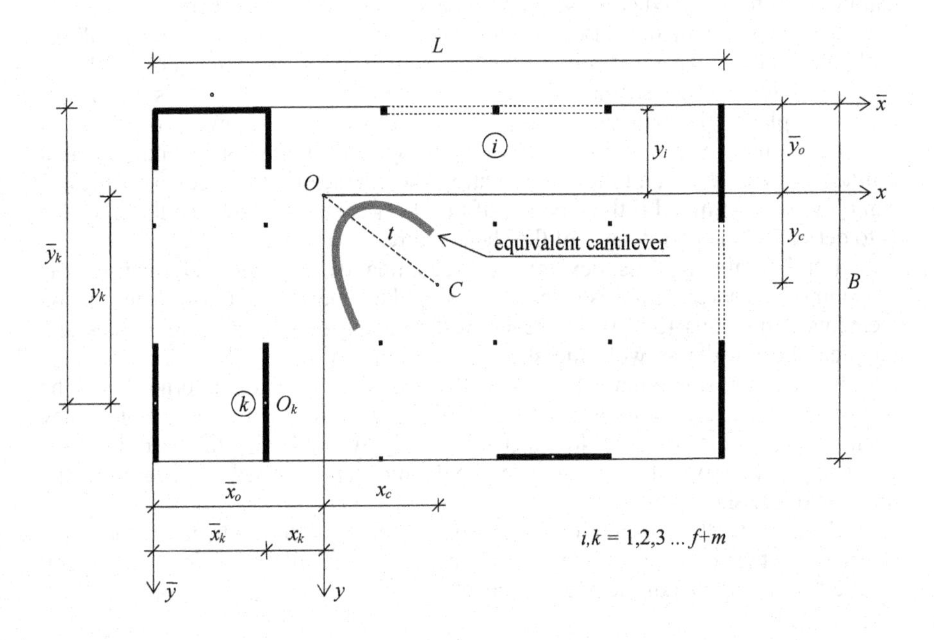

Figure 4.3 Layout with the equivalent column of open, thin-walled cross-section in the shear centre.

The effective Saint-Venant stiffness of the system may come from two sources: the Saint-Venant stiffness of the shear walls and cores and the effective shear stiffness of the frames as

$$(GJ)_e = \sum_{k=1}^{m} GJ_k + \sum_{i=1}^{f} \left((K_{e,i})_x y_i^2 + (K_{e,i})_y x_i^2\right) \tag{4.21}$$

In Equation (4.21)

J_k is the Saint-Venant constant of the kth wall/core
G is the modulus of elasticity in shear of the walls/cores
$(K_{e,i})_x$, $(K_{e,i})_y$ are the effective shear stiffnesses of ith frame/coupled shear walls in directions x and y, respectively
x_i, y_i are the perpendicular distances of the ith frame/coupled shear walls from the shear centre in directions x and y, respectively

If the bracing system consists of frames, (coupled) shear walls and cores of open cross-section, the first term in Equation (4.21) is normally small and negligible compared to the contribution of the frames.

The warping stiffness of the system may originate from three sources: the own warping stiffness of the cores, the bending stiffness of the walls and cores and the bending stiffness of the columns of the frames/wall sections of the coupled shear walls:

$$EI_\omega = E_w \sum_{k=1}^{m} \left(I_{\omega,k} + (I_{w,k})_x y_k^2 + (I_{w,k})_y x_k^2 \right) + \\ + E_c \sum_{i=1}^{f} \left((I_{c,i} r_i)_x y_i^2 + (I_{c,i} r_i)_y x_i^2 \right) \qquad (4.22)$$

with

$I_{\omega,k}$ the warping constant of the kth wall/core
$E_w(I_{w,k})_x$, $E_w(I_{w,k})_y$ the bending stiffnesses of the kth wall/core in directions x and y, respectively
$E_c(I_{c,i}r_i)_x$, $E_c(I_{c,i}r_i)_y$ the bending stiffnesses of the columns/wall sections of the ith frame/coupled shear walls in directions x and y, respectively
x_k, y_k the perpendicular distances of the kth wall/core from the shear centre in directions x and y, respectively
x_i, y_i the perpendicular distances of the ith frame/coupled shear walls from the shear centre in directions x and y, respectively

The warping stiffness of a well-balanced bracing system is normally dominated by the contribution of the shear walls and cores (if their perpendicular distance from the shear centre is great enough). The contribution of the cores through their own warping stiffness [first term in Equation (4.22)] tends to be much smaller and the effect of the columns of the frames (last two terms) is generally negligible.

To facilitate the easy calculation of the warping constant I_ω, closed-form formulae for cross-sections widely used for bracing cores are given in Tables 2.6, 2.7 and 2.8. More formulae are available in (Zalka, 2000). For bracing units of special (irregular) cross-sections where no closed-form solution exists, good computer procedures are available. One such computer program is PROSEC (1994), later Prokon, whose accuracy has been established and proved to be within the range required for structural engineering calculations.

The original load of the cantilever—concentrated masses M_i at floor levels—

is replaced by uniformly distributed mass m over the height (as shown in Figure 4.2). The governing differential equation of the cantilever is obtained by examining the equilibrium of its elementary section:

$$r_f^2 EI_\omega \varphi''''(z,t) - r_f^2 (GJ)_e \varphi''(z,t) + i_p^2 m\ddot{\varphi}(z,t) = 0$$

where primes and dots mark differentiation by z and t (time).

After seeking the solution in a product form, separating the variables and eliminating the time dependent functions, this equation results in the boundary value problem

$$\frac{r_f^2 EI_\omega}{i_p^2}\varphi_1''''(z) - \frac{r_f^2 (GJ)_e}{i_p^2}\varphi_1''(z) - \omega^2 m\varphi_1(z) = 0 \qquad (4.23)$$

where ω is the circular frequency, i_p is the radius of gyration and $\varphi_1(z)$ defines rotational motions.

As the derivation of Equation (4.23) demonstrates (Zalka, 1994), the radius of gyration is related to the distribution of the mass of the building. For regular multi-storey buildings of rectangular plan-shape and subjected to a uniformly distributed mass at floor levels, the radius of gyration is obtained from

$$i_p = \sqrt{\frac{L^2 + B^2}{12} + t^2} \qquad \text{with} \qquad t = \sqrt{x_c^2 + y_c^2} \qquad (4.24)$$

In the above equations, L and B are the plan length and breadth of the building, t is the distance between the geometrical centre of the plan of the building (C) and the shear centre of the bracing system (O) and x_c and y_c are the coordinates of the geometrical centre (Figure 4.3):

$$x_c = \frac{L}{2} - \bar{x}_o \qquad \text{and} \qquad y_c = \frac{B}{2} - \bar{y}_o \qquad (4.25)$$

For arbitrary plan-shapes and/or other types of mass distribution, formulae for the radius of gyration are available elsewhere (Kollár, 1999; Zalka, 2000). It is important to note that the value of i_p depends on the geometrical characteristics of the plan of the building, rather than the stiffness characteristics of the bracing system.

As the origin of the coordinate system is fixed at the bottom of the equivalent column (Figure 4.2/b), the boundary conditions are

$$\varphi_1(0) = 0, \qquad \varphi_1'(0) = 0$$

and

$$\varphi_1''(H) = 0, \qquad EI_\omega \varphi_1'''(H) - (GJ)_e \varphi_1'(H) = 0$$

The eigenvalue problem (4.23) is clearly analogous with the one defined by the governing differential equation of lateral vibration [Equation (4.16)] and its boundary conditions. Bending stiffness EI and the elastic support defined by the effective shear stiffness K_e in Equation (4.16) correspond to warping stiffness EI_ω and effective Saint-Venant stiffness $(GJ)_e$, divided by i_p^2 in Equation (4.23), respectively. It follows that Equation (4.16) can be used and converted to represent the solution of Equation (4.23). In doing so, the fundamental frequency for pure torsional vibration is obtained in the same manner as with Equation (4.18):

$$f_\varphi = \sqrt{f_\omega^2 + f_t^2 + \left(\frac{\eta_\varphi^2}{0.313} - \frac{k_\varphi^2}{5} - 1\right) s_\varphi f_\omega^2} \tag{4.26}$$

In the above equation, the pure torsional frequency associated with the warping stiffness is obtained from

$$f_\omega^2 = \frac{0.313 r_f^2 EI_\omega}{i_p^2 H^4 m} \tag{4.27}$$

and the formula for the pure torsional frequency associated with the Saint-Venant stiffness is

$$f_t^2 = \frac{r_f^2 (GJ)_e}{(4H)^2 i_p^2 m} \tag{4.28}$$

The effectiveness of the Saint-Venant stiffness is expressed by the factor

$$s_\varphi = \sqrt{\frac{(GJ)_e}{(GJ)}} \tag{4.29}$$

where the “original” Saint-Venant stiffness is

$$(GJ) = \sum_{k=1}^{m} GJ_k + \sum_{i=1}^{f} \left((K_i)_x y_i^2 + (K_i)_y x_i^2\right) \tag{4.30}$$

Values of frequency parameter η_φ are given in Table 4.2 as a function of torsion parameter k_φ:

$$k_\varphi = H\sqrt{\frac{(GJ)_e}{EI_\omega}} \tag{4.31}$$

Values for the second and third frequencies are given in (Zalka, 2000).

In the special case when the bracing system has no warping stiffness—e.g. the building is braced by a single core of closed or partially closed cross-section—Equation (4.26) simplifies to

$$f_\varphi = f_t = \frac{r_f}{4Hi_p}\sqrt{\frac{(GJ)_e}{m}} \tag{4.32}$$

4.3 COUPLED LATERAL-TORSIONAL VIBRATION

Multi-storey buildings tend to have an asymmetric bracing system arrangement when the shear centre of the bracing system and the centre of the mass of the building do not coincide. In such cases, interaction occurs among the two lateral and pure torsional modes. There are two possibilities to take into account the effect of interaction: "exactly" or approximately. The "exact" method automatically covers all the three coupling possibilities (triple, double and no coupling) with an error range of 0-2%. This method is given first.

When the basic frequencies f_x, f_y and f_φ are known, their coupling can be taken into account in a simple way by using the cubic equation

$$\left(f^2\right)^3 + a_2\left(f^2\right)^2 + a_1 f^2 - a_0 = 0 \tag{4.33}$$

whose smallest root yields the combined lateral-torsional frequency of the building. The coefficients in the above cubic equation are

$$a_0 = \frac{f_x^2 f_y^2 f_\varphi^2}{1-\tau_x^2-\tau_y^2}, \qquad a_1 = \frac{f_x^2 f_y^2 + f_\varphi^2 f_x^2 + f_\varphi^2 f_y^2}{1-\tau_x^2-\tau_y^2}$$

$$a_2 = \frac{f_x^2\tau_x^2 + f_y^2\tau_y^2 - f_x^2 - f_y^2 - f_\varphi^2}{1-\tau_x^2-\tau_y^2} \tag{4.34}$$

where τ_x and τ_y are eccentricity parameters:

$$\tau_x = \frac{x_c}{i_p} \quad \text{and} \quad \tau_y = \frac{y_c}{i_p} \tag{4.35}$$

Radius of gyration i_p and the coordinates of the geometrical centre x_c and y_c are given by Equations (4.24) and (4.25).

If a quick solution is needed or a cubic equation solver is not available or if one of the basic frequencies is much smaller than the others, the following approximate method based on the Föppl-Papkovich theorem (Tarnai, 1999) may be used.

For asymmetric bracing systems when the centroid of the mass of the building does not lie on either principal axis of the bracing system, triple coupling occurs and the resulting fundamental frequency is obtained using the reciprocal summation

$$\frac{1}{f^2} = \frac{1}{f_x^2} + \frac{1}{f_y^2} + \frac{1}{f_\varphi^2}$$

as

$$f = \left(\frac{1}{f_x^2} + \frac{1}{f_y^2} + \frac{1}{f_\varphi^2} \right)^{-\frac{1}{2}} \tag{4.36}$$

If the arrangement of the bracing system is monosymmetric and the centroid of the mass of the building lies on one of the principal axes of the bracing system (say, axis x), then two things may happen. Vibration may develop in direction x (defined by f_x) or vibration in direction y (f_y) couples with pure torsional vibration around axis z (f_φ). The frequency of this coupled vibration is obtained from

$$f_{y\varphi} = \left(\frac{1}{f_y^2} + \frac{1}{f_\varphi^2} \right)^{-\frac{1}{2}} \tag{4.37}$$

The fundamental frequency of the building is the smaller one of f_x and $f_{y\varphi}$, i.e.:

$$f = \min(f_x, f_{y\varphi}) \tag{4.38}$$

If the arrangement of the bracing system is doubly symmetric and the centroid of the mass of the building coincides with the shear centre of the bracing system, then no coupling occurs and the fundamental frequency of the building is the smallest one of f_x, f_y and f_φ, i.e.:

$$f = \min(f_x, f_y, f_\varphi) \tag{4.39}$$

The value of the coupled frequency of the building basically depends on two factors: the values of the basic frequencies (f_x, f_y and f_φ) and the eccentricities of the bracing system (τ_x and τ_y). The great disadvantage of using the summation equations for determining the coupled frequency may be that they totally ignore the eccentricity of the system. If the system has relatively small eccentricity, then the summation equations tend to result in very conservative estimates.

The natural frequencies of buildings are also affected by other factors, such as foundation flexibility, reduced stiffness due to cracking, damping, etc. The treatment of such "secondary" effects is outside the scope of this book; more detailed information is available elsewhere (Barkan, 1962; Rosman, 1973; Fintel, 1974; Ellis, 1986).

4.4 WORKED EXAMPLES

Two worked examples are presented in this section for the calculation of the fundamental frequency of buildings under uniformly distributed mass over the floors, braced by frames and shear walls. The calculations are based on the material presented earlier in Chapters 2 and 4, and the numbers of the equations used will be given on the right-hand side in curly brackets.

4.4.1 Fundamental frequency of twenty-five storey symmetric building

The twenty-five storey reinforced concrete building (Figure 4.4) is subjected to uniformly distributed mass over the floors.

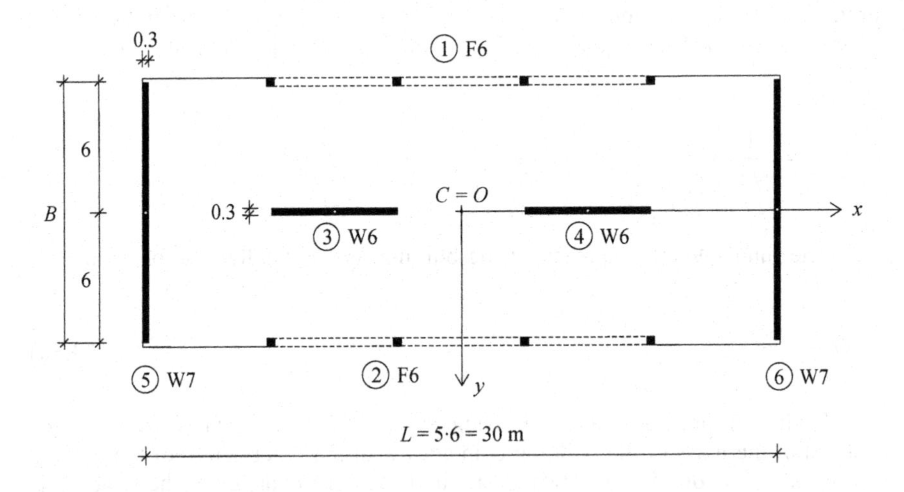

Figure 4.4 Layout of twenty-five storey doubly symmetric building.

The doubly symmetric bracing system consists of two frames (F6) and two shear walls (W6) in direction x and two shear walls (W7) in direction y (Figure 4.5). The modulus of elasticity is $E = 25000$ MN/m^2, the modulus of elasticity in shear is $G = 10400$ MN/m^2, the storey height is $h = 3$ m and the total height of the building is $H = 75$ m. The thickness of the shear walls is 0.30 m.

The weight per unit volume of the building is assumed to be $\gamma = 2.5$ kN/m^3. This leads to the mass density per unit length as

$$m = \rho A = \frac{\gamma}{g} LB = \frac{2.5}{9.81} 12 \cdot 30 = 91.74 \text{ kg/m} \qquad \{4.1\}$$

Mass distribution factor r_f is obtained from Table 4.1 as a function of the number of storeys:

$$r_f = 0.961 \qquad \{\text{Table 4.1}\}$$

Before the whole system of two frames and four shear walls is investigated, it is advantageous to establish the basic characteristics of the individual bracing units.

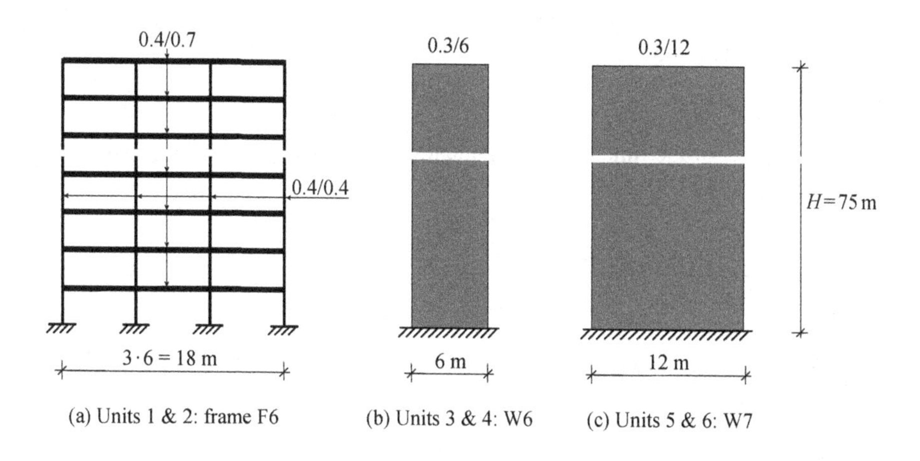

Figure 4.5 Bracing units for the twenty-five storey doubly symmetric building.

Individual bracing units

Mass density per unit length $m = 91.74$ kg/m will be used in connection with the individual bracing units. Although this value relates to the mass density of the *building*, it can be used here as the actual value of m is irrelevant at this stage as its value drops out of the equations.

Bracing Units 1 & 2 (frame F6, Figure 4.5/a)

The cross-sections of the columns and beams of frame F6 are 0.40/0.40 and 0.40/0.70 (metres), respectively.

The shear stiffness that is associated with the beams of the frame is

$$K_{b,1} = \sum_{j=1}^{n-1} \frac{12E_b I_{b,j}}{l_j h} = 3\frac{12 \cdot 25 \cdot 10^6 \cdot 0.4 \cdot 0.7^3}{12 \cdot 6 \cdot 3} = 571667 \text{ kN} \qquad \{2.28\}$$

The shear stiffness that is associated with the columns of the frame is

$$K_{c,1} = \sum_{j=1}^{n} \frac{12E_c I_{c,j}}{h^2} = 4\frac{12 \cdot 25 \cdot 10^6 \cdot 0.4^4}{12 \cdot 3^2} = 284444 \text{ kN} \qquad \{2.29\}$$

The combination of the two-part shear stiffnesses gives the "original" shear stiffness of the frame:

$$K_1 = K_{b,1}\frac{K_{c,1}}{K_{b,1}+K_{c,1}} = 571667\frac{284444}{571667+284444} = 189937 \text{ kN} \tag{2.27}$$

where the reduction factor

$$r_1 = \frac{K_{c,1}}{K_{b,1}+K_{c,1}} = \frac{284444}{571667+284444} = 0.3323 \tag{2.30}$$

is introduced.

The square of the fundamental frequency associated with the "original" shear stiffness of Bracing Unit 1 is

$$f_{s',1}^2 = \frac{1}{(4H)^2}\frac{r_f^2 K_1}{m} = \frac{0.961^2 \cdot 189937}{(4\cdot 75)^2 \cdot 91.74} = 0.02124 \text{ Hz}^2 \tag{2.42}$$

The global second moment of area of the cross-sections of the columns is

$$I_{g,1} = \sum_{j=1}^{n} A_{c,j} t_j^2 = 0.4\cdot 0.4(9^2+3^2)2 = 28.8 \text{ m}^4 \tag{2.32}$$

The square of the fundamental frequency that is associated with the global full-height bending vibration of the frame is

$$f_{g,1}^2 = \frac{0.313 r_f^2 E_c I_{g,1}}{H^4 m} = \frac{0.313\cdot 0.961^2 \cdot 25\cdot 10^6 \cdot 28.8}{75^4 \cdot 91.74} = 0.07170 \text{ Hz}^2 \tag{2.43}$$

There is an interaction between the "original" shear and global bending modes that reduces the effectiveness of the shear stiffness. The factor of effectiveness can be calculated using the squares of the two relevant frequencies as

$$s_{f,1}^2 = \frac{f_{g,1}^2}{f_{g,1}^2 + f_{s',1}^2} = \frac{0.0717}{0.0717+0.02124} = 0.7715 \tag{2.46}$$

The effective shear stiffness can now be obtained:

$$K_{e,1} = K_1 s_{f,1}^2 = 189937\cdot 0.7715 = 146536 \text{ kN} \tag{2.45}$$

Bracing Units 3 & 4 (shear wall W6, Figure 4.5/b)
The size of the shear wall is 6.0 metres with a thickness of 0.3 m. The second moment of area and the Saint-Venant constant of the shear wall are

$$I_{w,3} = I_{w,4} = \frac{0.3 \cdot 6^3}{12} = 5.4 \text{ m}^4, \qquad J_{w,3} = J_{w,4} = \frac{6 \cdot 0.3^3}{3} = 0.054 \text{ m}^4$$

Bracing Units 5 & 6 (shear wall W7, Figure 4.5/c)
The size of the shear wall is 12.0 metres with a thickness of 0.3 m. The second moment of area and the Saint-Venant constant of the shear wall are

$$I_{w,5} = I_{w,6} = \frac{0.3 \cdot 12^3}{12} = 43.2 \text{ m}^4, \qquad J_{w,5} = J_{w,6} = \frac{12 \cdot 0.3^3}{3} = 0.108 \text{ m}^4$$

Lateral vibration in direction x (Bracing Units 1, 2, 3 and 4)

The effective shear stiffness for the whole system (that contains the two frames marked F6) is obtained as

$$K_e = \sum_{i=1}^{f} s_{f,i}^2 K_i = 0.7715 \cdot 189937 \cdot 2 = 293073 \text{ kN} \tag{4.11}$$

The "original" shear stiffness is

$$K = \sum_{i=1}^{f} K_i = 189937 \cdot 2 = 379874 \text{ kN} \tag{4.6}$$

The effectiveness factor for the lateral system in direction x can now be established:

$$s_f = \sqrt{\frac{K_e}{K}} = \sqrt{\frac{293073}{379874}} = 0.8784 \tag{4.12}$$

The square of the frequency that belongs to the effective shear stiffness is

$$f_s^2 = \frac{1}{(4H)^2} \frac{r_f^2 K_e}{m} = \frac{0.961^2 \cdot 293073}{(4 \cdot 75)^2 \cdot 91.74} = 0.03278 \text{ Hz}^2 \tag{4.13}$$

Regarding the bending stiffness for the system (Bracing Units 1, 2, 3 and 4), the contribution of the columns of the two frames is small compared to the contribution of the two shear walls. After neglecting the first term in Equation (4.14), the bending stiffness is

$$EI = E_c I_c + E_w I_w \Rightarrow E_w \sum_{k=1}^{m} I_{w,k} = 25 \cdot 10^6 \cdot 5.4 \cdot 2 = 2.7 \cdot 10^8 \,\text{kNm}^2 \qquad \{4.14\}$$

The square of the lateral frequency that is associated with this bending stiffness is

$$f_b^2 = \frac{0.313 r_f^2 EI}{H^4 m} = \frac{0.313 \cdot 0.961^2 \cdot 2.7 \cdot 10^8}{75^4 \cdot 91.74} = 0.02689 \,\text{Hz}^2 \qquad \{4.15\}$$

With the non-dimensional parameter

$$k = H\sqrt{\frac{K_e}{EI}} = 75\sqrt{\frac{293073}{2.7 \cdot 10^8}} = 2.471 \qquad \{4.17\}$$

the frequency parameter is obtained from Table 4.2 as

$$\eta = 0.8628 + \frac{0.9809 - 0.8628}{0.50}(2.471 - 2.0) = 0.974 \qquad \{\text{Table 4.2}\}$$

Finally, the lateral frequency of the system in direction x is

$$f_x = \sqrt{f_b^2 + f_s^2 + \left(\frac{\eta^2}{0.313} - \frac{k^2}{5} - 1\right) s_f f_b^2} \qquad \{4.18\}$$

$$= \sqrt{0.02689 + 0.03278 + \left(\frac{0.974^2}{0.313} - \frac{2.471^2}{5} - 1\right) 0.8784 \cdot 0.02689} = 0.281 \,\text{Hz}$$

Lateral vibration in direction y (Bracing Units 5 and 6)

Compared to Bracing Units 5 and 6, the contribution of Bracing Units 1, 2, 3 and 4 is small and can safely be ignored. As there are no frames acting in direction y, Equation (4.18) is reduced to its first term and Equation (4.15) directly leads to the fundamental frequency as

$$f_y = \frac{0.56 r_f}{H^2}\sqrt{\frac{EI}{m}} = \frac{0.56 \cdot 0.961}{75^2}\sqrt{\frac{25 \cdot 10^6 \cdot 43.2 \cdot 2}{91.74}} = 0.464 \,\text{Hz} \qquad \{4.18; 4.15\}$$

Pure torsional vibration (with all bracing units participating)

The calculation is made fairly simple by the fact that the system is doubly symmetric and the location of the shear centre and the locations of the bracing units in relation to the shear centre are readily known.

The radius of gyration is

$$i_p = \sqrt{\frac{L^2 + B^2}{12} + t^2} = \sqrt{\frac{30^2 + 12^2}{12}} = \sqrt{87} = 9.33 \text{ m} \qquad \{4.24\}$$

The "original" Saint-Venant stiffness is

$$(GJ) = \sum_{k=1}^{m} GJ_k + \sum_{i=1}^{f} \left((K_i)_x y_i^2 + (K_i)_y x_i^2\right) \qquad \{4.30\}$$

$$= 2 \cdot 10.4 \cdot 10^6 (0.054 + 0.108) + 2 \cdot 189937 \cdot 6^2 = 17045064 \text{ kNm}^2$$

The effective Saint-Venant stiffness is

$$(GJ)_e = \sum_{k=1}^{m} GJ_k + \sum_{i=1}^{f} \left((K_{e,i})_x y_i^2 + (K_{e,i})_y x_i^2\right) \qquad \{4.21\}$$

$$= 2 \cdot 10.4 \cdot 10^6 (0.054 + 0.108) + 2 \cdot 146536 \cdot 6^2 = 13920192 \text{ kNm}^2$$

The effectiveness factor is the square root of the ratio of the effective and "original" Saint-Venant stiffnesses:

$$s_\varphi = \sqrt{\frac{(GJ)_e}{(GJ)}} = \sqrt{\frac{13920192}{17045064}} = 0.9037 \qquad \{4.29\}$$

The warping stiffness of the system originates from the bending stiffness of the four shear walls and the bending stiffness of the columns of the two frames (with the own warping stiffness of the bracing units being zero). The contribution of shear wall W6 and the columns of frame F6 is very small compared to the contribution of shear wall W7. After ignoring these contributions, the warping stiffness of the system is

$$EI_\omega = E_w \sum_{k=1}^{m} \left(I_{\omega,k} + (I_{w,k})_x y_k^2 + (I_{w,k})_y x_k^2\right) + E_c \sum_{i=1}^{f} \left((I_{c,i} r_i)_x y_i^2 + (I_{c,i} r_i)_y x_i^2\right)$$

$$= 25 \cdot 10^6 \cdot 2 \cdot 43.2 \cdot 15^2 = 486 \cdot 10^9 \text{ kNm}^4 \qquad \{4.22\}$$

The pure torsional frequency associated with the warping stiffness is needed:

$$f_\omega^2 = \frac{0.313 r_f^2 EI_\omega}{i_p^2 H^4 m} = \frac{0.313 \cdot 0.961^2 \cdot 486 \cdot 10^9}{87 \cdot 75^4 \cdot 91.74} = 0.5563 \text{ Hz}^2 \qquad \{4.27\}$$

and the formula for the pure torsional frequency associated with the Saint-Venant stiffness is

$$f_t^2 = \frac{r_f^2 (GJ)_e}{16 i_p^2 H^2 m} = \frac{0.961^2 \cdot 1.392 \cdot 10^7}{16 \cdot 87 \cdot 75^2 \cdot 91.74} = 0.01790\,\text{Hz}^2 \qquad \{4.28\}$$

With the non-dimensional parameter

$$k_\varphi = H\sqrt{\frac{(GJ)_e}{EI_\omega}} = 75\sqrt{\frac{1.392 \cdot 10^7}{4.86 \cdot 10^{11}}} = 0.4014 \qquad \{4.31\}$$

the torsional frequency parameter is obtained using Table 4.2 as

$$\eta_\varphi = 0.5761 + \frac{0.5851 - 0.5761}{0.5 - 0.4}(0.4014 - 0.4) = 0.5762 \qquad \{\text{Table 4.2}\}$$

The pure torsional frequency can now be determined:

$$f_\varphi = \sqrt{f_\omega^2 + f_t^2 + \left(\frac{\eta_\varphi^2}{0.313} - \frac{k_\varphi^2}{5} - 1\right) s_\varphi f_\omega^2} \qquad \{4.26\}$$

$$= \sqrt{0.5563 + 0.0179 + \left(\frac{0.5762^2}{0.313} - \frac{0.4014^2}{5} - 1\right) 0.9037 \cdot 0.5563} = 0.767\,\text{Hz}$$

Because of the doubly symmetric arrangement of the bracing system, there is no coupling among the two lateral and pure torsional modes, and the fundamental frequency of the building is the smallest one of the three "basic" frequencies:

$$f = \min(f_x, f_y, f_\varphi) = f_x = 0.281\,\text{Hz} \qquad \{4.18; 4.39\}$$

The Finite Element based computer program Axis VM X5 (2019) gives $f = 0.268$ Hz for the fundamental frequency of the building.

4.4.2 Fundamental frequency of twenty-storey asymmetric building

The fundamental frequency of the twenty-storey reinforced concrete asymmetric building (Figure 4.6) is determined in this section. The building is subjected to uniformly distributed mass over the floors. The bracing system consists of two frames (F5) and two shear walls (W5) in direction x and a frame (F7) and a shear wall (W7) in direction y (Figure 4.7). The contribution of the two columns marked

"C" to the overall resistance of the bracing system is negligible and is therefore ignored. The modulus of elasticity is $E = 25000$ MN/m^2, the modulus of elasticity in shear is $G = 10400$ MN/m^2, the storey height is $h = 3$ m and the total height of the building is $H = 60$ m. The thicknesses of shear walls W5 and W7 are 0.20 m and 0.30 m, respectively.

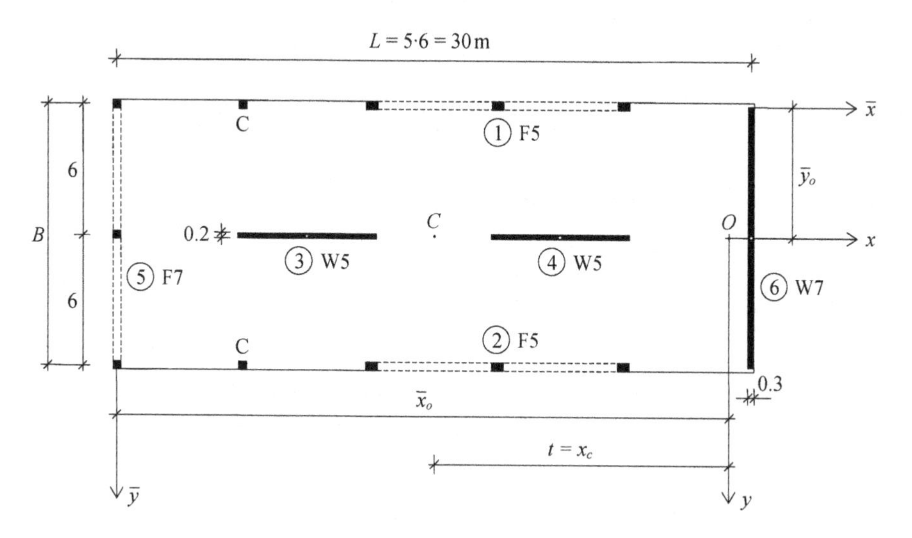

Figure 4.6 Layout of twenty-storey asymmetric building.

The weight per unit volume of the building is assumed to be $\gamma = 2.5$ kN/m^3. This leads to the mass density per unit length as

$$m = \rho A = \frac{\gamma}{g} LB = \frac{2.5}{9.81} 12 \cdot 30 = 91.74 \text{ kg/m} \qquad \{4.1\}$$

Mass distribution factor r_f will be needed throughout the calculations. Its value is obtained from Table 4.1 as a function of the number of storeys:

$$r_f = 0.952 \qquad \{\text{Table } 4.1\}$$

The calculation is carried out in three steps. The basic characteristics of the individual bracing units are determined first. The three basic vibration modes are then considered with the calculation of the frequencies that belong to the lateral and pure torsional motions. Finally, the question of the coupling of the three basic modes is addressed.

Basic characteristics of the individual bracing units

Mass density per unit length $m = 91.74$ kg/m will be used in connection with the individual bracing units. Although this value relates to the mass density of the *building*, it can be used here as the actual value of m is irrelevant at this stage as its value drops out of the equations.

Bracing Units 1 & 2 (frame F5, Figure 4.7/a)
The cross-sections of the columns and beams of frame F5 are 0.40/0.70 and 0.40/0.40 (metres), respectively.

The shear stiffness that is associated with the beams of the frame is

$$K_{b,1}=\sum_{j=1}^{n-1}\frac{12E_bI_{b,j}}{l_jh}=2\frac{12\cdot 25\cdot 10^6\cdot 0.4^4}{12\cdot 6\cdot 3}=71111\,\text{kN} \qquad \{2.28\}$$

The shear stiffness that is associated with the columns of the frame is

$$K_{c,1}=\sum_{j=1}^{n}\frac{12E_cI_{c,j}}{h^2}=3\frac{12\cdot 25\cdot 10^6\cdot 0.4\cdot 0.7^3}{12\cdot 3^2}=1143333\,\text{kN} \qquad \{2.29\}$$

The combination of the two part shear stiffnesses gives the "original" shear stiffness of the frame as

$$K_1=K_{b,1}\frac{K_{c,1}}{K_{b,1}+K_{c,1}}=71111\frac{1143333}{71111+1143333}=66947\,\text{kN} \qquad \{2.27\}$$

where

$$r_1=\frac{K_{c,1}}{K_{b,1}+K_{c,1}}=\frac{1143333}{71111+1143333}=0.9414 \qquad \{2.30\}$$

is the reduction factor.

The square of the fundamental frequency associated with the "original" shear stiffness of Bracing Unit 1 is

$$f_{s',1}^2=\frac{1}{(4H)^2}\frac{r_f^2K_1}{m}=\frac{0.952^2\cdot 66947}{(4\cdot 60)^2\cdot 91.74}=0.01148\,\text{Hz}^2 \qquad \{2.42\}$$

The global second moment of area of the cross-sections of the columns is

$$I_{g,1}=\sum_{j=1}^{n}A_{c,j}t_j^2=0.4\cdot 0.7\cdot 6^2\cdot 2=20.16\,\text{m}^4 \qquad \{2.32\}$$

The square of the fundamental frequency that is associated with the global full-height bending vibration of the frame is

$$f_{g,1}^2 = \frac{0.313 r_f^2 E_c I_{g,1}}{H^4 m} = \frac{0.313 \cdot 0.952^2 \cdot 25 \cdot 10^6 \cdot 20.16}{60^4 \cdot 91.74} = 0.1202\,\text{Hz}^2 \qquad \{2.43\}$$

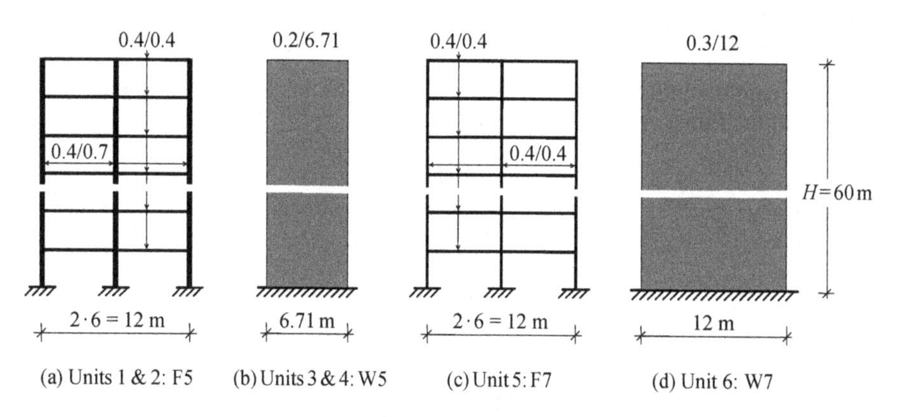

Figure 4.7 Bracing units of twenty-storey asymmetric building.

There is an interaction between the "original" shear and global bending modes that reduces the effectiveness of the shear stiffness. The factor of effectiveness can be calculated using the squares of the two relevant frequencies as

$$s_{f,1}^2 = \frac{f_{g,1}^2}{f_{g,1}^2 + f_{s',1}^2} = \frac{0.1202}{0.1202 + 0.01148} = 0.9128 \qquad \{2.46\}$$

The effective shear stiffness is now obtained as

$$K_{e,1} = s_{f,1}^2 K_1 = 0.9128 \cdot 66947 = 61109\,\text{kN} \qquad \{2.45\}$$

Bracing Units 3 & 4 (shear wall W5, Figure 4.7/b)
The size of the shear wall is 6.71 metres with a thickness of 0.2 m. The second moment of area and the Saint-Venant constant of the shear wall are

$$I_{w,3}^{\max} = I_{w,4}^{\max} = \frac{0.2 \cdot 6.71^3}{12} = 5.035\,\text{m}^4, \qquad J_{w,3} = J_{w,4} = \frac{6.71 \cdot 0.2^3}{3} = 0.0179\,\text{m}^4$$

Bracing Unit 5 (frame F7, Figure 4.7/c)
The cross-sections of both the columns and the beams of frame F7 are 0.40/0.40 (metres).

The shear stiffness that is associated with the beams of the frame is

$$K_{b,5} = \sum_{j=1}^{n-1} \frac{12 E_b I_{b,j}}{l_j h} = 2 \frac{12 \cdot 25 \cdot 10^6 \cdot 0.4^4}{12 \cdot 6 \cdot 3} = 71111\,\text{kN} \qquad \{2.28\}$$

The shear stiffness that is associated with the columns of the frame is

$$K_{c,5} = \sum_{j=1}^{n} \frac{12E_c I_{c,j}}{h^2} = 3\frac{12 \cdot 25 \cdot 10^6 \cdot 0.4^4}{12 \cdot 3^2} = 213333\,\text{kN} \qquad \{2.29\}$$

The combination of the two part shear stiffnesses leads to the "original" shear stiffness of the frame as

$$K_5 = K_{b,5}\frac{K_{c,5}}{K_{b,5}+K_{c,5}} = 71111\frac{213333}{71111+213333} = 53333\,\text{kN} \qquad \{2.27\}$$

where

$$r_5 = \frac{K_{c,5}}{K_{b,5}+K_{c,5}} = \frac{213333}{71111+213333} = 0.75 \qquad \{2.30\}$$

is the reduction factor.

The square of the fundamental frequency associated with the "original" shear stiffness of Bracing Unit 5 is

$$f_{s',5}^2 = \frac{1}{(4H)^2}\frac{r_f^2 K_5}{m} = \frac{0.952^2 \cdot 53333}{(4 \cdot 60)^2 \cdot 91.74} = 0.00915\,\text{Hz}^2 \qquad \{2.42\}$$

The global second moment of area of the cross-sections of the columns is

$$I_{g,5} = \sum_{j=1}^{n} A_{c,j} t_j^2 = 0.4 \cdot 0.4 \cdot 6^2 \cdot 2 = 11.52\,\text{m}^4 \qquad \{2.32\}$$

The square of the fundamental frequency that is associated with the global full-height bending vibration of the frame is

$$f_{g,5}^2 = \frac{0.313 r_f^2 E_c I_{g,5}}{H^4 m} = \frac{0.313 \cdot 0.952^2 \cdot 25 \cdot 10^6 \cdot 11.52}{60^4 \cdot 91.74} = 0.0687\,\text{Hz}^2 \qquad \{2.43\}$$

There is an interaction between the "original" shear and global bending modes that reduces the effectiveness of the shear stiffness. The factor of effectiveness can be calculated using the squares of the two relevant frequencies as

$$s_{f,5}^2 = \frac{f_{g,5}^2}{f_{g,5}^2 + f_{s',5}^2} = \frac{0.0687}{0.0687+0.00915} = 0.8825 \qquad \{2.46\}$$

The effective shear stiffness of the frame can now be determined:

$$K_{e,5} = s_{f,5}^2 K_5 = 0.8825 \cdot 53333 = 47066 \text{ kN} \qquad \{2.45\}$$

With the above effective shear stiffness, the square of the fundamental frequency associated with the effective shear stiffness of Bracing Unit 5 is

$$f_{s,5}^2 = \frac{1}{(4H)^2} \frac{r_f^2 K_{e,5}}{m} = \frac{0.952^2 \cdot 47066}{(4 \cdot 60)^2 \cdot 91.74} = 0.008072 \text{ Hz}^2 \qquad \{2.44\}$$

The lateral frequency of Bracing Unit 5 will also be needed later on with the torsional analysis.

With the local bending stiffness of the frame

$$EI_5 = Er_5 \sum_{j=1}^{n} I_{c,j} = 25 \cdot 10^6 \cdot 0.75 \cdot 3 \cdot \frac{0.4^4}{12} = 120000 \text{ kNm}^2 \qquad \{2.31\}$$

the fundamental frequency that is associated with the local bending stiffness is

$$f_{b,5}^2 = \frac{0.313 r_f^2 EI_5}{H^4 m} = \frac{0.313 \cdot 0.952^2 \cdot 120000}{60^4 \cdot 91.74} = 0.00002863 \text{ Hz}^2 \qquad \{2.47\}$$

As a function of the non-dimensional parameter

$$k = H\sqrt{\frac{K_e}{EI}} = 60\sqrt{\frac{47066}{120000}} = 37.58 \qquad \{2.51\}$$

the frequency parameter is obtained from Table 4.2 as

$$\eta = 9.016 + \frac{10.26 - 9.016}{40 - 35}(37.58 - 35) = 9.658 \qquad \{\text{Table 4.2}\}$$

The fundamental frequency of the frame is

$$f_5 = \sqrt{f_{b,5}^2 + f_{s,5}^2 + \left(\frac{\eta^2}{0.313} - \frac{k^2}{5} - 1\right) s_{f,5} f_{b,5}^2} \qquad \{2.52\}$$

$$= \sqrt{0.00002863 + 0.008072 + \left(\frac{9.658^2}{0.313} - \frac{37.58^2}{5} - 1\right) 0.9394 \cdot 0.00002863} = 0.0922 \text{ Hz}$$

Bracing Unit 6 (shear wall W7, Figure 4.7/d)
The size of the shear wall is 12.0 metres with a thickness of 0.3 m. The second moment of area and the Saint-Venant constant of the shear wall are

$$I_{w,6} = \frac{0.3 \cdot 12^3}{12} = 43.2\,\text{m}^4 \qquad \text{and} \qquad J_{w,6} = \frac{12 \cdot 0.3^3}{3} = 0.108\ \text{m}^4$$

The lateral frequency of the shear wall will also be needed with the torsional analysis:

$$f_6 = \frac{0.56 r_f}{H^2}\sqrt{\frac{EI_{w,6}}{m}} = \frac{0.56 \cdot 0.952}{60^2}\sqrt{\frac{25 \cdot 10^6 \cdot 43.2}{91.74}} = 0.5081\,\text{Hz} \qquad \{2.92\}$$

Lateral vibration in direction x (Bracing Units 1, 2, 3 and 4)
The effective shear stiffness for the whole system (that contains the two frames marked F5) is obtained as

$$K_e = \sum_{i=1}^{f} s_{f,i}^2 K_i = 0.9128 \cdot 66947 \cdot 2 = 122218\,\text{kN} \qquad \{4.11\}$$

The "original" shear stiffness is

$$K = \sum_{i=1}^{f} K_i = 66947 \cdot 2 = 133894\,\text{kN} \qquad \{4.6\}$$

The effectiveness factor for the lateral system can now be established:

$$s_f = \sqrt{\frac{K_e}{K}} = \sqrt{\frac{122218}{133894}} = 0.9554 \qquad \{4.12\}$$

The square of the frequency that belongs to the effective shear stiffness is

$$f_s^2 = \frac{1}{(4H)^2}\frac{r_f^2 K_e}{m} = \frac{0.952^2 \cdot 122218}{(4 \cdot 60)^2 \cdot 91.74} = 0.02096\,\text{Hz}^2 \qquad \{4.13\}$$

Regarding the bending stiffness for the system (Bracing Units 1, 2, 3 and 4), the contribution of the columns of the two frames is small compared to the contribution of the two shear walls. Neglecting the first term in Equation (4.14), the bending stiffness is

$$EI = E_c I_c + E_w I_w \Rightarrow E_w \sum_{1}^{m} I_{w,k} = 25 \cdot 10^6 \cdot 5.035 \cdot 2 = 2.52 \cdot 10^8\,\text{kNm}^2 \qquad \{4.14\}$$

The square of the lateral frequency that is associated with this bending stiffness is

$$f_b^2 = \frac{0.313 r_f^2 EI}{H^4 m} = \frac{0.313 \cdot 0.952^2 \cdot 2.52 \cdot 10^8}{60^4 \cdot 91.74} = 0.06013 \text{ Hz}^2 \qquad \{4.15\}$$

With the non-dimensional parameter

$$k = H\sqrt{\frac{K_e}{EI}} = 60\sqrt{\frac{122218}{2.52 \cdot 10^8}} = 1.32 \qquad \{4.17\}$$

the frequency parameter is obtained from Table 4.2 as

$$\eta = 0.71 + \frac{0.7302 - 0.71}{0.1}(1.32 - 1.3) = 0.714 \qquad \{\text{Table 4.2}\}$$

The lateral frequency of the system in direction x is

$$f_x = \sqrt{f_b^2 + f_s^2 + \left(\frac{\eta^2}{0.313} - \frac{k^2}{5} - 1\right) s_f f_b^2} \qquad \{4.18\}$$

$$= \sqrt{0.06013 + 0.02096 + \left(\frac{0.714^2}{0.313} - \frac{1.32^2}{5} - 1\right) 0.9554 \cdot 0.06013} = 0.312 \text{ Hz}$$

Lateral vibration in direction y (Bracing Units 5 and 6)

The effective shear stiffness for the whole system (that contains frame F7) is obtained as

$$K_e = \sum_{i=1}^{f} s_{f,i}^2 K_i = 0.8825 \cdot 53333 = 47066 \text{ kN} \qquad \{4.11\}$$

The "original" shear stiffness is

$$K = \sum_{i=1}^{f} K_i = 53333 \text{ kN} \qquad \{4.6\}$$

The effectiveness factor of the lateral system will also be needed for the determination of the lateral frequency. Its value is

$$s_f = \sqrt{\frac{K_e}{K}} = \sqrt{\frac{47066}{53333}} = 0.9394 \qquad \{4.12\}$$

The square of the frequency that belongs to the effective shear stiffness is

$$f_s^2 = \frac{1}{(4H)^2}\frac{r_f^2 K_e}{m} = \frac{0.952^2 \cdot 47066}{(4 \cdot 60)^2 \cdot 91.74} = 0.008072 \text{ Hz}^2 \qquad \{4.13\}$$

Regarding the bending stiffness for the system (Bracing Units 5 and 6), the contribution of the columns of the frame is small compared to the contribution of the shear wall. Neglecting the first term in Equation (4.14), the bending stiffness is

$$EI = E_c I_c + E_w I_w => E_w \sum_1^m I_{w,k} = 25 \cdot 10^6 \cdot 43.2 = 10.8 \cdot 10^8 \text{ kNm}^2 \qquad \{4.14\}$$

The square of the lateral frequency that is associated with this bending stiffness is

$$f_b^2 = \frac{0.313 r_f^2 EI}{H^4 m} = \frac{0.313 \cdot 0.952^2 \cdot 10.8 \cdot 10^8}{60^4 \cdot 91.74} = 0.2577 \text{ Hz}^2 \qquad \{4.15\}$$

With the non-dimensional parameter

$$k = H\sqrt{\frac{K_e}{EI}} = 60\sqrt{\frac{47066}{10.8 \cdot 10^8}} = 0.3961 \qquad \{4.17\}$$

the frequency parameter is obtained from Table 4.2 as

$$\eta = 0.569 + \frac{0.5761 - 0.569}{0.1}(0.3961 - 0.3) = 0.5758 \qquad \{\text{Table 4.2}\}$$

The lateral frequency of the system in direction y can now be determined:

$$f_y = \sqrt{f_b^2 + f_s^2 + \left(\frac{\eta^2}{0.313} - \frac{k^2}{5} - 1\right) s_f f_b^2} \qquad \{4.18\}$$

$$= \sqrt{0.2577 + 0.008072 + \left(\frac{0.5758^2}{0.313} - \frac{0.3961^2}{5} - 1\right) 0.9394 \cdot 0.2577} = 0.522 \text{ Hz}$$

Pure torsional vibration (with all bracing units participating)

The location of the shear centre is needed first. The two coordinates in the $\bar{x}-\bar{y}$ coordinate system (Figure 4.6) are

$$\bar{x}_o = \frac{\sum_{i=1}^{f+m} f_{y,i}^2 \bar{x}_i}{\sum_{i=1}^{f+m} f_{y,i}^2} = \frac{f_6^2 L}{f_5^2 + f_6^2} = \frac{0.5081^2 \cdot 30}{0.0922^2 + 0.5081^2} = 29.04 \text{ m}, \quad \bar{y}_o = 6 \text{ m} \qquad \{4.19\}$$

where the effect of shear wall W5, being small, is ignored.

The origin of the coordinate system is now moved to the shear centre. In this new coordinate system the radius of gyration is

$$i_p = \sqrt{\frac{L^2 + B^2}{12} + t^2} = \sqrt{\frac{30^2 + 12^2}{12} + 14.04^2} = \sqrt{284.1} = 16.86 \text{ m} \qquad \{4.24\}$$

The perpendicular distances of the bracing units from the shear centre are:

$$t_1 = t_2 = 6 \text{ m}, \qquad t_3 = \bar{x}_o - 9 = 20.04 \text{ m}, \qquad t_4 = \bar{x}_o - 21 = 8.04 \text{ m}$$

$$t_5 = \bar{x}_o = 29.04 \text{ m} \qquad \text{and} \qquad t_6 = L - \bar{x}_o = 30 - 29.04 = 0.96 \text{ m}$$

The "original" Saint-Venant stiffness is

$$(GJ) = \sum_{k=1}^{m} GJ_k + \sum_{i=1}^{f} \left((K_i)_x y_i^2 + (K_i)_y x_i^2 \right) \qquad \{4.30\}$$

$$= 10.4 \cdot 10^6 (2 \cdot 0.01789 + 0.108) + 2 \cdot 66947 \cdot 6^2 + 53333 \cdot 29.04^2 = 51292367 \text{ kNm}^2$$

The effective Saint-Venant stiffness is

$$(GJ)_e = \sum_{k=1}^{m} GJ_k + \sum_{i=1}^{f} \left((K_{e,i})_x y_i^2 + (K_{e,i})_y x_i^2 \right) \qquad \{4.21\}$$

$$= 10.4 \cdot 10^6 (2 \cdot 0.01789 + 0.108) + 2 \cdot 61109 \cdot 6^2 + 47066 \cdot 29.04^2 = 45586934 \text{ kNm}^2$$

The effectiveness factor is the square root of the ratio of the effective and "original" Saint-Venant stiffnesses:

$$s_\varphi = \sqrt{\frac{(GJ)_e}{(GJ)}} = \sqrt{\frac{45586934}{51292367}} = 0.9427 \qquad \{4.29\}$$

The warping stiffness of the system originates from the bending stiffness of the shear walls and the bending stiffness of the columns of the frames (with the own warping stiffness of the bracing units being zero):

$$EI_\omega = E_w \sum_{k=1}^{m} \left(I_{\omega,k} + (I_{w,k})_x y_k^2 + (I_{w,k})_y x_k^2 \right) + E_c \sum_{i=1}^{f} \left((I_{c,i} r_i)_x y_i^2 + (I_{c,i} r_i)_y x_i^2 \right)$$

$$= 25 \cdot 10^6 \left(0.00447(20.04^2 + 8.04^2) + 43.2 \cdot 0.96^2 \right) + \qquad \{4.22\}$$

$$+25 \cdot 10^6 \left(6 \frac{0.4 \cdot 0.7^3}{12} 0.9414 \cdot 6^2 + 3 \frac{0.4^4}{12} 0.75 \cdot 29.04^2 \right) = 1207 \cdot 10^6 \text{ kNm}^4$$

The square of the pure torsional frequency associated with the warping stiffness is

$$f_\omega^2 = \frac{0.313 r_f^2 EI_\omega}{i_p^2 H^4 m} = \frac{0.313 \cdot 0.952^2 \cdot 1207 \cdot 10^6}{284.1 \cdot 60^4 \cdot 91.74} = 0.001014 \text{ Hz}^2 \qquad \{4.27\}$$

and the formula for the pure torsional frequency associated with the Saint-Venant stiffness results in

$$f_t^2 = \frac{r_f^2 (GJ)_e}{16 i_p^2 H^2 m} = \frac{0.952^2 \cdot 45.59 \cdot 10^6}{16 \cdot 284.1 \cdot 60^2 \cdot 91.74} = 0.02752 \text{ Hz}^2 \qquad \{4.28\}$$

With the non-dimensional parameter

$$k_\varphi = H \sqrt{\frac{(GJ)_e}{EI_\omega}} = 60 \sqrt{\frac{45.59 \cdot 10^6}{12.07 \cdot 10^8}} = 11.66 \qquad \{4.31\}$$

the torsional frequency parameter is obtained using Table 4.2 as

$$\eta_\varphi = 3.172 + \frac{3.295 - 3.172}{12 - 11.5}(11.66 - 11.5) = 3.211 \qquad \{\text{Table 4.2}\}$$

Making use of the part torsional frequencies and the non-dimensional auxiliary parameters, the pure torsional frequency can now be determined:

$$f_\varphi = \sqrt{f_\omega^2 + f_t^2 + \left(\frac{\eta_\varphi^2}{0.313} - \frac{k_\varphi^2}{5} - 1 \right) s_\varphi f_\omega^2} = \qquad \{4.26\}$$

$$= \sqrt{0.001014 + 0.02752 + \left(\frac{3.211^2}{0.313} - \frac{11.66^2}{5} - 1\right) 0.9427 \cdot 0.001014} = 0.182\ \text{Hz}$$

Coupling of the basic modes

The bracing system is asymmetric to a large degree. In such cases, the Föppl-Papkovich theorem offers good estimates for taking the effect of coupling into account.

The centroid of the mass of the building lies on one of the principal axes of the bracing system (on axis x). Two things may happen. Vibration may develop in direction x (defined by f_x), or vibration in direction y (f_y) couples with pure torsional vibration around axis z (f_φ). The frequency of this coupled vibration is

$$f_{y\varphi} = \left(\frac{1}{f_y^2} + \frac{1}{f_\varphi^2}\right)^{-\frac{1}{2}} = \frac{1}{\sqrt{\dfrac{1}{0.522^2} + \dfrac{1}{0.182^2}}} = 0.172\ \text{Hz} \qquad \{4.37\}$$

As this frequency is smaller than $f_x = 0.312$ Hz, and the fundamental frequency of the building is the smaller one of f_x and $f_{y\varphi}$, this is also the fundamental frequency of the building:

$$f = \min(f_x, f_{y\varphi}) = f_{y\varphi} = 0.172\ \text{Hz} \qquad \{4.38\}$$

The computer-based "exact" solution by Axis VM X5 (2019) for the fundamental frequency of the structure is $f = 0.173$ Hz.

5

Stability analysis of buildings

The stability of a building can, and should, be assessed by looking at the stability of its individual elements as well as examining its stability as a whole. National codes have detailed instructions for the first case but the buckling analysis of whole structures is not so well regulated and therefore this chapter intends to address the second case. The designer basically has two possibilities to tackle whole building behaviour in either using finite element packages or relying on analytical methods. The analytical approach is used here.

A great number of methods have been developed for the stress analysis of individual frames, coupled shear walls and shear walls. Fewer methods are available to deal with a *system* of these bracing units. The availability of methods for the *stability* analysis of a system of frames, coupled shear walls, shear walls and cores is even more limited. This follows from the fact that the interaction among the elements (beams/lintels and columns/walls) of a single frame or coupled shear walls is complex enough but then the bracing units interact with one another not only in planar behaviour but normally also in a three-dimensional fashion. This is why the available analytical methods make one or more simplifying assumptions regarding the characteristic stiffnesses of the bracing units, the geometry of the building, or loading.

In using an equivalent Timoshenko-beam, Goschy (1970) developed a simple hand-method for the stability analysis of buildings under top-level load. Goldberg (1973) concentrated on plane buckling and presented two simple approximate formulae which can be used in the two extreme cases when the building undergoes pure shear mode or pure bending mode buckling. The interaction of the two modes is taken into account by applying the Föppl-Papkovich summation formula to the flexural and shear mode critical loads. Using the continuum approach (Gluck and Gellert, 1971; Rosman, 1974), Stafford Smith and Coull (1991) presented a more rigorous analysis for the sway and pure torsional buckling analysis of doubly symmetric multi-storey buildings whose vertical elements develop no or negligible axial deformations. Based on the top translation of the building (obtained from a plane frame analysis) and assuming a straight line deflection shape, MacLeod and Marshall (1983) derived a simple formula for the sway critical load of buildings. In using simple closed-form solutions for the critical loads of the individual bracing frames and coupled shear walls, Southwell's summation theorem results in a lower bound for the sway critical load of multi-storey buildings (Zalka and Armer, 1992). Even when the critical loads of the individual bracing units are not available, the repeated application of summation formulae leads to conservative estimates of the critical load in a simple manner (Kollár, 1999). In replacing the bracing units of a building with sandwich columns with thick faces, Hegedűs and Kollár (1999) developed a simple method for calculating the critical load of multi-storey buildings with bracing shear walls and frames in an arbitrary arrangement, subjected to concentrated top load. All these methods restrict the scope of analysis in one way or another and none were backed up with comprehensive accuracy analysis.

In taking into consideration all the characteristic stiffnesses of the bracing frames and shear walls as well as the interaction among the elements of the bracing structures and among the bracing units themselves (Zalka, 2002), the aim of this chapter is to introduce a simple analytical method for the calculation of the critical load of buildings braced by a system of frames, (coupled) shear walls and cores.

In addition to the general assumptions listed in Chapter 1, it is also assumed for the analysis that the load of the building is uniformly distributed over the floors. The critical load of the structures defines the bifurcation point.

The procedure for establishing the method for the determination of the global critical load of the building will be very similar to the way the method for the calculation of the fundamental frequency was developed in the previous chapter. First, the basic stiffness characteristics will be established for the analysis. The effective shear stiffness will be introduced, which, as in the previous chapter, makes it possible to create an equivalent column by the simple summation of the relevant stiffnesses. Second, based on the equivalent column, the eigenvalue problems characterising the sway buckling and pure torsional buckling problems will be presented and solved. Third, the coupling of the basic (sway and pure torsional) modes will be taken into account. Finally, two worked examples show how the method is used in structural engineering practice.

5.1 SWAY BUCKLING OF A SYSTEM OF FRAMES, (COUPLED) SHEAR WALLS AND CORES

Consider a *system* of frames and coupled shear walls ($i = 1...f$) and shear walls and cores ($k = 1...m$), shown in Figure 5.1/a. The whole bracing system can be characterised by the shear stiffness of the frames and coupled shear walls, the global bending stiffness of the frames and coupled shear walls, and the local bending stiffness of the individual columns/wall sections, shear walls and cores. These are the key characteristics of the system, which shall be combined into bending stiffness EI and effective shear stiffness K_e. The whole system shall be replaced with an equivalent cantilever with EI and K_e (Figure 5.1/b), enabling a relatively simple stability analysis.

At this stage, the load on the equivalent column consists of concentrated forces

$$F_i = QA$$

representing the total load at each floor level (Figure 5.2/a), where Q [kN/m^2] is the intensity of the uniformly distributed floor load and A is the floor area. To make the load on the equivalent column continuous, these concentrated forces are distributed over the height (Figure 5.2/b), leading to the uniformly distributed load on the substitute column as

$$q = \frac{F_i}{h} = \frac{QA}{h}$$

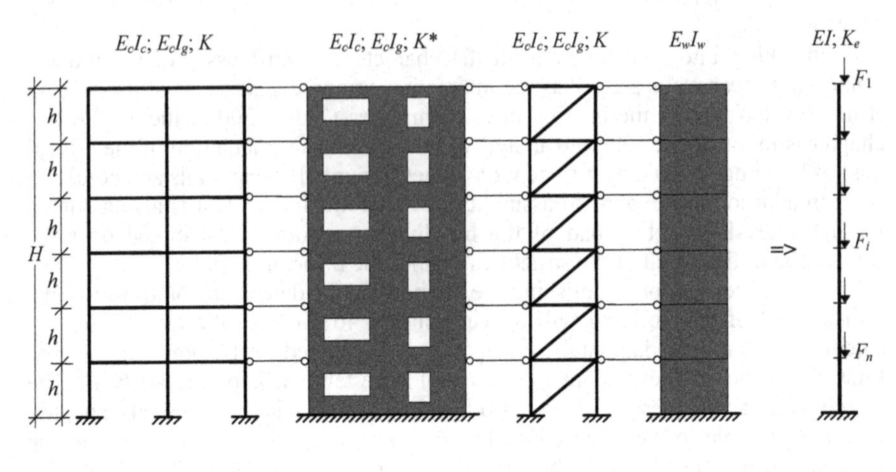

(a) a system of frames, (coupled) shear walls and cores

(b) equivalent cantilever

Figure 5.1 Model for the lateral stability analysis.

This procedure has a detrimental effect on the value of the critical load as the centroid of the original load moves downwards. However, this detrimental effect can be compensated for using a load distribution factor—see Equation (2.59) and the related explanation in Section 2.3.1.

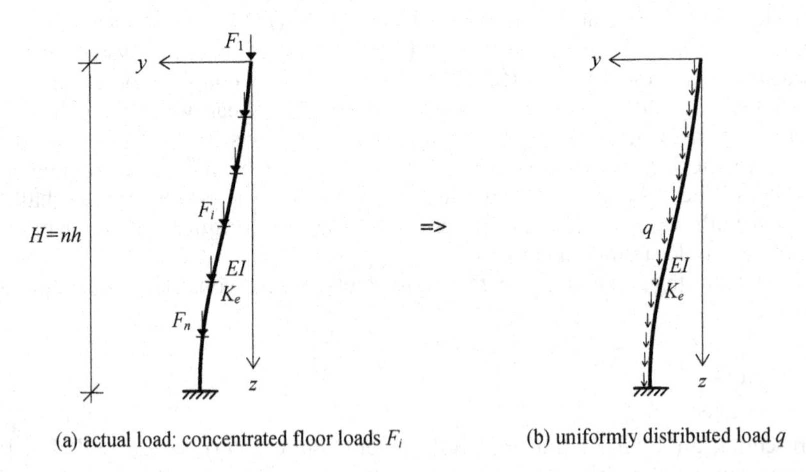

(a) actual load: concentrated floor loads F_i

(b) uniformly distributed load q

Figure 5.2 Load on the equivalent column.

The characteristic stiffnesses of the equivalent column shall be determined next.

The "original" shear stiffness originates from the frames and coupled shear walls. It consists of two parts. The *global* shear stiffness of the *i*th frame corresponds to the global (full-height) shear resistance of the frame and it is associated with the beams of the frame, assuming that the beams are continuously distributed over the height of the frame, resulting in full-height shear deformation (Figure 5.3/a). It is defined as with the deflection and frequency analyses, i.e. as

$$K_{b,i} = \sum_{j=1}^{n-1} \frac{12 E_b I_{b,j}}{l_j h} \tag{5.1}$$

with

E_b	the modulus of elasticity of the beams of the frame
$I_{b,j}$	the second moment of area of the *j*th beam of the *i*th frame
h	the storey height
l_j	the *j*th bay of the *i*th frame
n	the number of columns of the *i*th frame

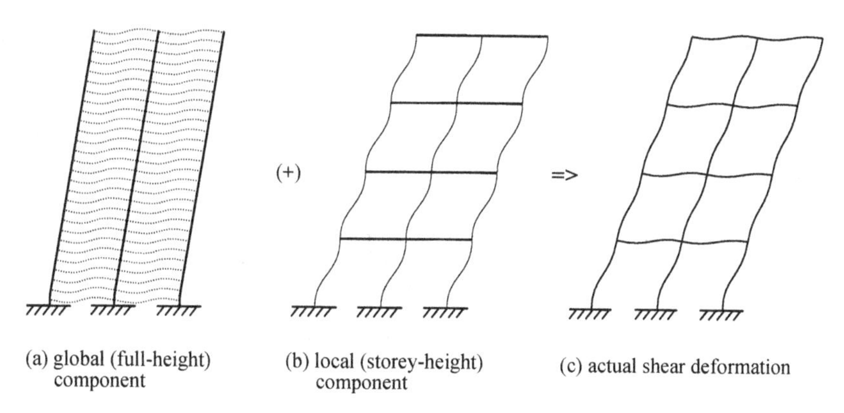

Figure 5.3 Shear deformation of the *i*th frame.

However, the beams are not distributed continuously over the height of the frame and only contribute to the shear resistance at floor levels. Between two floor levels, it is the responsibility of the storey-height columns to resist sway locally. It follows that the *local* shear stiffness is associated with the storey-height shear resistance of the structure (Figure 5.3/b) and—assuming fixed supports—is defined as

$$K_{c,i} = \sum_{j=1}^{n} \frac{\pi^2 E_c I_{c,j}}{h^2} \tag{5.2}$$

with

E_c	the modulus of elasticity of the columns of the frame

$I_{c,j}$ the second moment of area of the *j*th column of the *i*th frame

Note that $\pi^2 EI/h^2$ is the critical load of a column (with stiffness EI) of height h, with two built-in ends when the lateral movement of the upper end is not restricted. The fact that the local part of the shear stiffness is linked to the storey-height buckling makes it possible to handle frames with non-uniform storeys and also frames on pinned supports—see Section 2.4.

Using the above two components, the shear stiffness of the *i*th frame is obtained using the Föppl-Papkovich theorem in the form of

$$\frac{1}{K_i} = \frac{1}{K_{b,i}} + \frac{1}{K_{c,i}}$$

as

$$K_i = K_{b,i} r_i = K_{b,i} \frac{K_{c,i}}{K_{b,i} + K_{c,i}} \tag{5.3}$$

In the above equation, factor r_i is introduced (for later use) as a reduction factor:

$$r_i = \frac{K_{c,i}}{K_{b,i} + K_{c,i}} \tag{5.4}$$

The total shear stiffness of f bracing frames is

$$K = \sum_{i=1}^{f} K_i \tag{5.5}$$

The deformation that is associated with this "original" shear stiffness is shown in Figures 2.1/a and 5.3/c. It is called "original", as during buckling the system also develops bending type deformations that influence the shear deformation. In fact, the system develops two types of bending deformation, of which the global bending deformation (Figure 2.1/b) is worth paying attention to now, as it interacts with the above "original" shear deformation and tends to erode the original shear resistance of the system. [If coupled shear walls are also included in the system then, instead of Equation (5.1), Equations (2.72) and (2.73) should be used for the calculation of the global shear stiffness.]

By definition, K is also the shear critical load of the system.

Global bending deformation is resisted by the global second moment of area of the cross-sections of the columns. Its value for the *i*th frame is given by

$$I_{g,i} = \sum_{j=1}^{n} A_{c,j} t_j^2 \tag{5.6}$$

just like in previous chapters. In Equation (5.6)

$A_{c,j}$ is the cross-sectional area of the *j*th column of the *i*th frame
t_j is the distance of the *j*th column from the centroid of the cross-sections

With the global second moment of area, the global bending critical load of the *i*th frame is

$$N_{g,i} = \frac{7.837 r_s E_c I_{g,i}}{H^2} \tag{5.7}$$

where H is the height of the frame and r_s is the same load distribution factor that was introduced earlier in Section 2.3.1 where the stability of individual frames was investigated. It compensates for the detrimental approximation that the concentrated floor loads were distributed downwards when the uniformly distributed load of the equivalent column was created (Figure 5.2).

Values for r_s are given in Figure 2.15 for structures up to twenty storeys high; alternatively, if the structure is higher or more precise values are needed, Table 5.1 can be used.

Table 5.1 Load distribution factor r_s as a function of n (the number of storeys).

n	1	2	3	4	5	6	7	8	9	10	11
r_s	0.315	0.528	0.654	0.716	0.759	0.791	0.815	0.834	0.850	0.863	0.874
n	12	13	14	15	16	18	20	25	30	50	>50
r_s	0.883	0.891	0.898	0.904	0.910	0.919	0.926	0.940	0.950	0.969	$n/(n+1.588)$

During buckling, there is an interaction between the shear mode and the global bending mode. This interaction is detrimental as the resulting critical load is smaller than either the shear or the global bending critical load. This phenomenon can be taken into account using the Föppl-Papkovich theorem

$$\frac{1}{K_{e,i}} = \frac{1}{K_i} + \frac{1}{N_{g,i}}$$

and the role of global bending can be interpreted as an eroding effect which leads to a reduced shear stiffness. In doing so, this reduced shear stiffness is expressed as

$$K_{e,i} = K_i \frac{N_{g,i}}{K_i + N_{g,i}} = K_i s_i \tag{5.8}$$

where $K_{e,i}$ is defined as the *effective* shear stiffness of the *i*th frame and

$$s_i = \frac{N_{g,i}}{K_i + N_{g,i}} \tag{5.9}$$

is the effectiveness factor related to the shear stiffness of the *i*th frame.

It follows that the effective shear stiffness (and critical load) of the equivalent column (and of the whole system) is

$$K_e = \sum_{i=1}^{f} K_{e,i} = \sum_{i=1}^{f} K_i s_i \tag{5.10}$$

and the effectiveness factor for the whole system is

$$s = \frac{K_e}{K} \tag{5.11}$$

where K is the "original" shear stiffness [by Equation (5.5)].

The other characteristic stiffness of the system (and of the equivalent column) is the local bending stiffness. As all the columns of the frames, wall sections of the coupled shear walls, the shear walls and the cores have bending stiffness and all these structural items are made to work together by the floor slabs (and the beams of the frames), the total bending stiffness of the system is obtained by adding up the local bending stiffness of the vertical structural units:

$$EI = E_c I_c + E_w I_w = E_c \sum_{i=1}^{f} I_{c,i} r_i + E_w \sum_{k=1}^{m} I_{w,k} \tag{5.12}$$

where

E_w is the modulus of elasticity of the shear walls/cores
$I_{w,k}$ is the second moment of area of the *k*th shear wall/core

The equivalent column of the bracing system (Figure 5.2/b) has now been established with effective shear stiffness K_e, bending stiffness EI, and the uniformly distributed vertical load q.

When the system has mixed bracing units—both frames and shear walls/cores—the contribution of the columns of the frames [first term in Equation (5.12)] is small compared to that of the shear walls/cores and can normally be ignored. It is interesting to note that Equation (5.12) is formally identical to Equation (4.14) used for the frequency analysis. Careful investigation reveals that the two equations are in fact different—compare Equations (4.4) and (5.2). In most practical cases, however, the difference tends to be small.

In Equation (5.12), the bending stiffness of the columns of the frames are adjusted by combination factor r_i. Theoretical and numerical investigations (Hegedűs and Kollár, 1999; Zalka and Armer, 1992) demonstrate that this adjustment is necessary to prevent the over-representation of the second moments of area of the columns in the equivalent column where they are also represented in the shear stiffness.

With the above bending stiffness, the local bending critical load of the system of frames and shear walls/cores can now be presented as

$$N_l = N_f + N_w = \sum_{i=1}^{f} N_{f,i} r_i + \sum_{k=1}^{m} N_{w,k}$$

$$= \frac{7.837 r_s}{H^2}\left(E_c \sum_{i=1}^{f} I_{c,i} r_i + E_w \sum_{k=1}^{m} I_{w,k} \right) = \frac{7.837 r_s EI}{H^2} \tag{5.13}$$

where N_f represents the critical load of the full-height columns of the frames and N_w refers to the contribution of the shear walls.

With N_l and K_e (the shear critical load), an approximation of the critical load of the system can already be given. According to the Southwell theorem, the two part critical loads simply have to be added up:

$$N_{cr} = N_l + K_e \tag{5.14}$$

The beauty of this formula is that it is extremely simple and also it is conservative. However, it does not take into account the interaction between the (local) bending and shear modes.

This interaction can be taken into account automatically by setting up and solving the differential equation of the problem. The governing differential equation of the equivalent column is obtained by examining the equilibrium of an elementary section of the column. This leads to the eigenvalue problem

$$r_s EI y''''(z) + \left[\left(N(z) - K_e \right) y'(z) \right]' = 0 \tag{5.15}$$

where $N(z) = qz$ is the vertical load at z. For reasons of convenience, the origin of the coordinate system is placed at and fixed to the top of the equivalent column (Figure 5.2/b), so the boundary conditions are as follows:

$$y(0) = 0, \qquad y''(0) = 0$$

and

$$y'(H) = 0, \qquad y'''(H) = 0$$

This kind of eigenvalue problem can be solved relatively easily using the generalized power series method (Zalka and Armer, 1992). Critical load parameter (as the eigenvalue of the problem)

$$\alpha = \frac{N_{cr}}{N_l} \tag{5.16}$$

and the part critical load ratio

$$\beta = \frac{K_e}{N_l} \tag{5.17}$$

are introduced first. The application of the power series method follows. Finally, after some rearrangement, the solution for the sway buckling of the equivalent column is obtained as

$$N_{cr} = (\alpha - \beta)N_l + K_e$$

Values of critical load parameter α are given in Table 5.2 as a function of part critical load ratio β.

Table 5.2 Critical load parameters α and α_φ as a function of part critical load ratios β and β_φ.

β or β_φ	α or α_φ	β or β_φ	α or α_φ	β or β_φ	α or α_φ	β or β_φ	α or α_φ
0.0000	1.0000	0.06	1.1782	4	9.1001	200	246.3
0.0005	1.0015	0.07	1.2075	5	10.697	300	359.5
0.001	1.0030	0.08	1.2367	6	12.241	400	471.3
0.002	1.0060	0.09	1.2659	7	13.749	500	582.1
0.003	1.0090	0.10	1.2949	8	15.227	600	686.5
0.004	1.0120	0.20	1.5798	9	16.682	700	798.0
0.005	1.0150	0.30	1.8556	10	18.118	800	904.5
0.006	1.0180	0.40	2.1226	20	31.820	1000	1115
0.007	1.0210	0.50	2.3817	30	44.862	2000	2199
0.008	1.0240	0.60	2.6333	40	57.545	5000	5361
0.009	1.0270	0.70	2.8780	50	69.991	10000	10567
0.010	1.0300	0.80	3.1163	60	82.265	50000	51016
0.020	1.0598	0.90	3.3488	70	94.405	100000	102579
0.030	1.0896	1.00	3.5758	80	106.44	1000000	1011864
0.040	1.1192	2.00	5.6244	90	118.38	2000000	2018802
0.050	1.1487	3.00	7.4266	100	130.25	>2000000	β or β_φ

Before this solution is used for the sway buckling analysis of the whole bracing system, however, a small modification has to be made. The first term in the above equation stands for the bending contribution of the individual columns/wall sections, shear walls and cores in the system *and* it also represents the increase of the critical load of the system, due to the interaction between the bracing units in bending and the bracing units in shear. However, because of the fact that the effectiveness of the shear stiffness is normally smaller than 100% [c.f. Equation (5.11) where $s \leq 1$ holds], these two contributions have to be separated and the effectiveness factor should be applied to the part which is due to the interaction. This is easily done by first supplementing the above formula with the term $(N_l - N_l)$ of zero value, then separating $N_l + K_e$, and finally applying s to the rest of the formula. In so doing, the formula for the sway critical load then emerges as

$$N_{cr} = N_l + K_e + (\alpha - \beta - 1)sN_l \tag{5.18}$$

In the right-hand side of the above equation, the first two terms stand for the bending and shear critical loads of the system—compare Equation (5.18) and the approximate solution represented by Equation (5.14)—while the third term represents the effect of the interaction between the bending and shear deformations. As is the case with systems subjected to horizontal load, the interaction is beneficial. Bearing in mind that $(\alpha - \beta - 1) \geq 1$ always holds, the evaluation of the third term demonstrates that in practical cases the effect of interaction increases the critical load of the system. The evaluation of the data in Table 5.2 shows that the maximum increase is 87%, at $\beta = 2.1$.

The method can also be used when the building is subjected to a concentrated force on top of the building—e.g. a swimming pool on the top floor. In such a case, N_l and N_g in the relevant formulae are to be replaced by the corresponding Euler critical loads. It is interesting to note that in this load case there is no interaction between the bending and shear modes; the value of the term in brackets in Equation (5.18) becomes zero and Equation (5.14)—with the Euler critical loads—becomes the exact solution. See Section 5.5 for details.

A building may develop sway buckling in the two principal directions and both critical loads have to be calculated. These critical loads are obtained using Equation (5.18) where N_l, K_e, β and s are calculated by taking into account the bracing units in the relevant principal directions, say, in x and y.

5.2 SWAY BUCKLING: SPECIAL BRACING SYSTEMS

The following—idealised—special cases of bracing systems are worth considering (where the term “frame” refers to frames and coupled shear walls and the term “wall” covers both shear walls and cores).

5.2.1 Bracing systems consisting of shear walls only

In this special case, there is no shear stiffness in the sense it is used in this chapter. This translates to $K = 0$, $\beta = 0$ and $\alpha = 1$. As $N_f = 0$ in Equation (5.13), Equation (5.18) simplifies to

$$N_{cr} = N_l = N_w = \frac{7.837 r_s E_w I_w}{H^2} \tag{5.19}$$

which is the standard solution for the sway buckling of a bracing system in pure bending.

5.2.2 Bracing systems consisting of frames only

Equations (5.13) and (5.18) hold with $E_wI_w = 0$; everything else is unchanged.

If, furthermore, the beam/column stiffness ratio is very high, then the formula for the critical load further simplifies. In this special case, $K_b >> K_c$ and $N_g \approx \infty$ hold. Consequently, $r_i \approx 0$, $K \approx K_c$, $s_i \approx 1$, $K_e \approx K_c$ and $s \approx 1$. This leads to

$$N_{cr} = \sum_{i=1}^{f} K_{c,i} = \sum_{i=1}^{f}\left(\sum_{j=1}^{n}\frac{\pi^2 E_c I_{c,j}}{h^2}\right) \tag{5.20}$$

showing that the building loses stability through storey-height sway (shear failure from the point of view of the whole building), which is resisted by the bending stiffness of the columns. Equation (5.20) can also be used for checking stability when there is a loss of stiffness at a particular storey, making that storey vulnerable to local shear buckling (Zalka, 2000).

5.2.3 Shear walls and frames with very high beam/column stiffness ratio

When the bracing system consists of shear walls and frames with very high beam/column stiffness ratio, three sub-cases are worth considering.

First, the axial deformations of the columns are assumed to be negligible. The practical case that may belong here is the case of low-rise buildings. In this special case, $K_b >> K_c$ and $N_g \approx \infty$ hold. Consequently, $r_i \approx 0$, $K \approx K_c$, $s_i \approx 1$, $K_e \approx K_c$ and $s \approx 1$. $\beta \approx K_c/N_w > 0$ and $\alpha > 1$. This leads to

$$N_{cr} = (\alpha - \beta)N_w + K_c = \alpha N_w = \alpha\frac{7.837 E_w I_w r_s}{H^2} \tag{5.21}$$

showing that the critical load is based on the bending critical load of the shear walls and cores. This value is increased (through $\alpha > 1$) due to the interaction between the bracing units in shear (frames with stiff beams) and the bracing units in bending (walls and cores). The shear stiffness is characterised by the weakest link (i.e. by the stiffness of the columns).

Second, it is assumed that the axial deformations of the columns are not negligible. The practical case that may belong here is the case of medium-rise buildings. In this case, $K_b >> K_c$ and $N_g \neq \infty$ hold. Consequently, $r_i \approx 0$, $K \approx K_c$, $s_i < 1$, $K_e \approx \sum K_{c,i}s_i < K_c$ and $s < 1$. $\beta \approx sK_c/N_w > 0$ and $\alpha > 1$. This results in

$$N_{cr} = N_l + sK_c + (\alpha - \beta - 1)sN_l = [s(\alpha - \beta - 1) + 1 + \beta]\frac{7.837 E_w I_w r_s}{H^2} \tag{5.22}$$

As $[s(\alpha - \beta - 1) + 1 + \beta] > 1$ always holds, due to the supporting effect of the shear stiffness of the frames, the overall critical load is again greater than that of the shear walls/cores. However, the magnitude of the increase in this case is more difficult to estimate as, in addition to the effect of the columns as in the previous case, it also depends on the "eroding" effect of the axial deformations of the columns (through parameter s).

Third, the axial deformations of the columns are assumed to be very great. The practical case that may belong here is the case of medium/high-rise buildings with columns of relatively small cross-section. In this special case, $K_b >> K_c$ and $N_g \approx 0$ hold. Consequently, $r_i \approx 0$, $K \approx K_c$, $s_i \approx 0$, $K_e \approx 0$ and $s \approx 0$. $\beta \approx 0$ and $\alpha \approx 1$. This results in

$$N_{cr} = [s(\alpha - \beta) + 1 - s]N_w = N_w = \frac{7.837 E_w I_w r_s}{H^2} \tag{5.23}$$

Due to the excessive axial deformation of the columns, the shear capacity of the frames is practically eroded and the shear walls and cores act as individual bracing units in bending—c.f. Section 5.2.1 and Equation (5.19).

5.2.4 Shear walls and frames with very high column/beam stiffness ratio

The situation is similar to that of the previous case and, again, three characteristic cases can be distinguished.

First, the axial deformations of the columns are assumed to be negligible. Practical case: low/medium-rise buildings.

In this special case, $K_b << K_c$ and $N_g \approx \infty$ hold. Consequently, $r_i \approx 1$, $K \approx K_b$, $s_i \approx 1$, $K_e \approx K_b$ and $s \approx 1$. $\beta \approx K_b/N_l > 0$ and $\alpha > 1$. This leads to

$$N_{cr} = (\alpha - \beta)N_l + K_b = \alpha(N_f + N_w) = \alpha \frac{7.837 r_s (E_c I_c + E_w I_w)}{H^2} \tag{5.24}$$

showing that the critical load is based on the bending critical load of the full-height columns, shear walls and cores; this value is presumably slightly increased (through $\alpha > 1$) due to the interaction between the bracing units in shear (frames) and the bracing units in bending (walls and cores). The shear stiffness is characterised by the weakest link, i.e. by the stiffness of the beams.

Second, it is assumed that axial deformations of the columns are not negligible. Practical case: low/medium-rise buildings.

In this case, $K_b << K_c$ and $N_g \neq \infty$ hold. Consequently, $r_i \approx 1$, $K \approx K_b$, $s_i < 1$, $K_e \approx \sum K_{b,i} s_i < K_b$ and $s < 1$. $\beta \approx sK_b/N_l > 0$ and $\alpha > 1$. This results in

$$N_{cr} = [s(\alpha - \beta - 1) + 1 + \beta] \frac{7.837 r_s (E_c I_c + E_w I_w)}{H^2} \tag{5.25}$$

As $[s(\alpha - \beta - 1) + 1 + \beta] > 1$ always holds, due to the supporting effect of the shear stiffness of the frames, the overall critical load is greater than that of the shear walls/cores. However, the magnitude of the increase in this case is more difficult to estimate as, in addition to the effect of the columns as in the previous case, it also depends on the "eroding" effect of the axial deformations of the columns (through parameter s). The stiffness of the columns ($E_c I_c$) is in most practical cases negligible compared to the stiffness of the shear walls and cores (E_wI_w).

Third, the axial deformations of the columns are assumed to be very great. Practical case: high-rise buildings with frames of great global slenderness.

In this special case, $K_b << K_c$ and $N_g \approx 0$ hold. Consequently, $r_i \approx 1$, $K \approx K_b$, $s_i \approx 0$, $K_e \approx 0$ and $s \approx 0$. $\beta \approx 0$ and $\alpha \approx 1$. This results in

$$N_{cr} = [s(\alpha - \beta) + 1 - s](N_f + N_w) = \frac{7.837 r_s (E_c I_c + E_w I_w)}{H^2} \tag{5.26}$$

Due to the excessive axial deformation of the columns, all the shear capacity of the frames is eroded and the shear walls and cores work as individual bracing units in bending. The bending stiffness of the columns (E_cI_c) is in most practical cases negligible compared to that of the shear walls and cores (E_wI_w).

5.3 PURE TORSIONAL BUCKLING

Although the torsional buckling problem is more complex than that of sway buckling, the solution is obtained in a relatively simple way. An analogy can be used between the three-dimensional torsional problem and the two-dimensional buckling problem (discussed in the previous section). This analogy is well known in the stress analysis of thin-walled structures in bending and torsion (Vlasov, 1961; Kollbrunner and Basler, 1969). (The same analogy was used in Chapter 4 for the analysis of pure torsional vibration.) According to the analogy, translations, bending moments and shear forces correspond to rotations, warping moments and torsional moments, respectively. It will be demonstrated in the following that the analogy can be extended to the sway buckling of an elastically supported cantilever (discussed in the previous section) and the pure torsional buckling of a cantilever of thin-walled cross-section (to be investigated in this section).

The model which is used for the pure torsional buckling analysis of the building is an equivalent cantilever of thin-walled, open cross-section which has effective Saint-Venant stiffness $(GJ)_e$ and warping stiffness EI_ω. This equivalent column replaces the bracing system of the building for the torsional analysis (Figure 5.4). As the equivalent column is situated in the shear centre, and for the determination of the characteristic torsional stiffnesses the location of the shear centre is needed, the first step is to determine the location of the shear centre.

For bracing systems developing bending deformation only, the location of the shear centre is calculated using the bending stiffness of the bracing units. However, with bracing systems having frames and coupled shear walls as well, the shear deformation of some of the bracing units may be of considerable magnitude (in addition to their bending deformation). The behaviour of such systems is complex (and the location of the shear centre may even vary over the height). No exact solution is available for this case but, as a good approximation, the formulae given below can be used to determine the location of the shear centre.

As the critical load of a bracing unit reflects both its bending and shear stiffnesses, the location of the shear centre may be determined using the sway critical loads of the bracing units:

$$\bar{x}_o = \frac{\sum_{i=1}^{f+m} N_{y,i}\bar{x}_i}{\sum_{i=1}^{f+m} N_{y,i}}, \qquad \bar{y}_o = \frac{\sum_{i=1}^{f+m} N_{x,i}\bar{y}_i}{\sum_{i=1}^{f+m} N_{x,i}} \tag{5.27}$$

where $\bar{x}_i$ and $\bar{y}_i$ are the perpendicular distances of the bracing units from axes $\bar{y}$ and $\bar{x}$ and f and m are the number of frames/coupled shear walls and shear walls/cores, respectively (Figure 5.4). Any suitable method can be used for the calculation of the critical loads in Equations (5.27), including Equation (5.18) given in Section 5.1—c.f. special cases discussed in Sections 5.2.

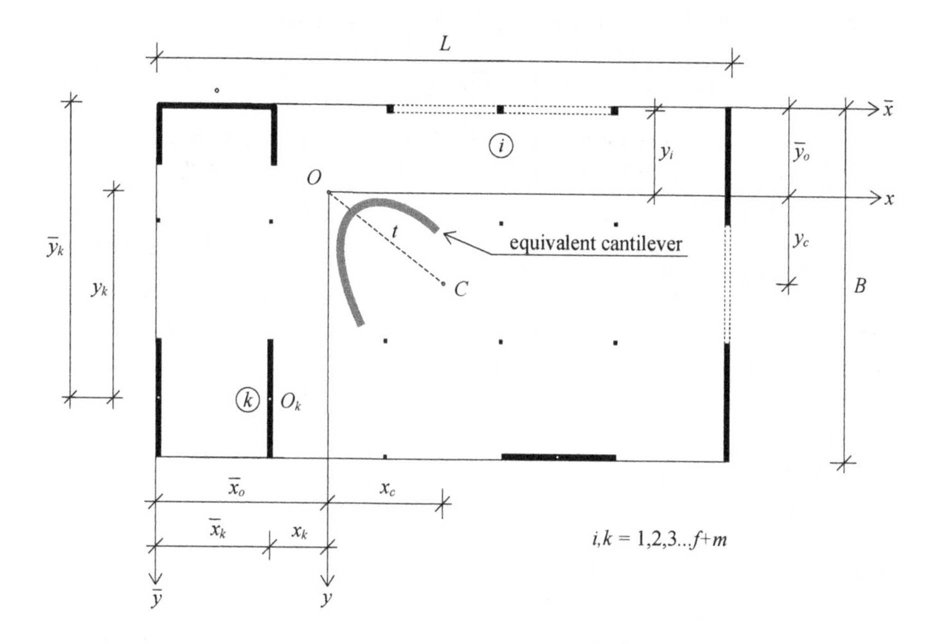

Figure 5.4 Layout with the equivalent cantilever of open, thin-walled cross-section in the shear centre.

When the bracing system consists of shear walls (and cores) only, Equations (5.27) for the location of the shear centre simplify and the shear centre coordinates can be determined using

$$\bar{x}_o = \frac{\sum_{k=1}^{m} I_{x,k}\bar{x}_k}{\sum_{k=1}^{m} I_{x,k}}, \qquad \bar{y}_o = \frac{\sum_{k=1}^{m} I_{y,k}\bar{y}_k}{\sum_{k=1}^{m} I_{y,k}} \tag{5.28}$$

Equations (5.28) can also be used for the frequency and deflection calculations for systems that only have bending stiffness.

With the location of the shear centre now established, the characteristic torsional stiffnesses of the equivalent column can be determined.

The effective Saint-Venant stiffness of the system may come from two sources: from the Saint-Venant stiffness of the shear walls and cores and from the effective shear stiffness of the frames as

$$(GJ)_e = \sum_{k=1}^{m} GJ_k + \sum_{i=1}^{f} \left((K_{e,i})_x y_i^2 + (K_{e,i})_y x_i^2 \right) \tag{5.29}$$

with

J_k	the Saint-Venant constant of the *k*th wall/core
G	the modulus of elasticity in shear of the walls/cores
$(K_{e,i})_x$, $(K_{e,i})_y$	the effective shear stiffnesses of the *i*th frame/coupled shear walls in directions x and y, respectively
x_i, y_i	the perpendicular distances of the *i*th frame/coupled shear walls from the shear centre in directions x and y, respectively

If the bracing system consists of frames, (coupled) shear walls and cores of open cross-section, the first term in Equation (5.29) is normally negligible compared to the contribution of the frames.

The warping stiffness of the system may originate from three sources: the own warping stiffness of the cores, the bending stiffness of the walls and the bending stiffness of the columns of the frames/wall sections of the coupled shear walls:

$$EI_\omega = E_w \sum_{k=1}^{m} \left(I_{\omega,k} + (I_{w,k})_x y_k^2 + (I_{w,k})_y x_k^2 \right) + E_c \sum_{i=1}^{f} \left((I_{c,i} r_i)_x y_i^2 + (I_{c,i} r_i)_y x_i^2 \right) \tag{5.30}$$

with

$I_{\omega,k}$	the warping constant of the *k*th wall/core
$E_w(I_{w,k})_x$, $E_w(I_{w,k})_y$	the bending stiffnesses of the *k*th wall/core in directions x and y, respectively
$E_c(I_{c,i}r_i)_x$, $E_c(I_{c,i}r_i)_y$	the bending stiffnesses of the columns/wall sections of the *i*th frame in directions x and y, respectively
x_k, y_k	the perpendicular distances of the *k*th wall/core from the shear centre in directions x and y, respectively
x_i, y_i	the perpendicular distances of the *i*th frame/coupled shear walls from the shear centre in directions x and y, respectively

The warping stiffness of a well-balanced bracing system is normally dominated by the contribution of the shear walls and cores (if their perpendicular distance from the shear centre is great enough). The contribution of the cores through their own warping stiffness [first term in Equation (5.30)] tends to be much smaller and the effect of the columns of the frames (last two terms) is normally negligible.

To facilitate the easy calculation of warping constant I_ω, closed-form formulae for cross-sections widely used for bracing cores are given in Tables 2.6, 2.7 and 2.8. More formulae are available in (Zalka, 2000). For bracing units of special (irregular) cross-sections where no closed-form solution is available, computer programs can be used. One such program is PROSEC (1994), later Prokon, whose accuracy has been established and proved to be within the range required for structural engineering calculations.

Everything is now ready for the torsional buckling analysis. After examining the equilibrium of an elementary section of the equivalent column with the above torsional stiffness characteristics, the governing differential equation of the torsional problem emerges as

$$\frac{r_s EI_\omega}{i_p^2}\varphi''''(z)+\left[\left(N(z)-\frac{(GJ)_e}{i_p^2}\right)\varphi'(z)\right]'=0 \tag{5.31}$$

where $\varphi(z)$ is the rotation of the cross-section, $N(z)$ is the vertical load at z, and i_p is the radius of gyration. As the derivation of an equation identical in structure to Equation (5.31) demonstrates (Zalka and Armer, 1992), the radius of gyration enters the picture when the role of the load of the structure is investigated. With multi-storey buildings, the load in question is the vertical load on the floors. For regular multi-storey buildings of rectangular plan-shape that are subjected to a uniformly distributed load over the floors, the radius of gyration is obtained from

$$i_p=\sqrt{\frac{L^2+B^2}{12}+t^2} \qquad \text{with} \qquad t=\sqrt{x_c^2+y_c^2} \tag{5.32}$$

In the above equations L and B are the plan length and breadth of the building, t is the distance between the geometrical centre of the plan of the building (C) and the shear centre of the bracing system (O) and x_c and y_c are the coordinates of the geometrical centre (Figure 5.4):

$$x_c=\frac{L}{2}-\bar{x}_o \qquad \text{and} \qquad y_c=\frac{B}{2}-\bar{y}_o \tag{5.33}$$

For arbitrary plan-shapes and/or other types of load distribution, formulae for the radius of gyration are available elsewhere (Kollár, 1999; Zalka, 2000). It is important to note that the value of i_p depends on the geometrical characteristics of the plan of the building, rather than the stiffness characteristics of the bracing system.

The origin of the coordinate system is placed at and fixed to the shear centre of the cross-section at the top of the equivalent column (Figure 5.5). The boundary conditions in this coordinate system are:

$$\varphi(0)=0, \qquad \varphi''(0)=0$$

$$\varphi'(H) = 0, \qquad \varphi'''(H) = 0$$

This eigenvalue problem is clearly analogous with the one defined by the governing differential equation of the sway buckling problem [Equation (5.15)] and its boundary conditions. Bending stiffness EI and the elastic support defined by the effective shear stiffness K_e in Equation (5.15) correspond to warping stiffness EI_ω and effective Saint-Venant stiffness $(GJ)_e$, divided by i_p^2 in Equation (5.31), respectively. It follows that the solution to Equation (5.15) can be used and converted to represent the solution of Equation (5.31). In doing so, the critical load of pure torsional buckling is obtained in the same manner as with Equation (5.18):

$$N_{cr,\varphi} = N_\omega + N_t + (\alpha_\varphi - \beta_\varphi - 1)s_\varphi N_\omega \tag{5.34}$$

where the warping critical load of the system is

$$N_\omega = \frac{7.837 r_s EI_\omega}{i_p^2 H^2} \tag{5.35}$$

and the Saint-Venant critical load is

$$N_t = \frac{(GJ)_e}{i_p^2} \tag{5.36}$$

It is interesting to note that the value of the Saint-Venant critical load does not depend on the height of the building.

The effectiveness of the Saint-Venant stiffness is expressed by the factor

$$s_\varphi = \frac{(GJ)_e}{(GJ)} \tag{5.37}$$

where the "original" Saint-Venant stiffness is

$$(GJ) = \sum_{k=1}^{m} GJ_k + \sum_{i=1}^{f} \left((K_i)_x y_i^2 + (K_i)_y x_i^2\right) \tag{5.38}$$

Values for the critical load parameter α_φ are given in Table 5.2 as a function of parameter β_φ:

$$\beta_\varphi = \frac{N_t}{N_\omega} \tag{5.39}$$

In making use of the analogy, special cases can be investigated in the same manner as in Section 5.2.

In the special case when the bracing system only consists of a single bracing unit with no warping stiffness—e.g. a closed or partially closed U-core—Equation (5.34) simplifies to

$$N_{cr,\varphi} = N_t = \frac{GJ}{i_p^2} \tag{5.40}$$

This case is covered in Section 2.7.4—see Equation (2.99)—which deals with the stability of a single core. Section 6.4 also deals with this case in great detail.

5.4 COUPLED SWAY-TORSIONAL BUCKLING

In structural engineering practice, bracing systems tend to be asymmetric when the shear centre does not coincide with the geometrical centre of the building (Figure 5.5/a). In such cases, interaction occurs between the two lateral and pure torsional modes and sway buckling in the two principal directions combines with pure torsional buckling, resulting in triple coupling (Figure 5.5/b).

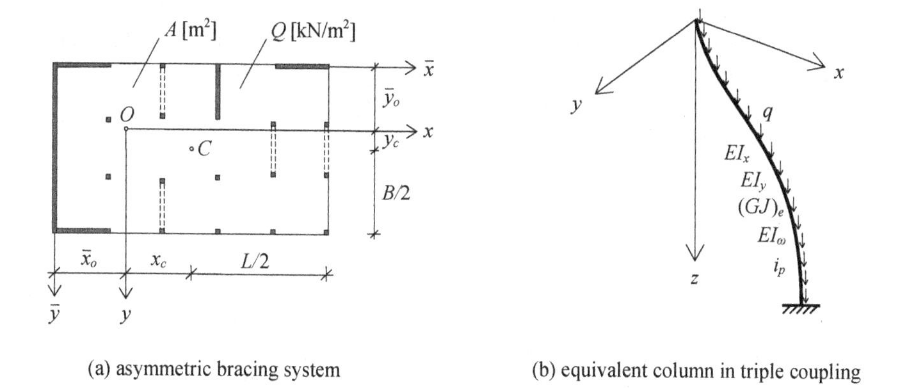

(a) asymmetric bracing system

(b) equivalent column in triple coupling

Figure 5.5 Combined sway-torsional buckling.

The situation is very similar to that of the vibration problem. Accordingly, there are two possibilities to take into account the effect of interaction: exactly and approximately. The exact method automatically covers all the three possibilities (triple, double, and no coupling). This method is presented first.

The critical load is obtained by solving the cubic equation

$$N^3 + b_2 N^2 + b_1 N - b_0 = 0 \tag{5.41}$$

whose smallest root yields the combined global critical load of the building. In the case of buildings subjected to uniformly distributed floor load, Equation (5.41) is exact, as far as the effect of the coupling of the three modes is concerned.

The coefficients in Equation (5.41) are

$$b_0 = \frac{N_{cr,x} N_{cr,y} N_{cr,\varphi}}{1-\tau_x^2-\tau_y^2}, \qquad b_1 = \frac{N_{cr,x} N_{cr,y} + N_{cr,\varphi} N_{cr,x} + N_{cr,\varphi} N_{cr,y}}{1-\tau_x^2-\tau_y^2}$$

$$b_2 = \frac{N_{cr,x}\tau_x^2 + N_{cr,y}\tau_y^2 - N_{cr,x} - N_{cr,y} - N_{cr,\varphi}}{1-\tau_x^2-\tau_y^2} \tag{5.42}$$

where τ_x and τ_y are eccentricity parameters:

$$\tau_x = \frac{x_c}{i_p} \quad \text{and} \quad \tau_y = \frac{y_c}{i_p} \tag{5.43}$$

The radius of gyration i_p and the coordinates of the geometrical centre x_c and y_c are given by Equations (5.32) and (5.33).

If a quick solution is needed or a cubic equation solver is not available or if one of the basic critical loads is much smaller than the others, the Föppl-Papkovich theorem (Tarnai, 1999) offers a simple albeit approximate solution.

If the bracing system is asymmetric and the centroid of the vertical load of the building does not lie on either principal axis of the bracing system, triple coupling occurs and an approximation of the resulting combined critical load is obtained using the reciprocal summation as

$$N_{cr} = \left(\frac{1}{N_{cr,x}} + \frac{1}{N_{cr,y}} + \frac{1}{N_{cr,\varphi}}\right)^{-1} \tag{5.44}$$

If the arrangement of the bracing system is monosymmetric and the centroid of the vertical load of the building lies on one of the principal axes of the bracing system (say, axis x), then two things may happen. Sway buckling may develop in direction x (defined by $N_{cr,x}$) or buckling in direction y (defined by $N_{cr,y}$) couples with pure torsional buckling (defined by $N_{cr,\varphi}$). The critical load of this coupled buckling is obtained from

$$N_{y\varphi} = \left(\frac{1}{N_{cr,y}} + \frac{1}{N_{cr,\varphi}}\right)^{-1} \tag{5.45}$$

The global critical load of the building is the smaller one of $N_{cr,x}$ and $N_{y\varphi}$, i.e.:

$$N_{cr} = \min(N_{cr,x}, N_{y\varphi}) \tag{5.46}$$

In the special case when the arrangement of the bracing system is doubly symmetric and the centroid of the vertical load of the building coincides with the shear centre of the bracing system, no coupling occurs and the global critical load of the building is the smallest one of $N_{cr,x}$, $N_{cr,y}$ and $N_{cr,\varphi}$, i.e.:

$$N_{cr} = \min(N_{cr,x}, N_{cr,y}, N_{cr,\varphi}) \tag{5.47}$$

Simplicity and the fact that the above Föppl-Papkovich equations are always conservative may justify the use of the approximate solutions. However, their application may lead to rather uneconomic structural solutions as they cannot take into consideration the degree of eccentricity of the bracing system. The (conservative) error of Equation (5.44) can be as much as 67%.

When the global critical load of the system is calculated, the global critical load ratio (critical load/actual load) can be used to assess the effectiveness of the bracing system. It also indicates whether or not a more sophisticated second-order analysis needs to be carried out. The application of the global critical load ratio is discussed in the next chapter in detail and it is only mentioned here that the greater the global critical load ratio, the greater the level of safety against buckling.

5.5 CONCENTRATED TOP LOAD

In certain cases, concentrated load on top of the building may need to be taken into account. A panorama restaurant, a swimming pool or a water tank may represent some extra load that cannot be covered by the uniformly distributed floor load, considered to be the same at each floor level over the height of the building. The critical load for the concentrated top load case can be determined relatively easily following the procedure presented in the previous sections but with using the corresponding Euler critical loads as shown below.

The critical concentrated load for sway buckling, based on Equation (5.18), assumes the form

$$F_{cr} = F_l + K_e \tag{5.48}$$

where F_l is the local bending critical load and K_e is the effective shear stiffness for concentrated top load case. The effective shear stiffness is determined as

$$K_e = \left(\frac{1}{K} + \frac{1}{F_g}\right)^{-1} \tag{5.49}$$

where F_g is the global bending critical load. Note that for the part critical loads F_l and F_g, the corresponding Euler critical loads should be used:

$$F_l = \frac{\pi^2 EI}{4H^2} \tag{5.50}$$

and

$$F_g = \frac{\pi^2 EI_g}{4H^2} \tag{5.51}$$

To calculate sway critical loads $F_{cr,x}$ and $F_{cr,y}$ in principal directions x and y, the relevant bracing units (with their stiffnesses) should be used.

For pure torsional buckling, Equation (5.34) leads to

$$F_{cr,\varphi} = F_\omega + F_t \tag{5.52}$$

where the warping critical load of the system is

$$F_\omega = \frac{\pi^2 EI_\omega}{4H^2 i_p^2} \tag{5.53}$$

and the warping stiffness is given by Equation (5.30).

The Saint-Venant critical load is obtained from

$$F_t = \frac{(GJ)_e}{i_p^2} \tag{5.54}$$

where effective Saint-Venant stiffness $(GJ)_e$ is determined using Equation (5.29) which consists of two parts. The first term stands for the Saint-Venant stiffness of the cores/walls and the second term represents the contribution of the frames. This term is calculated using K_e according to Equation (5.49).

It is interesting to note that when the load is concentrated on top of the building, there is no interaction between the bending and shear modes (that would increase the critical load).

Once the basic critical loads $F_{cr,x}$, $F_{cr,y}$ and $F_{cr,\varphi}$ are available, the coupling of the basic modes is taken into account exactly as with $N_{cr,x}$, $N_{cr,y}$ and $N_{cr,\varphi}$ in Section 5.4 above, i.e. applying Equations (5.41), (5.42), (5.43) and (5.33) for the exact analysis or Equations (5.44), (5.45), (5.46) and (5.47) for the approximate analysis, using the Euler basic critical loads.

5.6 WORKED EXAMPLES

Two worked examples are presented in this chapter for the calculation of the global critical load of buildings under uniformly distributed vertical load over the floors. Both buildings braced by frames and shear walls (Systems S26 and S29) are part of the collection used for a comprehensive accuracy analysis in Chapter 7 (see Figure 7.23). The calculations are based on the material presented earlier in Chapters 2 and 5, and the numbers of the equations used will be given on the right-hand side in curly brackets.

5.6.1 Critical load of twenty-storey monosymmetric building

The twenty-storey building is braced by two reinforced concrete frames (F1) and two shear walls (W6) in direction x, a frame (F5) in direction y and a core (U) which acts in both directions (Figure 5.6). The core also has its own warping stiffness. This bracing system is used in Section 7.3 for a comprehensive accuracy analysis (as system S26 in Figure 7.23/d). The building is subjected to a uniformly distributed vertical floor load. The modulus of elasticity is $E = 25000$ MN/m^2, the modulus of elasticity in shear is $G = 10400$ MN/m^2, the storey height is $h = 3$ m and the total height of the building is $H = 60$ m.

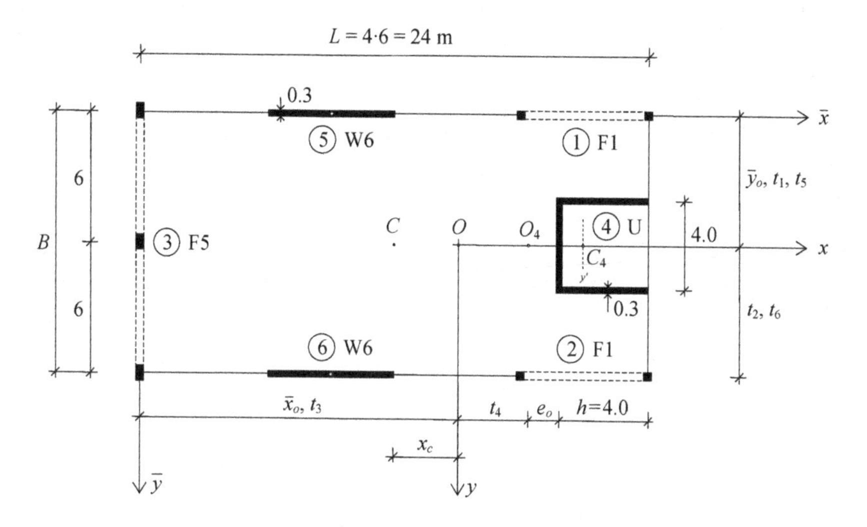

Figure 5.6 Layout of twenty-storey building.

Load distribution factor r_s will be needed throughout the calculations. Its value is obtained from Table 5.1 as a function of the number of storeys: $r_s = 0.926$.

The procedure for the determination of the global critical load of the building is carried out in three steps. The basic characteristics of the bracing units are calculated first. The basic critical loads are determined in the second step when it is tacitly assumed that sway buckling in direction x, sway buckling in direction y and pure torsional buckling happen independently of each other. Finally, the interaction among the basic critical loads is taken into account.

Basic characteristics of the individual bracing units

Before the whole system is investigated, it is advantageous to establish the basic characteristics of the individual bracing units for later use in steps 2 and 3.

Bracing Units 1 and 2 (frame F1, Figure 5.7/a)
The cross-sections of both the columns and the beams of frame F1 are 0.40/0.40 (metres).

The shear stiffness that is associated with the beams of the frame is

$$K_{b,1} = \sum_{j=1}^{n-1} \frac{12E_b I_{b,j}}{l_j h} = \frac{12 \cdot 25 \cdot 10^3 \cdot 0.4^4}{12 \cdot 6 \cdot 3} = 35.556 \text{ MN} \qquad \{2.53\}$$

The shear stiffness that is associated with the columns of the frame is

$$K_{c,1} = \sum_{j=1}^{n} \frac{\pi^2 E_c I_{c,j}}{h^2} = 2\frac{\pi^2 \cdot 25 \cdot 10^3 \cdot 0.4^4}{12 \cdot 3^2} = 116.97 \text{ MN} \qquad \{2.54\}$$

The combination of the two part shear stiffnesses gives the "original" shear stiffness of the frame:

$$K_1 = K_{b,1} \frac{K_{c,1}}{K_{b,1} + K_{c,1}} = 35.556 \frac{116.97}{35.556 + 116.97} = 27.27 \text{ MN} \qquad \{2.55\}$$

where

$$r_1 = \frac{K_{c,1}}{K_{b,1} + K_{c,1}} = \frac{116973}{35556 + 116973} = 0.7669 \qquad \{2.56\}$$

is the reduction factor.

The global second moment of area is

$$I_{g,1} = \sum_{j=1}^{n} A_{c,j} t_j^2 = 0.4 \cdot 0.4 \cdot 3^2 \cdot 2 = 2.88 \text{ m}^4 \qquad \{2.57\}$$

With the global second moment of area, the global bending critical load of the frame is

$$N_{g,1} = \frac{7.837 r_s E_c I_{g,1}}{H^2} = \frac{7.837 \cdot 0.926 \cdot 25 \cdot 10^3 \cdot 2.88}{60^2} = 145.1 \text{ MN} \qquad \{5.7\}$$

Because of the different buckling shape of the shear and global bending modes, there is an interaction between the two modes and the original shear stiffness is reduced by the effectiveness factor

$$s_1 = \frac{N_{g,1}}{K_1 + N_{g,1}} = \frac{145.1}{27.27 + 145.1} = 0.8418 \qquad \{5.9\}$$

The effective shear stiffness can now be determined:

$$K_{e,1} = K_1 s_1 = 27.27 \cdot 0.8418 = 22.96 \text{ MN} \qquad \{5.8\}$$

The local second moment of area of the frame, amended by r and making use of the fact that the two columns have the same cross-section, is

$$I_1 = r_1 \sum_{j=1}^{n} I_{c,j} = 0.7669 \cdot 2 \frac{0.4^4}{12} = 0.003272 \text{ m}^4 \qquad \{2.58\}$$

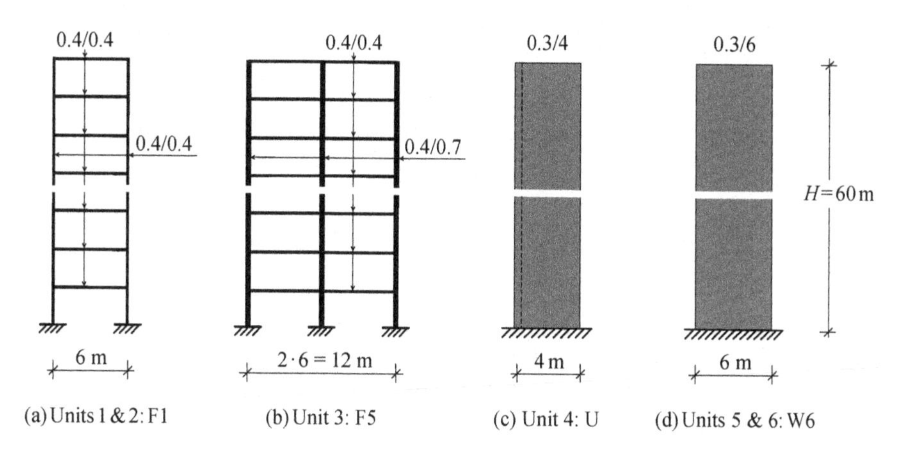

Figure 5.7 Bracing units for the twenty-storey building.

Bracing Unit 3 (frame F5, Figure 5.7/b)
The cross-sections of the columns and beams of frame F5 are 0.40/0.70 and 0.40/0.40 (metres), respectively.

The shear stiffness that is associated with the beams of the frame is

$$K_{b,3} = \sum_{j=1}^{n-1} \frac{12 E_b I_{b,j}}{l_j h} = 2 \frac{12 \cdot 25 \cdot 10^3 \cdot 0.4^4}{12 \cdot 6 \cdot 3} = 71.11 \text{ MN} \qquad \{2.53\}$$

The shear stiffness that is associated with the columns of the frame is

$$K_{c,3} = \sum_{j=1}^{n} \frac{\pi^2 E_c I_{c,j}}{h^2} = 3 \frac{\pi^2 \cdot 25 \cdot 10^3 \cdot 0.4 \cdot 0.7^3}{12 \cdot 3^2} = 940.4 \text{ MN} \qquad \{2.54\}$$

The combination of the two part shear stiffnesses gives the "original" shear stiffness of the frame:

$$K_3 = K_{b,3} \frac{K_{c,3}}{K_{b,3} + K_{c,3}} = 71.11 \frac{940.4}{71.11 + 940.4} = 66.11 \text{ MN} \qquad \{2.55\}$$

where

$$r_3 = \frac{K_{c,3}}{K_{b,3} + K_{c,3}} = \frac{940.4}{71.11 + 940.4} = 0.9297 \qquad \{2.56\}$$

is the reduction factor.

The global second moment of area is

$$I_{g,3} = \sum_{j=1}^{n} A_{c,j} t_j^2 = 0.4 \cdot 0.7 \cdot 6^2 \cdot 2 = 20.16 \text{ m}^4 \qquad \{2.57\}$$

With the global second moment of area, the global bending critical load of the frame is

$$N_{g,3} = \frac{7.837 r_s E_c I_{g,3}}{H^2} = \frac{7.837 \cdot 0.926 \cdot 25 \cdot 10^3 \cdot 20.16}{60^2} = 1016 \text{ MN} \qquad \{5.7\}$$

Because of the different buckling shape of the shear and global bending modes, there is an interaction between the two modes and the original shear stiffness is reduced by the effectiveness factor

$$s_3 = \frac{N_{g,3}}{K_3 + N_{g,3}} = \frac{1016}{66.11 + 1016} = 0.9389 \qquad \{5.9\}$$

The effective shear stiffness can now be determined:

$$K_{e,3} = K_3 s_3 = 66.11 \cdot 0.9389 = 62.07 \text{ MN} \qquad \{5.8\}$$

The local second moment of area of the frame, amended by r and making use of the fact that the three columns have the same cross-section, is

$$I_3 = r_3 \sum_{j=1}^{n} I_{c,j} = 0.9297 \cdot 3 \frac{0.4 \cdot 0.7^3}{12} = 0.03189 \text{ m}^4 \qquad \{2.58\}$$

As a function of the two coefficients given by Equation {2.62}

$$\frac{I_3}{I_{f,3}} = \frac{0.03189}{0.03189 + 20.16} = 0.00158, \quad \frac{K_3 H^2}{E I_{f,3} r_s} = \frac{66.11 \cdot 60^2}{25 \cdot 10^3 \cdot 20.1919 \cdot 0.926} = 0.509$$

parameter c_1 is obtained from Table 2.1

	0.001	0.005
0.5	0.665	0.815
1.0	1.222	1.403

using the relevant values:

$$1:\quad 0.665+\frac{0.815-0.665}{0.005-0.001}(0.00158-0.001)=0.68675$$

$$2:\quad 1.222+\frac{1.403-1.222}{0.005-0.001}(0.00158-0.001)=1.248245$$

$$3:\quad c_1=0.68675+\frac{1.248245-0.68675}{1-0.5}(0.509-0.5)=0.697$$

The critical load of the frame can now be calculated:

$$N_{cr,3}=c_1\frac{EI_{f,3}r_s}{H^2}=0.697\frac{25\cdot 10^3\cdot 20.1919\cdot 0.926}{60^2}=90.5\ \text{MN} \qquad \{2.60\}$$

Bracing Unit 4 (U-core, Figures 5.6 and 5.7/c)
The U-core contributes both to the lateral and the torsional resistance of the bracing system. Its basic stiffness characteristics are determined using the formulae given in Tables 2.6 and 2.7. These are collected in the following.

$$J=0.108\ \text{m}^4,\qquad I_\omega=18.29\ \text{m}^6,\qquad e_o=1.714\ \text{m} \qquad \{\text{Table 2.6}\}$$

$$I_x=11.245\ \text{m}^4,\qquad I_y=6.427\ \text{m}^4 \qquad \{\text{Table 2.7}\}$$

The critical load of the U-core in direction y will also be needed with the torsional analysis:

$$N_{cr,4}=\frac{7.837EI_xr_s}{H^2}=\frac{7.837\cdot 25\cdot 10^3\cdot 11.245\cdot 0.926}{60^2}=566.7\ \text{MN} \qquad \{2.96\}$$

Bracing Units 5 and 6 (shear wall W6; Figure 5.7/d)
The size of the shear wall is 6.0 metres with a thickness of 0.3 m. The second moments of area of the shear wall are

$$I_{w,5}^{\max}=I_{w,6}^{\max}=\frac{0.3\cdot 6^3}{12}=5.4\ \text{m}^4 \quad\text{and}\quad I_{w,5}^{\min}=I_{w,6}^{\min}=\frac{6\cdot 0.3^3}{12}=0.0135\ \text{m}^4$$

The Saint-Venant constant is

$$J_{w,5}=J_{w,6}=\frac{6\cdot 0.3^3}{3}=0.054\ \text{m}^4$$

The critical load of the shear wall in direction y will also be needed with the torsional analysis:

$$N_{cr,5} = \frac{7.837 E I_x r_s}{H^2} = \frac{7.837 \cdot 25 \cdot 10^3 \cdot 0.0135 \cdot 0.926}{60^2} = 0.68 \text{ MN} \qquad \{2.76\}$$

Sway buckling in direction x (Bracing Units 1, 2, 4, 5 and 6)

The original shear stiffness of the system originates from Bracing Units 1 and 2, the two identical frames F1. Using the data above, the original shear stiffness is

$$K = \sum_{i=1}^{f} K_i = 2 \cdot 27.27 = 54.54 \text{ MN} \qquad \{5.5\}$$

Using the effectiveness factor, the effective shear stiffness of the system is

$$K_e = \sum_{i=1}^{f} K_i s_i = 2 \cdot 27.27 \cdot 0.8418 = 45.91 \text{ MN} \qquad \{5.10\}$$

As the two frames of the system in direction x are identical, the effectiveness factor of the system equals that of frame F1, or using the above values

$$s = \frac{K_e}{K} = \frac{45.91}{54.54} = 0.8418 \qquad \{5.11\}$$

The local bending stiffness is also needed. It consists of the bending stiffness of the columns of the frames, the bending stiffness of the two shear walls and the relevant stiffness of the U-core. The contribution of the columns of the frames is small compared to that of the shear walls and the U-core and is therefore neglected:

$$EI = E_c \sum_{i=1}^{f} I_{c,i} r_i + E_w \sum_{k=1}^{m} I_{w,k} = 25 \cdot 10^3 (2 \cdot 5.4 + 6.427) = 4.31 \cdot 10^5 \text{ MNm}^2 \qquad \{5.12\}$$

With the local bending stiffness, the local bending critical load is

$$N_l = \frac{7.837 r_s EI}{H^2} = \frac{7.837 \cdot 0.926 \cdot 4.31 \cdot 10^5}{60^2} = 868.8 \text{ MN} \qquad \{5.13\}$$

The two main components (N_l and K_e) of the sway critical load are now available. The third component—responsible for the interaction between the bending and shear modes—is obtained as a function of the dimensionless ratio

$$\beta = \frac{K_e}{N_l} = \frac{45.91}{868.8} = 0.0528 \qquad \{5.17\}$$

According to Table 5.2, the critical load parameter is

$$\alpha = 1.1487 + \frac{1.1782 - 1.1487}{0.06 - 0.05}(0.0528 - 0.05) = 1.157 \qquad \{\text{Table 5.2}\}$$

Finally, the sway critical load in direction x is

$$N_{cr,x} = N_l + K_e + (\alpha - \beta - 1)sN_l \qquad \{5.18\}$$

$$= 868.8 + 45.91 + (1.157 - 0.0528 - 1)0.8418 \cdot 868.8 = 990.9 \text{ MN}$$

Sway buckling in direction y (Bracing Units 3 and 4)

The situation in direction y is slightly simpler as only one frame (F5) and the U-core have to be considered. The effective shear stiffness and the effectiveness factor are already available at frame F5 as

$K_e = 62.07$ MN and $s = 0.9389$

The local bending stiffness is also needed. It consists of the bending stiffness of the columns of the frames, the bending stiffness of the two shear walls and the relevant stiffness of the U-core. The contribution of the columns of the frames and that of the two shear walls, being small, is ignored:

$$EI = E_c \sum_{i=1}^{f} I_{c,i} r_i + E_w \sum_{k=1}^{m} I_{w,k} = 25 \cdot 10^3 \cdot 11.245 = 2.81 \cdot 10^5 \text{ MNm}^2 \qquad \{5.12\}$$

With the local bending stiffness, the local bending critical load is

$$N_l = \frac{7.837 r_s EI}{H^2} = \frac{7.837 \cdot 0.926 \cdot 2.81 \cdot 10^5}{60^2} = 566.5 \text{ MN} \qquad \{5.13\}$$

As a function of the dimensionless ratio

$$\beta = \frac{K_e}{N_l} = \frac{62.07}{566.5} = 0.1096 \qquad \{5.17\}$$

the critical load parameter is

$$\alpha = 1.2949 + \frac{1.5798 - 1.2949}{0.2 - 0.1}(0.1096 - 0.1) = 1.322 \qquad \{\text{Table 5.2}\}$$

Finally, the sway critical load in direction y is

$$N_{cr,y} = N_l + K_e + (\alpha - \beta - 1)sN_l \qquad \{5.18\}$$

$$= 566.5 + 62.07 + (1.322 - 0.1096 - 1)0.9389 \cdot 566.5 = 741.5 \text{ MN}$$

Pure torsional buckling (all bracing units participating)

The coordinates of the shear centre of the bracing system are needed first. Because of symmetry with regard to direction x, only coordinate $\bar{x}_o$ needs calculation:

$$\bar{x}_o = \frac{\sum_{i=1}^{f+m} N_{y,i}\bar{x}_i}{\sum_{i=1}^{f+m} N_{y,i}} = \frac{566.7(24-4-1.714)+2\cdot 0.68\cdot 9}{566.7+2\cdot 0.68+90.5} = 15.75 \text{ m}, \quad \bar{y}_o = 6 \text{ m} \qquad \{5.27\}$$

With the coordinates of the shear centre, a new x-y coordinate system is established whose origin is in the shear centre. The calculation is carried out in this new coordinate system.

The coordinates of the centroid of the floor area are

$$x_c = \frac{L}{2} - \bar{x}_o = \frac{24}{2} - 15.75 = -3.75 \qquad \text{and} \qquad y_c = \frac{B}{2} - \bar{y}_o = 0 \qquad \{5.33\}$$

The radius of gyration is

$$i_p = \sqrt{\frac{L^2+B^2}{12} + t^2} = \sqrt{\frac{24^2+12^2}{12} + 3.75^2} = \sqrt{74.06} = 8.61 \text{ m} \qquad \{5.32\}$$

The distances of the bracing units from the shear centre are

$$t_1 = t_2 = t_5 = t_6 = 6 \text{ m}, \qquad t_3 = \bar{x}_o = 15.75 \text{ m}, \qquad t_4 = L - \bar{x}_o - e_o - 4 = 2.536 \text{ m}$$

The "original" Saint-Venant stiffness consists of two parts. In addition to the Saint-Venant stiffness of the two shear walls and the U-core, the three frames also have contributions:

$$(GJ) = \sum_{k=1}^{m} GJ_k + \sum_{i=1}^{f}\left((K_i)_x y_i^2 + (K_i)_y x_i^2\right) \qquad \{5.38\}$$

$$= 10400(2\cdot 0.054 + 0.108) + 2\cdot 27.27\cdot 6^2 + 66.11\cdot 15.75^2 = 20609 \text{ MNm}^2$$

The first term represents the "own" contribution of the shear walls and the core and it amounts to 11% of the total Saint-Venant stiffness.

The real (effective) Saint-Venant stiffness is always smaller than the "original" one as the effect of the frames is limited by their effectiveness factor:

$$(GJ)_e = \sum_{k=1}^{m} GJ_k + \sum_{i=1}^{f} \left((K_{e,i})_x y_i^2 + (K_{e,i})_y x_i^2 \right) \qquad \{5.29\}$$

$$=10400(2\cdot 0.054+0.108)+2\cdot 0.8418\cdot 27.27\cdot 6^2+0.9389\cdot 66.11\cdot 15.75^2=19297 \text{ MNm}^2$$

The effectiveness of the Saint-Venant stiffness for the whole system can now be determined:

$$s_\varphi = \frac{(GJ)_e}{(GJ)} = \frac{19297}{20609} = 0.9363 \qquad \{5.37\}$$

The warping stiffness of the system comes from three sources: the own warping stiffness of the U-core, the bending stiffness of the two shear walls and the bending stiffness of the columns of the frames [Equation (5.30)]. When the first two items exist, the contribution of the third is normally negligible. This is the case now and the third source is neglected in the following equation:

$$EI_\omega = E_w \sum_{k=1}^{m} \left(I_{\omega,k} + (I_{w,k})_x y_k^2 + (I_{w,k})_y x_k^2 \right) \qquad \{5.30\}$$

$$= 25\cdot 10^3 (18.29 + 2\cdot 5.4\cdot 6^2 + 11.245\cdot 2.536^2) = 11.99\cdot 10^6 \text{ MNm}^4$$

The first term represents the "own" contribution of the cores and it amounts to 4% of the total warping stiffness.

With the above stiffnesses, the part torsional critical loads can now be determined.

The warping critical load of the system is

$$N_\omega = \frac{7.837 r_s EI_\omega}{i_p^2 H^2} = \frac{7.837\cdot 0.926\cdot 11.99\cdot 10^6}{74.06\cdot 60^2} = 326.4 \text{ MN} \qquad \{5.35\}$$

and the Saint-Venant critical load is

$$N_t = \frac{(GJ)_e}{i_p^2} = \frac{19297}{74.06} = 260.6 \text{ MN} \qquad \{5.36\}$$

As a function of the ratio of the above part critical loads

$$\beta_\varphi = \frac{N_t}{N_\omega} = \frac{260.6}{326.4} = 0.798 \qquad \{5.39\}$$

the critical load parameter is obtained from Table 5.2:

$$\alpha_\varphi = 2.878 + \frac{3.1163 - 2.878}{0.8 - 0.7}(0.798 - 0.7) = 3.112 \qquad \{\text{Table 5.2}\}$$

Finally, making use of the part critical loads and the dimensionless parameters, the critical load of pure torsional buckling is

$$N_{cr,\varphi} = N_\omega + N_t + (\alpha_\varphi - \beta_\varphi - 1)s_\varphi N_\omega \qquad \{5.34\}$$

$$= 326.4 + 260.6 + (3.112 - 0.798 - 1)0.9363 \cdot 326.4 = 988.6 \text{ MN}$$

The third term represents the effect of the interaction between the Saint-Venant and warping torsional modes. It amounts to a considerable 41% of the total pure torsional critical load.

Coupling of the basic critical loads

The centroid of the layout does not coincide with the shear centre of the bracing system so there is an interaction among the basic critical loads. As the interaction reduces the value of the global critical load, its effect must be taken into account. The centroid lies on axis x and it follows that interaction is realized as a double coupling: sway buckling in direction y couples with pure torsional buckling and sway buckling in direction x may happen independently of the combined mode. The interaction is best taken into account by solving the cubic equation

$$N^3 + b_2 N^2 + b_1 N - b_0 = 0 \qquad \{5.41\}$$

which automatically handles the different coupling possibilities.

With eccentricity parameters τ_x and τ_y

$$\tau_x = \frac{x_c}{i_p} = \frac{-3.75}{8.61} = -0.4355 \quad \text{and} \quad \tau_y = \frac{y_c}{i_p} = \frac{0.0}{8.61} = 0 \qquad \{5.43\}$$

the coefficients are

$$b_0 = \frac{N_{cr,x} N_{cr,y} N_{cr,\varphi}}{1 - \tau_x^2 - \tau_y^2} = \frac{990.9 \cdot 741.5 \cdot 988.6}{1 - (-0.4355)^2} = 8.964 \cdot 10^8$$

$$b_1 = \frac{N_{cr,x} N_{cr,y} + N_{cr,\varphi} N_{cr,x} + N_{cr,\varphi} N_{cr,y}}{1 - \tau_x^2 - \tau_y^2} = 3.020 \cdot 10^6 \qquad \{5.42\}$$

$$b_2 = \frac{N_{cr,x}\tau_x^2 + N_{cr,y}\tau_y^2 - N_{cr,x} - N_{cr,y} - N_{cr,\varphi}}{1-\tau_x^2-\tau_y^2} = -3126$$

The critical load is the smallest root of the above cubic equation:

$$N_{cr} = 583 \text{ MN}$$

The exact critical load by Axis VM X5 (2019) is $N_{cr}(\text{FE}) = 612$ MN.

5.6.2 Critical load of fifteen-storey asymmetric building

The fifteen-storey reinforced concrete building is braced by one frame (F3) and one shear wall (W6) in direction x, and three frames (F1, F5, F6) and two shear walls (W6) in direction y (Figure 5.8). This bracing system is among those used in Section 7.3 for a comprehensive accuracy analysis (as system S29 in Figure 7.23/g).

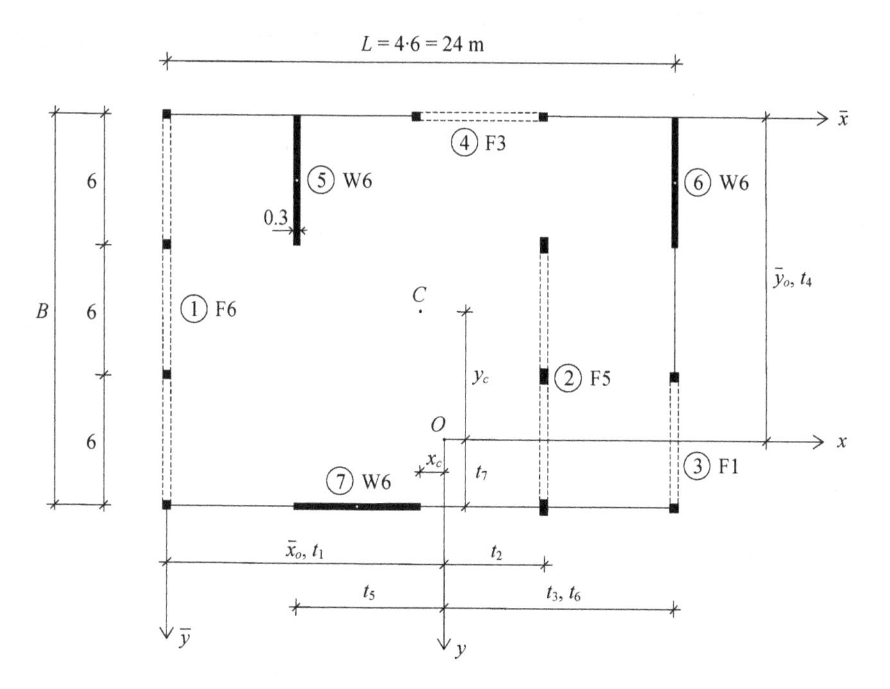

Figure 5.8 Layout of fifteen-storey asymmetric building.

The building is subjected to a uniformly distributed vertical floor load. The modulus of elasticity is $E = 25000$ MN/m^2, the modulus of elasticity in shear is $G = 10400$ MN/m^2, the storey height is $h = 3$ m and the total height of the building is $H = 45$ m.

Load distribution factor r_s will be needed throughout the calculations. Its value is obtained from Table 5.1 as a function of the number of storeys: $r_s = 0.904$.

The procedure for the determination of the critical load of the building is carried out in three steps. The basic characteristics of the bracing units are calculated first. The basic critical loads are determined in the second step when it is tacitly assumed that sway buckling in direction x, sway buckling in direction y and pure torsional buckling happen independently of each other. Finally, the interaction among the basic critical loads is taken into account.

Basic characteristics of the individual bracing units

Before the whole system of four frames and three shear walls is investigated, it is advantageous to establish the basic characteristics of the individual bracing units for later use in steps 2 and 3.

Bracing Unit 1 (frame F6, Figure 5.9/a)
The cross-sections of the columns and the beams of three-bay frame F6 are 0.40/0.40 and 0.40/0.70 (metres), respectively.

The shear stiffness that is associated with the beams of the frame is

$$K_{b,1} = \sum_{j=1}^{n-1} \frac{12 E_b I_{b,j}}{l_j h} = 3\frac{12 \cdot 25 \cdot 10^3 \cdot 0.4 \cdot 0.7^3}{12 \cdot 6 \cdot 3} = 571.7 \text{ MN} \qquad \{2.53\}$$

The shear stiffness that is associated with the columns of the frame is

$$K_{c,1} = \sum_{j=1}^{n} \frac{\pi^2 E_c I_{c,j}}{h^2} = 4\frac{\pi^2 \cdot 25 \cdot 10^3 \cdot 0.4^4}{12 \cdot 3^2} = 233.9 \text{ MN} \qquad \{2.54\}$$

The combination of the two part shear stiffnesses gives the "original" shear stiffness of the frame:

$$K_1 = K_{b,1} \frac{K_{c,1}}{K_{b,1} + K_{c,1}} = 571.7 \frac{233.9}{571.7 + 233.9} = 166.0 \text{ MN} \qquad \{2.55\}$$

where

$$r_1 = \frac{K_{c,1}}{K_{b,1} + K_{c,1}} = \frac{233.9}{571.7 + 233.9} = 0.2903 \qquad \{2.56\}$$

is the reduction factor.

The global second moment of area is

$$I_{g,1} = \sum_{j=1}^{n} A_{c,j} t_j^2 = 0.4 \cdot 0.4(3^2 + 9^2)2 = 28.8 \text{ m}^4 \qquad \{2.57\}$$

With the global second moment of area, the global bending critical load of the frame is

$$N_{g,1} = \frac{7.837 r_s E_c I_{g,1}}{H^2} = \frac{7.837 \cdot 0.904 \cdot 25 \cdot 10^3 \cdot 28.8}{45^2} = 2519\,\text{MN} \qquad \{5.7\}$$

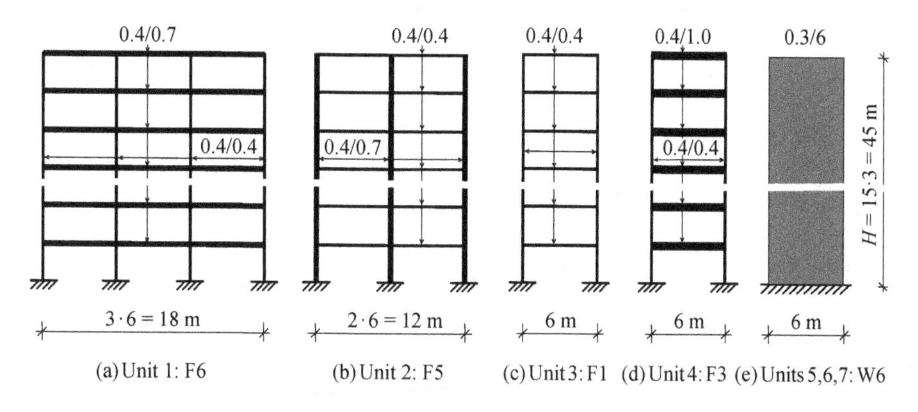

Figure 5.9 Bracing units for the fifteen-storey asymmetric building.

Because of the different buckling shape of the shear and global bending modes, there is an interaction between the two modes and the original shear stiffness is reduced by the effectiveness factor

$$s_1 = \frac{N_{g,1}}{K_1 + N_{g,1}} = \frac{2519}{166 + 2519} = 0.9382 \qquad \{5.9\}$$

The effective shear stiffness can now be determined:

$$K_{e,1} = K_1 s_1 = 166 \cdot 0.9382 = 155.7\ \text{MN} \qquad \{5.8\}$$

The local second moment of area of the frame, amended by r and making use of the fact that the four columns have the same cross-section, is

$$I_1 = r_1 \sum_{j=1}^{n} I_{c,j} = 0.2903 \cdot 4 \frac{0.4^4}{12} = 0.002477\ \text{m}^4 \qquad \{2.58\}$$

The critical load of the frame can be determined by the comprehensive method (Section 2.3.1) and the simple method (2.4.1). As condition

$$\frac{I_1}{I_{f,1}} = \frac{0.002477}{0.002477 + 28.8} = 0.000086 \le 0.0003 \qquad \{2.66\}$$

is fulfilled, the simple method will be used.

As a function of

$$\beta_s = \frac{K_1}{N_{g,1}} = \frac{166}{2519} = 0.066 \qquad \{2.64\}$$

the critical load parameter is obtained from Table 2.2 as

$$\alpha_s = 1 \qquad \{\text{Table 2.2}\}$$

The critical load of the frame is

$$N_{cr,1} = \alpha_s K_1 = 1 \cdot 166 = 166 \text{ MN} \qquad \{2.63\}$$

Bracing Unit 2 (frame F5, Figure 5.9/b)

The cross-sections of the columns and beams of frame F5 are 0.40/0.70 and 0.40/0.40 (metres), respectively.

The shear stiffness that is associated with the beams of the frame is

$$K_{b,2} = \sum_{j=1}^{n-1} \frac{12 E_b I_{b,j}}{l_j h} = 2\frac{12 \cdot 25 \cdot 10^3 \cdot 0.4^4}{12 \cdot 6 \cdot 3} = 71.11 \text{ MN} \qquad \{2.53\}$$

The shear stiffness that is associated with the columns of the frame is

$$K_{c,2} = \sum_{j=1}^{n} \frac{\pi^2 E_c I_{c,j}}{h^2} = 3\frac{\pi^2 \cdot 25 \cdot 10^3 \cdot 0.4 \cdot 0.7^3}{12 \cdot 3^2} = 940.4 \text{ MN} \qquad \{2.54\}$$

The combination of the two part shear stiffnesses gives the "original" shear stiffness of the frame:

$$K_2 = K_{b,2} \frac{K_{c,2}}{K_{b,2} + K_{c,2}} = 71.11 \frac{940.4}{71.11 + 940.4} = 66.11 \text{ MN} \qquad \{2.55\}$$

where

$$r_2 = \frac{K_{c,2}}{K_{b,2} + K_{c,2}} = \frac{940.4}{71.11 + 940.4} = 0.9297 \qquad \{2.56\}$$

is the reduction factor.

The global second moment of area is

$$I_{g,2} = \sum_{j=1}^{n} A_{c,j} t_j^2 = 0.4 \cdot 0.7 \cdot 6^2 \cdot 2 = 20.16 \text{ m}^4 \qquad \{2.57\}$$

With the global second moment of area, the global bending critical load of the frame is

$$N_{g,2} = \frac{7.837 r_s E_c I_{g,2}}{H^2} = \frac{7.837 \cdot 0.904 \cdot 25 \cdot 10^3 \cdot 20.16}{45^2} = 1763 \text{ MN} \qquad \{5.7\}$$

Because of the different buckling shape of the shear and global bending modes, there is an interaction between the two modes and the original shear stiffness is reduced by the effectiveness factor

$$s_2 = \frac{N_{g,2}}{K_2 + N_{g,2}} = \frac{1763}{66.11 + 1763} = 0.9639 \qquad \{5.9\}$$

The effective shear stiffness can now be determined:

$$K_{e,2} = K_2 s_2 = 66.11 \cdot 0.9639 = 63.72 \text{ MN} \qquad \{5.8\}$$

The local second moment of area of the frame, amended by r and making use of the fact that the three columns have the same cross-section, is

$$I_2 = r_2 \sum_{j=1}^{n} I_{c,j} = 0.9297 \cdot 3 \frac{0.4 \cdot 0.7^3}{12} = 0.03189 \text{ m}^4 \qquad \{2.58\}$$

Parameter c_1 is needed for the calculation of the critical load of the frame. Its value is obtained from Table 2.1 as a function of the two parameters {2.62}:

$$\frac{I_2}{I_{f,2}} = \frac{0.03189}{0.03189 + 20.16} = 0.00158, \qquad \frac{K_2 H^2}{E I_{f,2} r_s} = \frac{66.11 \cdot 45^2}{25 \cdot 10^3 \cdot 20.1919 \cdot 0.904} = 0.2934$$

With the relevant values from Table 2.1

	0.001	0.005
0.2	0.304	0.412
0.5	0.665	0.815

parameter c_1 is obtained after three interpolations:

$$1: \quad 0.304 + \frac{0.412 - 0.304}{0.005 - 0.001}(0.00158 - 0.001) = 0.31966$$

$$2: \quad 0.665 + \frac{0.815 - 0.665}{0.005 - 0.001}(0.00158 - 0.001) = 0.68675$$

$$3: \quad c_1 = 0.31966 + \frac{0.68675 - 0.31966}{0.5 - 0.2}(0.2934 - 0.2) = 0.434$$

The critical load of the frame can now be calculated:

$$N_{cr,2} = c_1 \frac{EI_{f,2} r_s}{H^2} = 0.434 \frac{25 \cdot 10^3 \cdot 20.1919 \cdot 0.904}{45^2} = 97.8 \text{ MN} \qquad \{2.60\}$$

Bracing Unit 3 (frame F1, Figure 5.9/c)
The cross-sections of both the columns and the beams of frame F1 are 0.40/0.40 (metres).

The shear stiffness that is associated with the beams of the frame is

$$K_{b,3} = \sum_{j=1}^{n-1} \frac{12 E_b I_{b,j}}{l_j h} = \frac{12 \cdot 25 \cdot 10^3 \cdot 0.4^4}{12 \cdot 6 \cdot 3} = 35.556 \text{ MN} \qquad \{2.53\}$$

The shear stiffness that is associated with the columns of the frame is

$$K_{c,3} = \sum_{j=1}^{n} \frac{\pi^2 E_c I_{c,j}}{h^2} = 2 \frac{\pi^2 \cdot 25 \cdot 10^3 \cdot 0.4^4}{12 \cdot 3^2} = 116.97 \text{ MN} \qquad \{2.54\}$$

The combination of the two part shear stiffnesses gives the “original” shear stiffness of the frame:

$$K_3 = K_{b,3} \frac{K_{c,3}}{K_{b,3} + K_{c,3}} = 35.556 \frac{116.97}{35.556 + 116.97} = 27.27 \text{ MN} \qquad \{2.55\}$$

where

$$r_3 = \frac{K_{c,3}}{K_{b,3} + K_{c,3}} = \frac{116.97}{35.556 + 116.97} = 0.7669 \qquad \{2.56\}$$

is the reduction factor.

The global second moment of area is

$$I_{g,3} = \sum_{j=1}^{n} A_{c,j} t_j^2 = 0.4 \cdot 0.4 \cdot 3^2 \cdot 2 = 2.88 \text{ m}^4 \qquad \{2.57\}$$

With the global second moment of area, the global bending critical load of the frame is

$$N_{g,3} = \frac{7.837 r_s E_c I_{g,3}}{H^2} = \frac{7.837 \cdot 0.904 \cdot 25 \cdot 10^3 \cdot 2.88}{45^2} = 251.9 \text{ MN} \qquad \{5.7\}$$

Because of the different buckling shape of the shear and global bending modes, there is an interaction between the two modes and the original shear stiffness is reduced by the effectiveness factor

$$s_3 = \frac{N_{g,3}}{K_3 + N_{g,3}} = \frac{251.9}{27.27 + 251.9} = 0.9023 \qquad \{5.9\}$$

The effective shear stiffness can now be determined:

$$K_{e,3} = K_3 s_3 = 27.27 \cdot 0.9023 = 24.61 \text{ MN} \qquad \{5.8\}$$

The local second moment of area of the frame, amended by r and making use of the fact that the two columns have the same cross-section, is

$$I_3 = r_3 \sum_{j=1}^{n} I_{c,j} = 0.7669 \cdot 2 \frac{0.4^4}{12} = 0.003272 \text{ m}^4 \qquad \{2.58\}$$

Parameter c_1 is needed for the calculation of the critical load of the frame. Its value is obtained from Table 2.1 as a function of the two parameters {2.62}

$$\frac{I_3}{I_{f,3}} = \frac{0.003272}{0.003272 + 2.88} = 0.00113, \quad \frac{K_3 H^2}{E I_{f,3} r_s} = \frac{27.27 \cdot 45^2}{25 \cdot 10^3 \cdot 2.883272 \cdot 0.904} = 0.847$$

With the relevant values from Table 2.1

	0.001	0.005
0.5	0.665	0.815
1.0	1.222	1.403

parameter c_1 is obtained after three interpolations:

$$1: \quad 0.665 + \frac{0.815 - 0.665}{0.005 - 0.001}(0.00113 - 0.001) = 0.6699$$

$$2: \quad 1.222 + \frac{1.403 - 1.222}{0.005 - 0.001}(0.00113 - 0.001) = 1.2279$$

$$3: \quad c_1 = 0.6699 + \frac{1.2279 - 0.6699}{1 - 0.5}(0.847 - 0.5) = 1.057$$

The critical load of the frame can now be calculated:

$$N_{cr,3} = c_1 \frac{EI_{f,3} r_s}{H^2} = 1.057 \frac{25 \cdot 10^3 \cdot 2.883272 \cdot 0.904}{45^2} = 34.01 \text{ MN} \qquad \{2.60\}$$

Bracing Unit 4 (frame F3, Figure 5.9/d)
The cross-sections of the columns and the beams of frame F3 are 0.40/0.40 and 0.40/1.00 (metres), respectively.

The shear stiffness that is associated with the beams of the frame is

$$K_{b,4} = \sum_{j=1}^{n-1} \frac{12 E_b I_{b,j}}{l_j h} = \frac{12 \cdot 25 \cdot 10^3 \cdot 0.4 \cdot 1.0^3}{12 \cdot 6 \cdot 3} = 555.55 \text{ MN} \qquad \{2.53\}$$

The shear stiffness that is associated with the columns of the frame is

$$K_{c,4} = \sum_{j=1}^{n} \frac{\pi^2 E_c I_{c,j}}{h^2} = 2 \frac{\pi^2 \cdot 25 \cdot 10^3 \cdot 0.4^4}{12 \cdot 3^2} = 116.97 \text{ MN} \qquad \{2.54\}$$

The combination of the two part shear stiffnesses gives the “original” shear stiffness of the frame:

$$K_4 = K_{b,4} \frac{K_{c,4}}{K_{b,4} + K_{c,4}} = 555.55 \frac{116.97}{555.55 + 116.97} = 96.63 \text{ MN} \qquad \{2.55\}$$

where

$$r_4 = \frac{K_{c,4}}{K_{b,4} + K_{c,4}} = \frac{116.97}{555.55 + 116.97} = 0.1739 \qquad \{2.56\}$$

is the reduction factor.

The global second moment of area is

$$I_{g,4} = \sum_{j=1}^{n} A_{c,j} t_j^2 = 0.4 \cdot 0.4 \cdot 3^2 \cdot 2 = 2.88 \text{ m}^4 \qquad \{2.57\}$$

With the global second moment of area, the global bending critical load of the frame is

$$N_{g,4} = \frac{7.837 r_s E_c I_{g,4}}{H^2} = \frac{7.837 \cdot 0.904 \cdot 25 \cdot 10^3 \cdot 2.88}{45^2} = 251.9 \text{ MN} \qquad \{5.7\}$$

Because of the different buckling shape of the shear and global bending modes, there is an interaction between the two modes and the original shear stiffness is reduced by the effectiveness factor

$$s_4 = \frac{N_{g,4}}{K_4 + N_{g,4}} = \frac{251.9}{96.63 + 251.9} = 0.7228 \qquad \{5.9\}$$

The effective shear stiffness can now be determined:

$$K_{e,4} = K_4 s_4 = 96.63 \cdot 0.7228 = 69.84 \text{ MN} \qquad \{5.8\}$$

The local second moment of area of the frame, amended by r and making use of the fact that the two columns have the same cross-section, is

$$I_4 = r_4 \sum_{j=1}^{n} I_{c,j} = 0.1739 \cdot 2 \frac{0.4^4}{12} = 0.000742 \text{ m}^4 \qquad \{2.58\}$$

The critical load of the frame can be determined by the comprehensive method (Section 2.3.1) and the simple method (2.4.1). As condition

$$\frac{I_4}{I_{f,4}} = \frac{0.000742}{0.000742 + 2.88} = 0.00026 \le 0.0003 \qquad \{2.62\}$$

is fulfilled, the simple method can be used.

As a function of

$$\beta_s = \frac{K_4}{N_{g,4}} = \frac{96.63}{251.9} = 0.384 \qquad \{2.64\}$$

the critical load parameter is obtained from Table 2.2 as

$$\alpha_s = 1 - \frac{1 - 0.9972}{0.4 - 0.3}(0.384 - 0.3) = 0.9976 \qquad \{\text{Table 2.2}\}$$

The critical load of the frame is

$$N_{cr,4} = \alpha_s K_4 = 0.9976 \cdot 96.63 = 96.4 \text{ MN} \qquad \{2.63\}$$

Bracing Units 5, 6 and 7 (shear wall W6; Figure 5.9/e)

The size of the shear wall is 6.0 metres with a thickness of 0.3 m. The second moments of area of the shear wall are

$$I_w^{max} = \frac{0.3 \cdot 6^3}{12} = 5.4\,\text{m}^4 \qquad \text{and} \qquad I_w^{min} = \frac{6 \cdot 0.3^3}{12} = 0.0135\ \text{m}^4$$

The Saint-Venant constant is

$$J_w = \frac{6 \cdot 0.3^3}{3} = 0.054\ \text{m}^4$$

The critical loads of the shear walls in directions x and y will also be needed with the torsional analysis:

$$N_{cr,w}^{min} = \frac{7.837 E I_w^{min} r_s}{H^2} = \frac{7.837 \cdot 25 \cdot 10^3 \cdot 0.0135 \cdot 0.904}{45^2} = 1.18\ \text{MN} \qquad \{2.76\}$$

$$N_{cr,w}^{max} = \frac{7.837 E I_w^{max} r_s}{H^2} = \frac{7.837 \cdot 25 \cdot 10^3 \cdot 5.4 \cdot 0.904}{45^2} = 472.3\ \text{MN} \qquad \{2.76\}$$

Sway buckling in direction x (Bracing Units 4 and 7)

Frame F3 (Bracing Unit 4) is the only frame in direction x. It follows that its original shear stiffness, effective shear stiffness and effectiveness factor are also the original shear stiffness, effective shear stiffness and effectiveness factor of the system:

$$K = \sum_{i=1}^{f} K_i = 96.63\ \text{MN} \qquad \{5.5\}$$

$$K_e = \sum_{i=1}^{f} K_i s_i = 96.63 \cdot 0.7228 = 69.84\ \text{MN} \qquad \{5.10\}$$

$$s = \frac{K_e}{K} = \frac{69.84}{96.63} = 0.7228 \qquad \{5.11\}$$

The local bending stiffness is also needed. Shear wall W6 dominates the system and the contribution of the other units is negligible:

$$EI = E_c I_c + E_w I_w = E_w \sum_{k=1}^{m} I_{w,k} = 25 \cdot 10^3 \cdot 5.4 = 135 \cdot 10^3\ \text{MNm}^2 \qquad \{5.12\}$$

With the local bending stiffness, the local bending critical load is

$$N_l = \frac{7.837 r_s EI}{H^2} = \frac{7.837 \cdot 0.904 \cdot 135 \cdot 10^3}{45^2} = 472.3 \text{ MN} \qquad \{5.13\}$$

The two main components (N_l and K_e) of the sway critical load are now available. The third component—responsible for the interaction between the bending and shear modes—is obtained using the dimensionless ratio

$$\beta = \frac{K_e}{N_l} = \frac{69.84}{472.3} = 0.148 \qquad \{5.17\}$$

According to Table 5.2, the critical load parameter is

$$\alpha = 1.2949 + \frac{1.5798 - 1.2949}{0.2 - 0.1}(0.148 - 0.1) = 1.432 \qquad \{\text{Table } 5.2\}$$

Finally, the sway critical load in direction x is

$$N_{cr,x} = N_l + K_e + (\alpha - \beta - 1) s N_l \qquad \{5.18\}$$

$$= 472.3 + 69.84 + (1.432 - 0.148 - 1)0.7228 \cdot 472.3 = 639.1 \text{ MN}$$

Sway buckling in direction y (Bracing Units 1, 2, 3, 5 and 6)

The original and effective stiffnesses of the system originate from frames F1, F5 and F6 (Bracing Units 1, 2 and 3):

$$K = \sum_{i=1}^{f} K_i = 166 + 66.11 + 27.27 = 259.38 \text{ MN} \qquad \{5.5\}$$

$$K_e = \sum_{i=1}^{f} K_{e,i} = 155.7 + 63.72 + 24.61 = 244.03 \text{ MN} \qquad \{5.10\}$$

The effectiveness factor is the ratio of the effective and original stiffnesses:

$$s = \frac{K_e}{K} = \frac{244.03}{259.38} = 0.9408 \qquad \{5.11\}$$

The local bending stiffness is also needed. The two shear walls (W6 and W6) dominate the system and the contribution of the other units is negligible:

$$EI = E_c I_c + E_w I_w = E_w \sum_{k=1}^{m} I_{w,k} = 2 \cdot 25 \cdot 10^3 \cdot 5.4 = 270 \cdot 10^3 \text{ MNm}^2 \qquad \{5.12\}$$

With the local bending stiffness, the local bending critical load is

$$N_l = \frac{7.837 r_s EI}{H^2} = \frac{7.837 \cdot 0.904 \cdot 270 \cdot 10^3}{45^2} = 944.6 \text{ MN} \qquad \{5.13\}$$

The two main components (N_l and K_e) of the sway critical load are now available. The third component—responsible for the interaction between the bending and shear modes—is obtained using the dimensionless ratio

$$\beta = \frac{K_e}{N_l} = \frac{244.03}{944.6} = 0.258 \qquad \{5.17\}$$

According to Table 5.2, the critical load parameter is

$$\alpha = 1.5798 + \frac{1.8556 - 1.5798}{0.3 - 0.2}(0.258 - 0.2) = 1.74 \qquad \{\text{Table 5.2}\}$$

Finally, the sway critical load in direction y is

$$N_{cr,y} = N_l + K_e + (\alpha - \beta - 1) s N_l \qquad \{5.18\}$$

$$= 944.6 + 244.03 + (1.74 - 0.258 - 1)0.9408 \cdot 944.6 = 1617 \text{ MN}$$

Pure torsional buckling (all seven bracing units participating)

The critical loads of the bracing units determined in the "Basic characteristics of the bracing units" above are used to calculate the coordinates of the shear centre in the coordinate system $\bar{x} - \bar{y}$:

$$\bar{x}_o = \frac{\sum_{i=1}^{f+m} N_{y,i}\bar{x}_i}{\sum_{i=1}^{f+m} N_{y,i}} = \frac{472.3(6+24) + 97.8 \cdot 18 + 34.01 \cdot 24}{166 + 472.3 \cdot 2 + 97.8 + 34.01} = 13.48 \text{ m} \qquad \{5.27\}$$

$$\bar{y}_o = \frac{\sum_{i=1}^{f+m} N_{x,i}\bar{y}_i}{\sum_{i=1}^{f+m} N_{x,i}} = \frac{472.3 \cdot 18}{472.3 + 96.4} = 14.95 \text{ m} \qquad \{5.27\}$$

With the coordinates of the shear centre, a new x-y coordinate system is established whose origin is in the shear centre. From now on, the calculation is carried out in this new coordinate system.

The coordinates of the centroid of the floor area are

$$x_c = \frac{L}{2} - \bar{x}_o = \frac{24}{2} - 13.48 = -1.48\,\text{m},\; y_c = \frac{B}{2} - \bar{y}_o = \frac{18}{2} - 14.95 = -5.95\,\text{m} \quad \{5.33\}$$

The radius of gyration is

$$i_p = \sqrt{\frac{L^2 + B^2}{12} + t^2} = \sqrt{\frac{24^2 + 18^2}{12} + 1.48^2 + 5.95^2} = \sqrt{112.59} = 10.61\,\text{m} \quad \{5.32\}$$

The perpendicular distances of the bracing units from the shear centre are

$$t_1 = \bar{x}_o = 13.48\,\text{m}, \qquad t_2 = 18 - \bar{x}_o = 4.52\,\text{m}, \qquad t_3 = t_6 = L - \bar{x}_o = 10.52\,\text{m}$$

$$t_4 = \bar{y}_o = 14.95\,\text{m}, \qquad t_5 = \bar{x}_o - 6 = 7.48\,\text{m}, \qquad t_7 = B - \bar{y}_o = 3.05\,\text{m}$$

The "original" Saint-Venant stiffness consists of two parts. Both the Saint-Venant stiffness of the three shear walls and the four frames have contributions:

$$(GJ) = \sum_{k=1}^{m} GJ_k + \sum_{i=1}^{f}\left((K_i)_x y_i^2 + (K_i)_y x_i^2\right) = 10400 \cdot 0.054 \cdot 3 + \quad \{5.38\}$$

$$+166 \cdot 13.48^2 + 66.11 \cdot 4.52^2 + 27.27 \cdot 10.52^2 + 96.63 \cdot 14.95^2 = 57814\,\text{MNm}^2$$

The real (effective) Saint-Venant stiffness is always smaller than the "original" one as the effect of the frames is limited by their effectiveness factor:

$$(GJ)_e = \sum_{k=1}^{m} GJ_k + \sum_{i=1}^{f}\left((K_{e,i})_x y_i^2 + (K_{e,i})_y x_i^2\right) = 10400 \cdot 0.054 \cdot 3 + \quad \{5.29\}$$

$$+155.7 \cdot 13.48^2 + 63.72 \cdot 4.52^2 + 24.61 \cdot 10.52^2 + 69.84 \cdot 14.95^2 = 49612\,\text{MNm}^2$$

The first term represents the "own" contribution of the shear walls and it amounts to 3% of the total Saint-Venant stiffness.

The effectiveness of the Saint-Venant stiffness for the whole system can now be determined:

$$s_\varphi = \frac{(GJ)_e}{(GJ)} = \frac{49612}{57814} = 0.8581 \quad \{5.37\}$$

As for the warping stiffness of the system, the bracing units have no own warping stiffness, the contributions of the frames are negligible and that leaves the contribution of the three shear walls:

$$EI_{\omega} = E_w \sum_{k=1}^{m} \left((I_{w,k})_x y_k^2 + (I_{w,k})_y x_k^2 \right) \qquad \{5.30\}$$

$$= 25 \cdot 10^3 \cdot 5.4 \cdot (7.48^2 + 10.52^2 + 3.05^2) = 23.75 \cdot 10^6 \text{ MNm}^4$$

With the above stiffnesses, the part torsional critical loads can now be determined.

The warping torsional critical load of the system is

$$N_{\omega} = \frac{7.837 r_s EI_{\omega}}{i_p^2 H^2} = \frac{7.837 \cdot 0.904 \cdot 23.75 \cdot 10^6}{112.59 \cdot 45^2} = 738.0 \text{ MN} \qquad \{5.35\}$$

and the Saint-Venant torsional critical load is

$$N_t = \frac{(GJ)_e}{i_p^2} = \frac{49612}{112.59} = 440.6 \text{ MN} \qquad \{5.36\}$$

As a function of the ratio of the above part critical loads

$$\beta_{\varphi} = \frac{N_t}{N_{\omega}} = \frac{440.6}{738.0} = 0.597 \qquad \{5.39\}$$

the critical load parameter is obtained from Table 5.2:

$$\alpha_{\varphi} = 2.3817 + \frac{2.6333 - 2.3817}{0.6 - 0.5}(0.597 - 0.5) = 2.625 \qquad \{\text{Table } 5.2\}$$

Finally, the critical load of pure torsional buckling is

$$N_{cr,\varphi} = N_{\omega} + N_t + (\alpha_{\varphi} - \beta_{\varphi} - 1) s_{\varphi} N_{\omega} \qquad \{5.34\}$$

$$= 738.0 + 440.6 + (2.625 - 0.597 - 1)0.8581 \cdot 738.0 = 1830 \text{ MN}$$

The third term represents the effect of the interaction between the Saint-Venant and warping torsional modes. It amounts to 35% of the total pure torsional critical load.

Coupling of the basic critical loads

The centroid of the layout does not coincide with the shear centre so there is an interaction among the basic critical loads. As the interaction reduces the value of the global critical load, its effect must be taken into account. The interaction is best taken into account by solving the cubic equation

$$N^3 + b_2 N^2 + b_1 N - b_0 = 0 \qquad \{5.41\}$$

which automatically handles the different coupling possibilities.

With eccentricity parameters τ_x and τ_y

$$\tau_x = \frac{x_c}{i_p} = \frac{-1.48}{10.61} = -0.1395 \quad \text{and} \quad \tau_y = \frac{y_c}{i_p} = \frac{-5.95}{10.61} = -0.5608 \qquad \{5.43\}$$

the coefficients are

$$b_0 = \frac{N_{cr,x} N_{cr,y} N_{cr,\varphi}}{1 - \tau_x^2 - \tau_y^2} = \frac{639.1 \cdot 1617 \cdot 1830}{1 - (-0.1395)^2 - (-0.5608)^2} = 2.839 \cdot 10^9 \qquad \{5.42\}$$

$$b_1 = \frac{639.1 \cdot 1617 + 1830 \cdot 639.1 + 1830 \cdot 1617}{1 - (-0.1395)^2 - (-0.5608)^2} = 7.75 \cdot 10^6 \qquad \{5.42\}$$

$$b_2 = \frac{639.1 \cdot (-0.1395)^2 + 1617 \cdot (-0.5608)^2 - 639.1 - 1617 - 1830}{1 - (-0.1395)^2 - (-5608)^2} = -5353$$

The critical load is the smallest root of the above cubic equation:

$$N_{cr} = 561 \text{ MN}$$

The “exact” critical load by Axis VM X5 (2019) is N_{cr}(FE) = 556 MN.

6

Global structural analysis

Stability problems concerning multi-storey buildings can be investigated on two levels. An element-based "local" analysis can be carried out, step by step, aimed at certain key structural members. Codes of practice normally follow this avenue and have detailed instructions for the analysis. This approach makes it possible to carry out the analysis in a relatively simple way. However, this approach has disadvantages. It leaves the designer with the task of identifying all the key members and it cannot address the full-height, three-dimensional global behaviour of the multi-storey building. The "local" approach may also lead to uneconomic solutions as the elements of the whole structure tend to work together and, with the local approach, the possibility to take into account the effects of interaction is normally limited to the neighbouring members only.

The other approach is the global approach. The concept of the global critical load ratio has been around for some time. Around but not in use. Or at least not in use to such an extent as it should have been. As the results of an illustrative example given in this chapter will demonstrate, the global critical load ratio is far more than a stability parameter: it is a generic characteristic with which the designer can monitor the overall performance of the whole bracing system. It also links the three important areas of analysis: the stress, stability and dynamic analyses. The way the structure responds to the loads—in two- or three-dimensional manner—is automatically taken into account and made clear to the designer.

Around the middle of the last century, Chwalla (1959) emphasized the importance of the global approach and recommended the introduction of a "global factor". Halldorsson and Wang (1968) suggested that a "general safety factor" should be used for building structures, and its importance was comparable to that of the "overturning factor" used in the design of dams. Dowrick (1976) drew attention to the importance of the overall stability of structures. Dealing with plane structures, Stevens (1983) linked theory and practice and underlined the importance of the critical load in the design of frames. The idea of a global safety factor also surfaced in connection with the structural design of large structures (Zalka and Armer, 1992). MacLeod and Zalka (1996) and MacLeod (2005) advocated the use of the critical load ratio emphasizing its ability to handle torsional behaviour in a relatively simple way. The importance of torsional behaviour cannot be overemphasized, especially considering the fact that up to the emergence of the personal computer, relatively little attention had been paid to the three-dimensional behaviour of complex structures in university textbooks and in national and international codes of practice.

However, the situation seems to have changed. More and more powerful computers, sophisticated software packages and advanced guidelines on modelling complex structures (MacLeod, 1990 and 2005) make it easier to carry out true three-dimensional analyses. The global approach and methods developed with the global aspect in mind have also been emerging in structural designer handbooks and in codes of practice themselves (EN 1992, 2004; EN 1993, 2004; Martin and Purkiss, 2008). Applying the global approach, the structural engineer can rely on two types of technique: full-blown, albeit time-consuming analyses can be performed using advanced computer modelling, or quick, less accurate, but more descriptive investigations may be carried out that use specialized but simpler models (Howson, 2006). The latter technique is employed in this book.

6.1 THE GLOBAL CRITICAL LOAD RATIO

Depending on which direction the situation is looked at from, the global critical load ratio can be defined in two ways. First, it can be defined as

$$\frac{N}{N_{cr}} \tag{6.1}$$

where

$$N = LBQn \tag{6.2}$$

is the total vertical load of the multi-storey building with

N_{cr}	the global elastic critical load for buildings subjected to uniformly distributed floor load
L, B	the plan length and breadth of the building
Q	the intensity of the uniformly distributed floor load [kN/m^2]
n	the number of storeys

Practicing structural engineers may prefer the reciprocal definition when the global critical load ratio is defined as the ratio of the global elastic critical load and the total vertical load

$$\lambda = \frac{N_{cr}}{N} \tag{6.3}$$

as it carries a practical meaning that is easy to relate to the safety of the structure.

Somewhat confusingly, codes of practice use both definitions. In this book, from now on, the reciprocal definition [Equation (6.3)] will be used.

When there is significant extra load at the top floor level (e.g. a swimming pool), its detrimental effect cannot be ignored. For such cases, Equation (6.3) can be amended and the global critical load ratio can be obtained using

$$\frac{1}{\lambda} = \frac{N}{N_{cr}} + \frac{F}{F_{cr}} \tag{6.4}$$

In the above equation

F is the extra concentrated load at top floor level
F_{cr} is the global critical load for the concentrated top load case

The global critical loads can be determined carrying out a full-blown analysis using a computer program or by approximate analytical solutions, e.g. the ones presented in Chapter 5.

The global critical load ratio can be used in different ways. Codes of practice normally concentrate on its use as an indicator whether or not a second-order analysis is needed. If the condition

$$\lambda \geq 10 \tag{6.5}$$

is satisfied, then the suitability of the bracing system is proved and the vertical load bearing elements can be considered as braced (by the bracing system). Neglecting the second-order effects (due to sway and torsion) may result in a maximum 10% error.

If condition (6.5) is not satisfied (but $\lambda > 1$ holds), the stability of the building may still be acceptable but it must be demonstrated using a second-order analysis. However, there is a warning here: it is widely accepted in practicing structural engineering communities that the absolute minimum for a critical load ratio is four.

Another simple use of the global critical load ratio may be as a global safety factor: the greater the value of the global critical load ratio, the greater the safety of the multi-storey building against buckling.

The global critical load ratio can also be used as a performance indicator. As its value is calculated using the basic (sway and pure torsional) critical loads and taking into account the coupling of the basic modes, any weakness either in the bending/shear and torsional stiffnesses or in the geometrical arrangement of the bracing units (on which the detrimental coupling depends) is picked up automatically. As it happens, any weakness detected during the course of the stability analysis also points at unfavourable behaviour when the fundamental frequency and the maximum deflection of the building are calculated. This very useful quality is demonstrated in the next section where the structural performance of a building is monitored using the global critical load ratio.

6.2 ILLUSTRATIVE EXAMPLE

This case study uses an eight-storey building whose detailed global analysis is presented in Section 6.3 and only the main results are summarised here. The height of the building is $H = 8 \cdot 3 = 24$ metres, the plan length of the building is $L = 24$ metres and the breadth is $B = 18$ metres (Figure 6.1).

Four bracing units are available for making the building stable enough: two 3-bay reinforced concrete frames (F6) and two shear walls (W6) shown in Figure 6.2. Their location and orientation inside the layout are arbitrary.

Three different arrangements are considered. Three analyses are carried out for each arrangement: for the deflection analysis it is assumed that the building is subjected to a uniformly distributed wind load of intensity $q = 30$ kN/m, making

50° with axis x. The uniformly distributed floor load for the global stability analysis is $Q = 10$ kN/m^2. The weight per unit volume for the determination of the fundamental frequency of the building is assumed to be $\gamma = 3$ kN/m^3. In the three cases, the following characteristics are determined:

φ_{max}	maximum rotation of the building at top level [°]
d_{max}	maximum deflection of the building at top level [m]
d_{max}/d_{ASCE}	ratio of the maximum and recommended maximum deflection [–]
f	fundamental frequency of the building [Hz]
N_{cr}	global critical load of the building [MN]
λ	global critical load ratio of the building [–]

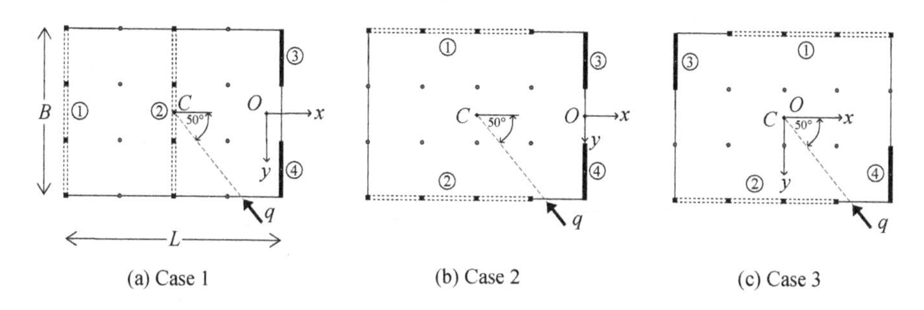

Figure 6.1 Eight-storey building braced by four bracing units in three arrangements.

The recommended maximum deflection of the building is $H/500 = 0.048$ m.

The main results are collected in Table 6.1.

Case 1 is an arrangement that is obviously unacceptable (Figure 6.1/a). This fact is spectacularly picked up by the global critical load ratio which (being smaller than 1.0) shows the unstable nature of the system. In line with the very small critical load ratio, the calculated maximum deflection is huge, the fundamental frequency and the critical load are very small. The fatal weakness of the bracing system is due to the lack of sufficient bracing in direction x.

By rotating the two frames by 90 degrees, the bracing system of Case 2 addresses the main problem with the previous arrangement and provides the building with sufficient bracing in direction x (Figure 6.1/b). The results clearly show the improvement. The global critical load ratio defines a theoretically stable structure (albeit the margin may not be considered sufficient enough in structural engineering practice); the maximum deflection is much smaller and the fundamental frequency and the critical load are much greater. However, as the details in Section 6.3 show, the torsional resistance is relatively small. This is due to the fact that the perpendicular distance of the two shear walls from the shear centre is zero.

By exchanging bracing units 1 and 3 for Case 3 (without changing their orientation), the shear centre of the system moves to the geometrical centre of the building without changing the value of the lateral stiffness in directions x and y (Figure 6.1/c).

The hugely beneficial consequence of this alteration is that all four bracing units can now contribute to the torsional resistance of the system as all four units now "have" perpendicular distance from the shear centre. In addition, as the bracing units are now placed along the sides of the layout *and* the farthest from the shear centre, their torsion arms are the longest possible and their efficiency is the greatest against torsion. The results reflect the beneficial change in the arrangement. The system is not only stable but the global critical load ratio now exceeds the recommended value ($\lambda = 13.6 > 10$). The maximum deflection of the building is much smaller than with the previous arrangement and it is now below the recommended value ($H/500$). The value of the fundamental frequency has also increased.

Table 6.1 The global critical load ratio as a performance indicator.

Characteristics	Case 1	Case 2	Case 3
φ_{max} [°]	0.03	0.15	0
d_{max} [m]	0.725	0.066	0.014
d_{max}/d_{ASCE}	15.1	1.4	0.3
f [Hz]	0.079	0.328	0.525
N_{cr} [MN]	12.5	196	471
λ	0.36	5.7	13.6

The results in Table 6.1 (together with the details of the calculations given in Section 6.3) and the results of hundreds of other examples show that the global critical load ratio is a reliable and sensitive indicator regarding the overall performance of the structural system. It may be advantageous to determine the value of the global critical load ratio for different arrangements and the arrangement that belongs to the greatest value is normally the best arrangement also when the maximum deflection and the fundamental frequency of the building are calculated. However, the global critical load ratio should not be too big—compared to 10—as it could easily lead to uneconomic structural arrangements.

Section 6.1 introduced the global critical load ratio as the ratio of the global critical load to the total vertical load of the building. Through three sets of comprehensive series of worked examples, it will be shown in the following sections how the global critical load ratio is calculated in different situations. In addition, it will also be demonstrated that the global critical load ratio is more than a simple ratio of two quantities; it also gives a strong indication regarding the performance of the structure. The greater the global critical load ratio, the better the performance of the bracing system, as far as the maximum top deflection, the fundamental frequency and the stability of the building are concerned.

In the three worked examples in the next section, the numbers of the equations used for the calculations are given in curly brackets on the right-hand side.

6.3 PRACTICAL APPLICATION NO. 1: ILLUSTRATIVE EXAMPLE

The eight-storey building braced by two reinforced concrete shear walls and two frames (Figure 6.1) was used in Section 6.2 to illustrate how the global critical load ratio links the three main areas of structural analysis: stability, dynamic and deflection analyses. The results are summarized in Table 6.1. The details of the calculations are given in this section.

With a storey height of $h = 3$ m, the height of the building is $H = 24$ m. The length and breadth of the building are $L = 24$ m and $B = 18$ m. Two reinforced concrete frames and two shear walls are available for providing the building with sufficient stability. The modulus of elasticity for the frames and shear walls is $E = 25000$ MN/m^2. The modulus of elasticity in shear is $G = 10400$ MN/m^2. In addition to the four bracing units, eight concrete columns are also part of the vertical load carrying system but, because of their pinned supports, they have no contribution to the lateral and torsional stiffnesses of the building. It is assumed for the analysis that the shear walls only develop bending deformation.

The arrangement of the bracing units is up to the designer. Three different arrangements will be examined in order to show how the critical load ratio is able to monitor the performance of the bracing system and to find an optimum solution. The weight per unit volume of the building (for the dynamic analysis) is $\gamma = 3$ kN/m^3. A uniformly distributed vertical floor load of $Q = 10$ kN/m^2 is assumed for the stability analysis and for the determination of the global critical load ratio. When the structures are subjected to lateral load and the top rotation and deflection are calculated, a uniformly distributed horizontal load of intensity $q = 30.0$ kN/m is considered, whose resultant makes 50° with axis x. The components of this wind load are

$$q_x = 30 \cdot \cos 50^\circ = 19.28\,\text{kN/m} \qquad \text{and} \qquad q_y = 30 \cdot \sin 50^\circ = 22.98\,\text{kN/m}$$

Frame F6 and shear wall W6 (Figure 6.2) will be used as bracing units. Both structures are part of a collection of bracing units that are used for a comprehensive accuracy analysis in Chapter 7. Their cross-sectional characteristics are collected in Table 6.2. As both bracing units will be used repeatedly in the following investigations, it is advantageous to summarize their basic characteristics for later use.

Table 6.2 Cross-sectional characteristics of the bracing units.

Characteristics	Bracing units 1 & 2: frame F6		Bracing units 3 & 4: shear wall W6
	Columns	Beams	
Cross-section	0.40×0.40	0.40×0.70	0.30 m×6.0 m
A [m^2]	0.16	0.28	1.8
I_{max} [m^4]	$0.0021\dot{3}$	$0.0114\dot{3}$	5.4
I_{min} [m^4]	$0.0021\dot{3}$	n/a	0.0135
J [m^4]	–	–	0.054

6.3.1 Basic characteristics

For the stability analysis, the load distribution factor is obtained from Table 5.1 as

$$r_s = 0.834 \qquad \{\text{Table 5.1}\}$$

Bracing Units 1 & 2: frame F6
The critical load of the frame is determined first. The calculation below follows the procedures presented in Sections 2.3.1 and 2.4.1.

The shear stiffness that is associated with the beams of the frame is

$$K_{b,1} = \sum_{j=1}^{n-1} \frac{12E_b I_{b,j}}{l_j h} = 3\frac{12 \cdot 25 \cdot 10^3 \cdot 0.4 \cdot 0.7^3}{12 \cdot 6 \cdot 3} = 571.7\,\text{MN} \qquad \{2.53\}$$

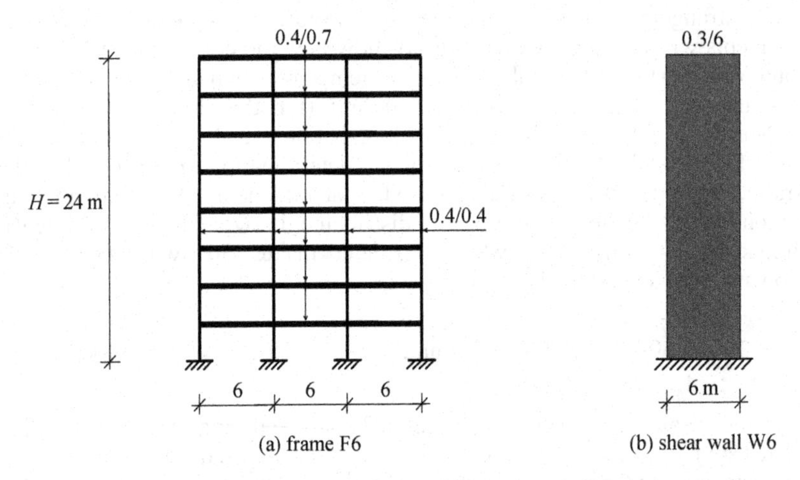

Figure 6.2 Bracing units for the eight-storey building.

The shear stiffness that is associated with the columns of the frame is

$$K_{c,1} = \sum_{j=1}^{n} \frac{\pi^2 E_c I_{c,j}}{h^2} = 4\frac{\pi^2 \cdot 25 \cdot 10^3 \cdot 0.4^4}{12 \cdot 3^2} = 233.9\,\text{MN} \qquad \{2.54\}$$

The combination of the two part shear stiffnesses gives the "original" shear stiffness of the frame:

$$K_1 = K_{b,1}\frac{K_{c,1}}{K_{b,1}+K_{c,1}} = 571.7\frac{233.9}{571.7+233.9} = 166.0\,\text{MN} \qquad \{2.55\}$$

In the above equation

$$r_1 = \frac{K_{c,1}}{K_{b,1} + K_{c,1}} = \frac{233.9}{571.7 + 233.9} = 0.2903 \qquad \{2.56\}$$

is the reduction factor.

The global second moment of area

$$I_{g,1} = \sum_{j=1}^{n} A_{c,j} t_j^2 = 2 \cdot 0.4 \cdot 0.4(3^2 + 9^2) = 28.8\,\mathrm{m}^4 \qquad \{2.57\}$$

is needed for the full-height global bending buckling of the frame. The corresponding critical load is

$$N_{g,1} = \frac{7.837 r_s E I_{g,1}}{H^2} = \frac{7.837 \cdot 0.834 \cdot 25 \cdot 10^3 \cdot 28.8}{24^2} = 8170\ \mathrm{MN} \qquad \{2.65\}$$

The local second moment of area of the frame, amended by r and making use of the fact that the four columns have the same cross-section, is

$$I_1 = r \sum_{j=1}^{n} I_{c,j} = 0.2903 \cdot 4 \frac{0.4^4}{12} = 0.002477\,\mathrm{m}^4 \qquad \{2.58\}$$

As condition

$$\frac{I_1}{I_{f,1}} = \frac{0.002477}{28.802477} = 0.000086 \le 0.0003 \qquad \{2.62\}$$

is fulfilled, the simpler method—the thin-walled sandwich model—can be used for the determination of the critical load.

With

$$\beta_s = \frac{K_1}{N_{g,1}} = \frac{166}{8170} = 0.02 \qquad \{2.64\}$$

the critical load parameter is obtained from Table 2.2 as

$$\alpha_s = 1.0 \qquad \{\text{Table 2.2}\}$$

and the critical load of the frame is

$$N_{cr,1} = \alpha_s K_1 = 1.0 \cdot 166.0 = 166\ \mathrm{MN} \qquad \{2.63\}$$

The effectiveness factor of the shear stiffness will also be needed later on:

$$s_1 = \frac{N_{g,1}}{K_1 + N_{g,1}} = \frac{8170}{166 + 8170} = 0.9801 \qquad \{5.9\}$$

The fundamental frequency and the maximum deflection of the frame are determined next. For practical reasons, the forces in these calculations are given in kilonewton (instead of meganewton). Some of the stiffness characteristics that were used above with the stability analysis have to be recalculated as the shear stiffness that is associated with the columns of the frame is now different:

$$K_{c,1} = \sum_{j=1}^{n} \frac{12 E_c I_{c,j}}{h^2} = 4 \frac{12 \cdot 25 \cdot 10^6 \cdot 0.4^4}{12 \cdot 3^2} = 284444 \text{ kN} \qquad \{2.29\}$$

The shear stiffness that is associated with the beams of the frame is unchanged

$$K_{b,1} = \sum_{j=1}^{n-1} \frac{12 E_b I_{b,j}}{l_j h} = 3 \frac{12 \cdot 25 \cdot 10^6 \cdot 0.4 \cdot 0.7^3}{12 \cdot 6 \cdot 3} = 571667 \text{ kN} \qquad \{2.53\}$$

With K_c and K_b, the original shear stiffness of the frame is

$$K_1 = K_{b,1} \frac{K_{c,1}}{K_{b,1} + K_{c,1}} = 571667 \frac{284444}{571667 + 284444} = 189937 \text{ kN} \qquad \{2.27\}$$

where

$$r_1 = \frac{K_{c,1}}{K_{b,1} + K_{c,1}} = \frac{284444}{571667 + 284444} = 0.3323 \qquad \{2.56\}$$

is the reduction factor.

Mass distribution factor r_f and the mass density per unit length are also needed for the calculations.

Mass distribution factor r_f is obtained from Table 4.1 as a function of the number of storeys as

$$r_f = 0.892 \qquad \{\text{Table 4.1}\}$$

The mass density per unit length is calculated with regard to the whole building as, eventually, the frequency values of the individual bracing units will be used for the determination of the fundamental frequency of the whole building:

$$m = \rho A = \frac{\gamma}{g} A = \frac{3}{9.81} 24 \cdot 18 = 132.1 \text{ kg/m} \qquad \{4.1\}$$

e "auxiliary" frequencies are needed before the value of the fundamental is determined. For practical reasons, instead of the basic frequencies s, their squares will be used.
frequencies that belong to the original shear stiffness and the global iffness are defined as

$$\frac{1}{(4H)^2}\frac{r_f^2 K_1}{m} = \frac{0.892^2 \cdot 189937}{(4\cdot 24)^2 \cdot 132.1} = 0.1241\,\text{Hz}^2 \qquad \{2.42\}$$

$$\frac{0.313 r_f^2 E_c I_{g,1}}{H^4 m} = \frac{0.313\cdot 0.892^2 \cdot 25\cdot 10^6 \cdot 28.8}{24^4 \cdot 132.1} = 4.091\,\text{Hz}^2 \qquad \{2.43\}$$

effectiveness factor is now obtained as

$$\sqrt{\frac{f_{g,1}^2}{f_{g,1}^2 + f_{s',1}^2}} = \sqrt{\frac{4.091}{4.091+0.1241}} = \sqrt{0.9706} = 0.9852 \qquad \{2.46\}$$

effectiveness factor leads to the effective shear stiffness:

$$s_{f,1}^2 K_1 = 0.9706\cdot 189937 = 184353\,\text{kN} \qquad \{2.45\}$$

frequency that belongs to the effective shear stiffness is given by

$$\frac{1}{(4H)^2}\frac{r_f^2 K_{e,1}}{m} = \frac{0.892^2 \cdot 184353}{(4\cdot 24)^2 \cdot 132.1} = 0.1205\,\text{Hz}^2 \qquad \{2.44\}$$

ocal second moment of area of the frame also has to be recalculated:

$$\sum_{j=1}^{n} I_{c,j} = 0.3323\cdot 4\frac{0.4^4}{12} = 0.002835\,\text{m}^4 \qquad \{2.31\}$$

fundamental frequency which is associated with the local bending defined by

$$\frac{0.313 r_f^2 EI_1}{H^4 m} = \frac{0.313\cdot 0.892^2 \cdot 25\cdot 10^6 \cdot 0.002835}{24^4 \cdot 132.1} = 0.0004027\,\text{Hz}^2 \qquad \{2.47\}$$

$$= H\sqrt{\frac{K_{e,1}}{EI_1}} = 24\sqrt{\frac{184353}{25\cdot10^6\cdot0.002835}} = 38.71 \qquad \{2.51\}$$

equency parameter is obtained using Table 4.2 as

$$= 9.016 + \frac{10.26 - 9.016}{40 - 35}(38.71 - 35) = 9.939 \qquad \{\text{Table 4.2}\}$$

Finally, the fundamental frequency is

$$_1 = \sqrt{f_{b,1}^2 + f_{s,1}^2 + \left(\frac{\eta^2}{0.313} - \frac{k^2}{5} - 1\right) s_{f,1} f_{b,1}^2} \qquad \{2.52\}$$

$$\sqrt{0.0004027 + 0.1205 + \left(\frac{9.939^2}{0.313} - \frac{38.71^2}{5} - 1\right) 0.9852\cdot0.0004027} = 0.3561 \text{ Hz}$$

The auxiliary constants needed for the determination of the maximum ction of the frame are

$$= 1 + \frac{I_1}{I_{g,1}} = 1 + \frac{0.002835}{28.8} = 1.000098 \cong 1.0 \qquad \{2.14\}$$

$$= \sqrt{\frac{K_1 s}{EI_1}} = \sqrt{\frac{189937}{25\cdot10^6\cdot0.002835}} = 1.637 \qquad \text{and} \qquad \kappa H = 39.29$$

The sum of the local and global second moments of area is

$$_{f,1} = I_1 + I_{g,1} = 28.802835 \text{ m}^4 \qquad \{2.23\}$$

The maximum deflection of the frame—under $q_y = 22.98$ kN/m—is

$$_1 = y(H) = \frac{22.98\cdot24^4}{8\cdot25\cdot10^6\cdot28.802835} + \frac{22.98\cdot24^2}{2\cdot189937} - \qquad \{2.24\}$$

$$-\frac{22.98\cdot25\cdot10^6\cdot0.002835}{189937^2}\left(\frac{1+39.29\sinh 39.29}{\cosh 39.29} - 1\right) = 0.0344 \text{ m}$$

'nits 3 and 4: shear wall W6

al load, the fundamental frequency and the maximum deflection of the are

$$= \frac{7.837EIr_s}{H^2} = \frac{7.837 \cdot 25 \cdot 10^3 \cdot 5.4 \cdot 0.834}{24^2} = 1532 \text{ MN} \qquad \{2.76\}$$

$$\frac{.56 r_f}{H^2}\sqrt{\frac{EI}{m}} = \frac{0.56 \cdot 0.892}{24^2}\sqrt{\frac{25 \cdot 10^6 \cdot 5.4}{132.1}} = 0.8767 \text{ Hz} \qquad \{2.75\}$$

$$\frac{wH^4}{EI_{max}} = \frac{22.98 \cdot 24^4}{8 \cdot 25 \cdot 10^6 \cdot 5.4} = 0.00706 \text{ m} \qquad \{2.74\}$$

maximum deflection of the shear walls in their perpendicular direction $_{min}$) will also be needed with the torsional analysis:

$$\frac{wH^4}{8EI_{min}} = \frac{22.98 \cdot 24^4}{8 \cdot 25 \cdot 10^6 \cdot 0.0135} = 2.824 \text{ m} \qquad \{2.74\}$$

1: an unacceptable bracing system arrangement

ng arrangement shown in Figure 6.3 is deliberately chosen as one not suitable for an eight-storey building. Nevertheless, it is included in igation as it spectacularly illustrates how the critical load ratio can e weakness(es) of a bracing system.

bility analysis

, the three basic (sway in directions x and y and pure torsional) critical to be calculated first, then the coupling of the basic modes has to be . Although it is obvious in this case that $N_{cr,x}$ will be the critical load of ng, for the sake of completeness, the complete procedure will be

:

ve bracing is provided for lateral stability in direction x and only the s and the full-height columns of the frames (normally ignored in the erpendicular to their plane) offer nominal resistance:

$$\frac{37EIr_s}{H^2} = \frac{7.837 \cdot 25 \cdot 10^3 \cdot 2(0.0135 + 4 \cdot 0.0021\dot{3})0.834}{24^2} = 12.5 \text{ MN} \qquad \{5.19\}$$

tion y
ffective shear stiffness of the system (with two frames) is

$$_e = \sum_{i=1}^{f} K_i s_i = 2 \cdot 166.0 \cdot 0.9801 = 325.4 \text{ MN} \qquad \{5.10\}$$

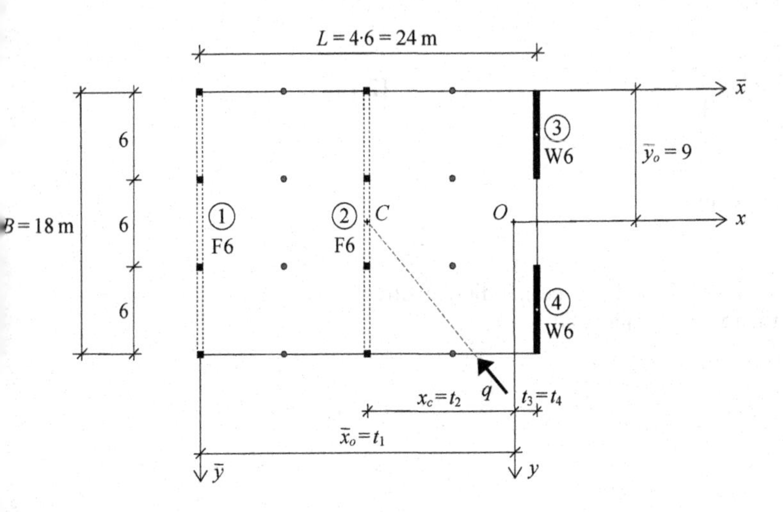

Figure 6.3 Case 1: an unacceptable bracing system arrangement.

The effectiveness factor for the whole system is

$$= \frac{K_e}{K} = \frac{325.4}{2 \cdot 166.0} = 0.9801 \qquad \{5.11\}$$

The total bending stiffness of the system is obtained by adding up the local ing stiffness of the vertical structural units (where the contribution of the nns of the frames, being negligible, is ignored):

$$I = 2 \cdot 25 \cdot 10^3 \cdot 5.4 = 270000 \text{ MNm}^2 \qquad \{5.12\}$$

The local bending critical load of the system is

$$_l = \frac{7.837 r_s EI}{H^2} = \frac{7.837 \cdot 0.834 \cdot 270000}{24^2} = 3064 \text{ MN} \qquad \{5.13\}$$

With the part critical load ratio

$$\frac{K_e}{} \quad 325.4$$

l load parameter is

312 {Table 5.2}

critical load in direction y is now obtained as

$$= N_l + K_e + (\alpha - \beta - 1)sN_l \qquad \{5.18\}$$

$$= 3064 + 325.4 + (1.312 - 0.106 - 1)0.9801 \cdot 3064 = 4008 \text{ MN}$$

onal buckling

critical loads of the individual bracing units now available (in Section first step is to establish the location of the shear centre. Because of only direction x requires calculation:

$$\frac{\sum_{i=1}^{f+m} N_{y,i}\bar{x}_i}{\sum_{i=1}^{f+m} N_{y,i}} = \frac{166 \cdot 12 + 2 \cdot 1532 \cdot 24}{2 \cdot 166 + 2 \cdot 1532} = 22.24 \text{ m}, \qquad \bar{y}_o = 9 \text{ m} \qquad \{5.27\}$$

the distance between the shear centre and the centroid of the layout

$$= \frac{L}{2} - \bar{x}_o = \frac{24}{2} - 22.24 = -10.24 \text{ m} \qquad \{5.32\}$$

of gyration is:

$$\sqrt{\frac{L^2 + B^2}{12} + t^2} = \sqrt{\frac{24^2 + 18^2}{12} + 10.24^2} = \sqrt{179.86} = 13.41 \text{ m} \qquad \{5.32\}$$

erpendicular distances of the bracing units from the shear centre are

$= 22.24$ m, $t_2 = x_c = 10.24$ m and $t_3 = t_4 = L - \bar{x}_o = 24 - 22.24 = 1.76$ m

"original" Saint-Venant stiffness originates from the shear walls and the

$$\sum_{k=1}^{m} GJ_k + \sum_{i=1}^{f}\left((K_i)_x y_i^2 + (K_i)_y x_i^2\right) \qquad \{5.38\}$$

The contribution of the own Saint-Venant stiffness of the shear walls (first is very little at 1.1%.
The effective Saint-Venant stiffness is

$$GJ)_e = \sum_{k=1}^{m} GJ_k + \sum_{i=1}^{f}\left((K_{e,i})_x y_i^2 + (K_{e,i})_y x_i^2\right) \qquad \{5.29\}$$

$$= 2 \cdot 10.4 \cdot 10^3 \cdot 0.054 + 166 \cdot 0.9801(22.24^2 + 10.24^2) = 98656 \text{ MNm}^2$$

The effectiveness factor is

$$_\varphi = \frac{(GJ)_e}{(GJ)} = \frac{98656}{100638} = 0.9803 \qquad \{5.37\}$$

The Saint-Venant torsional critical load that is associated with this stiffness is

$$V_t = \frac{(GJ)_e}{i_p^2} = \frac{98656}{179.86} = 548.5 \text{ MN} \qquad \{5.36\}$$

The warping stiffness of the system originates from the two shear walls and olumns of the frames:

$$EI_\omega = E_w \sum_{k=1}^{m}\left(I_{\omega,k} + (I_{w,k})_x y_k^2 + (I_{w,k})_y x_k^2\right) + E_c \sum_{i=1}^{f}\left((I_{c,i} r_i)_x y_i^2 + (I_{c,i} r_i)_y x_i^2\right)$$

$$= 25000\left(5.4 \cdot 1.76^2 \cdot 2 + 0.002477(10.24^2 + 22.24^2)\right) = 873475 \text{ MNm}^4 \qquad \{5.30\}$$

The warping torsional critical load of the system is

$$V_\omega = \frac{7.837 r_s EI_\omega}{i_p^2 H^2} = \frac{7.837 \cdot 0.834 \cdot 873475}{179.86 \cdot 24^2} = 55.1 \text{ MN} \qquad \{5.35\}$$

With the part critical load ratio

$$3_\varphi = \frac{N_t}{N_\omega} = \frac{548.5}{55.1} = 9.955 \qquad \{5.39\}$$

ritical load parameter is obtained as

$$18.118 - 16.682$$

With the critical load parameter, the critical load for pure torsional buckling is

$$N_{cr,\varphi} = N_\omega + N_t + (\alpha_\varphi - \beta_\varphi - 1)s_\varphi N_\omega \qquad \{5.34\}$$

$$= 55.1 + 548.5 + (18.05 - 9.955 - 1)0.9803 \cdot 55.1 = 987 \text{ MN}$$

With the three basic modes available, their coupling must be considered.

Mode coupling

The arrangement of the bracing system is monosymmetric and the centroid of the vertical load of the building lies on axis x. Two things may happen in such cases. Sway buckling may develop in direction x (defined by $N_{cr,x}$) or buckling in direction y ($N_{cr,y}$) couples with pure torsional buckling ($N_{cr,\varphi}$). As foreseen in the beginning, basic critical load $N_{cr,x}$ is much smaller than either of $N_{cr,y}$ and $N_{cr,\varphi}$. It follows that it is sufficient to use the much simpler approximate formula for taking into account the effect of coupling, as the actual value of the coupled critical load, being obviously greater than that of $N_{cr,x}$, is irrelevant. The critical load of this coupled buckling is obtained approximately as

$$N_{y\varphi} = \left(\frac{1}{N_{cr,y}} + \frac{1}{N_{cr,\varphi}} \right)^{-1} = \left(\frac{1}{4008} + \frac{1}{987} \right)^{-1} = 792 \text{ MN} \qquad \{5.45\}$$

The critical load of the building is the smaller one of $N_{cr,x}$ and $N_{y\varphi}$, i.e.:

$$N_{cr} = \min(N_{cr,x}, N_{y\varphi}) = N_{cr,x} = 12.5 \text{ MN} \qquad \{5.46\}$$

The total load on the building is

$$N = LBQn = 24 \cdot 18 \cdot 0.01 \cdot 8 = 34.56 \text{ MN} \qquad \{6.2\}$$

The global critical load ratio

$$\lambda = \frac{N_{cr}}{N} = \frac{12.5}{34.56} = 0.36 \qquad \{6.3\}$$

as expected, shows a totally unacceptable, unstable bracing system.

In a practical situation, this would be end of story: in the case of an unstable building, no further investigation is needed. For the sake of completeness, however, both the frequency and the deflection analyses will be carried out.

6.3.2.2 Frequency analysis

As with the stability analysis, the three basic (lateral in directions x and y and pure torsional) modes will be investigated first, then the coupling of the modes will be considered.

Direction x

As seen with the stability analysis, the system is very weak in direction x and the lateral frequency is governed by the small stiffness of the shear walls and the columns of the frames (acting as shear walls), perpendicular to their plane (which is normally neglected in practical structural engineering calculations):

$$EI_y = E_c I_c + E_w I_w = 25 \cdot 10^6 (8 \cdot 0.0021\dot{3} + 2 \cdot 0.0135) = 1102 \cdot 10^3 \text{ kNm}^2 \quad \{4.14\}$$

The lateral frequency is

$$f_x = \frac{0.56 r_f}{H^2}\sqrt{\frac{EI_y}{m}} = \frac{0.56 \cdot 0.892}{24^2}\sqrt{\frac{1102 \cdot 10^3}{132.1}} = 0.0792 \text{ Hz} \quad \{2.75\}$$

Direction y

With the bending stiffness of the two shear walls (as the contribution of the columns of the frames, being negligible compared to that of the shear walls, is ignored) the lateral frequency of the system in bending is obtained from

$$f_b^2 = \frac{0.313 r_f^2 EI}{H^4 m} = \frac{0.313 \cdot 0.892^2 \cdot 25 \cdot 10^6 \cdot 2 \cdot 5.4}{24^4 \cdot 132.1} = 1.534 \text{ Hz}^2 \quad \{4.15\}$$

The lateral frequency which is associated with shear deformation is determined using the effective shear stiffness of the two frames:

$$f_s^2 = \frac{1}{(4H)^2}\frac{r_f^2 K_e}{m} = \frac{0.892^2 \cdot 2 \cdot 184353}{(4 \cdot 24)^2 \cdot 132.1} = 0.2410 \text{ Hz}^2 \quad \{4.13\}$$

With the non-dimensional parameter

$$k = H\sqrt{\frac{K_e}{EI}} = 24\sqrt{\frac{2 \cdot 184353}{25 \cdot 10^6 \cdot 2 \cdot 5.4}} = 0.8869 \quad \{4.17\}$$

the frequency parameter is

$$\eta = 0.6223 + \frac{0.6376 - 0.6223}{0.9 - 0.8}(0.8869 - 0.8) = 0.6356 \quad \{\text{Table 4.2}\}$$

Using the above frequency parameter, the lateral frequency in direction y is obtained as

$$f_y = \sqrt{f_b^2 + f_s^2 + \left(\frac{\eta^2}{0.313} - \frac{k^2}{5} - 1\right) s_f f_b^2} = \quad \{4.18\}$$

$$= \sqrt{1.534 + 0.241 + \left(\frac{0.6356^2}{0.313} - \frac{0.8869^2}{5} - 1\right)0.9852 \cdot 1.534} = 1.406 \text{ Hz}$$

Pure torsional vibration

With the fundamental frequencies of the individual bracing units now available (in Section 6.3.1), the first step is to establish the location of the shear centre. Because of symmetry, only direction x requires calculation:

$$\bar{x}_o = \frac{\sum_{i=1}^{f+m} f_{y,i}^2 \bar{x}_i}{\sum_{i=1}^{f+m} f_{y,i}^2} = \frac{0.3561^2 \cdot 12 + 2 \cdot 0.8767^2 \cdot 24}{2(0.3561^2 + 0.8767^2)} = 21.45 \text{ m}, \qquad \bar{y}_o = 9 \text{ m} \quad \{4.19\}$$

With the distance between the shear centre and the centroid of the layout

$$t = x_c = \bar{x}_o - \frac{L}{2} - = 21.45 - \frac{24}{2} = 9.45 \text{ m} \qquad \{4.25\}$$

the radius of gyration is

$$i_p = \sqrt{\frac{L^2 + B^2}{12} + t^2} = \sqrt{\frac{24^2 + 18^2}{12} + 9.45^2} = \sqrt{164.3} = 12.82 \text{ m} \qquad \{4.24\}$$

The perpendicular distances of the bracing units from the shear centre are

$$t_1 = \bar{x}_o = 21.45 \text{ m}, \quad t_2 = x_c = 9.45 \text{ m} \quad \text{and} \quad t_3 = t_4 = L - \bar{x}_o = 24 - 21.45 = 2.55 \text{ m}$$

The “original” Saint-Venant stiffness originates from the shear walls and the frames as

$$(GJ) = \sum_{k=1}^{m} GJ_k + \sum_{i=1}^{f} \left((K_i)_x y_i^2 + (K_i)_y x_i^2\right) \qquad \{4.30\}$$

$$= 2 \cdot 10.4 \cdot 10^6 \cdot 0.054 + 189937(21.45^2 + 9.45^2) = 105.5 \cdot 10^6 \text{ kNm}^2$$

The contribution of the own Saint-Venant stiffness of the shear walls (first term) is very little: 1.1%.

The effective Saint-Venant stiffness is

$$(GJ)_e = \sum_{k=1}^{m} GJ_k + \sum_{i=1}^{f} \left((K_{e,i})_x y_i^2 + (K_{e,i})_y x_i^2\right) = \qquad \{4.21\}$$

$$= 2 \cdot 10.4 \cdot 10^6 \cdot 0.054 + 184353(21.45^2 + 9.45^2) = 102.4 \cdot 10^6 \text{ kNm}^2$$

The Saint-Venant torsional frequency that is associated with this stiffness is

$$f_t^2 = \frac{r_f^2 (GJ)_e}{16 i_p^2 H^2 m} = \frac{0.892^2 \cdot 102.4 \cdot 10^6}{16 \cdot 164.3 \cdot 24^2 \cdot 132.1} = 0.4073 \text{ Hz}^2 \qquad \{4.28\}$$

The effectiveness of the Saint-Venant stiffness is expressed by the factor

$$s_\varphi = \sqrt{\frac{(GJ)_e}{(GJ)}} = \sqrt{\frac{102.4}{105.5}} = 0.9852 \qquad \{4.29\}$$

The warping stiffness of the system originates from the two shear walls and the columns of the frames:

$$EI_\omega = E_w \sum_{k=1}^{m} \left(I_{\omega,k} + (I_{w,k})_x y_k^2 + (I_{w,k})_y x_k^2 \right) + E_c \sum_{i=1}^{f} \left((I_{c,i} r_i)_x y_i^2 + (I_{c,i} r_i)_y x_i^2 \right)$$

$$= 25 \cdot 10^6 \left(5.4 \cdot 2.55^2 \cdot 2 + 0.002835(21.45^2 + 9.45^2) \right) = 17.95 \cdot 10^8 \text{ kNm}^4 \qquad \{4.22\}$$

The warping torsional frequency of the system is

$$f_\omega^2 = \frac{0.313 r_f^2 EI_\omega}{i_p^2 H^4 m} = \frac{0.313 \cdot 0.892^2 \cdot 17.95 \cdot 10^8}{164.3 \cdot 24^4 \cdot 132.1} = 0.0621 \text{ Hz}^2 \qquad \{4.27\}$$

With torsion parameter

$$k_\varphi = H \sqrt{\frac{(GJ)_e}{EI_\omega}} = 24 \sqrt{\frac{102.4 \cdot 10^6}{17.95 \cdot 10^8}} = 5.73 \qquad \{4.31\}$$

the frequency parameter can now be obtained using Table 4.2:

$$\eta_\varphi = 1.7065 + \frac{1.827 - 1.7065}{6 - 5.5}(5.73 - 5.5) = 1.76 \qquad \{\text{Table 4.2}\}$$

The frequency for pure torsional vibration is

$$f_\varphi = \sqrt{f_\omega^2 + f_t^2 + \left(\frac{\eta_\varphi^2}{0.313} - \frac{k_\varphi^2}{5} - 1 \right) s_\varphi f_\omega^2} = \qquad \{4.26\}$$

$$= \sqrt{0.0621 + 0.4073 + \left(\frac{1.76^2}{0.313} - \frac{5.73^2}{5} - 1\right) 0.9852 \cdot 0.0621} = 0.782 \text{ Hz}$$

Coupling of the vibration modes
Pure torsional vibration and lateral vibration in direction y combine. The value of the combined frequency can be approximated as

$$f_{y\varphi} = \left(\frac{1}{f_y^2} + \frac{1}{f_\varphi^2}\right)^{-\frac{1}{2}} = \left(\frac{1}{1.406^2} + \frac{1}{0.782^2}\right)^{-\frac{1}{2}} = 0.683 \text{ Hz} \qquad \{4.37\}$$

As the frequency of lateral vibration in direction x is (much) smaller, the fundamental frequency of the building is this smaller value:

$$f = \min(f_x, f_{y\varphi}) = f_x = 0.0792 \text{ Hz} \qquad \{4.38\}$$

6.3.2.3 Maximum deflection

Wind load component q_y is balanced by all four bracing units acting in their plane while wind load component q_x is resisted by the four bracing units acting perpendicular to their plane.

Direction x
The sum of the second moments of area of the vertical elements that can brace the building in direction x is

$$I_y = I_c + I_w = 8 \cdot 0.0021\dot{3} + 2 \cdot 0.0135 = 0.0441 \text{ m}^4$$

The maximum deflection of the building is calculated as

$$u_{\max} = \frac{q_x H^4}{8EI_y} = \frac{19.28 \cdot 24^4}{8 \cdot 25 \cdot 10^6 \cdot 0.0441} = 0.725 \text{ m} \qquad \{2.74\}$$

There is wind load in direction of y as well. However, the bracing system acting in direction y is much more effective and the above value can be considered as a good approximation of the maximum deflection of the building. The maximum deflection of the building already far exceeds the recommended maximum:

$$d_{\text{ASCE}} = \frac{H}{500} = \frac{24}{500} = 0.048 \text{ m} \qquad \{3.67\}$$

In summary, the bracing system is clearly unacceptable: the maximum deflection is far too great, the fundamental frequency is very small and, above all, the building is not stable. The inadequacy of the system is clearly indicated by the value of the critical load ratio ($\lambda = 0.34$). For a theoretically stable bracing system, this value must be greater than 1.0 and, preferably, it should be greater than 10.

The main weakness of the bracing system lies in the lack of bracing in direction x. To improve the performance of the building, the two frames are rotated by 90 degrees and are moved towards the left-hand side of the building to create a more balanced arrangement (Figure 6.4).

6.3.3 Case 2: a more balanced bracing system arrangement

The arrangement shown in Figure 6.4 remedies a fatal flaw with Case 1, namely, this time the system has considerable stiffness in direction x as well. It remains to be seen if the improvement is great enough to result in an adequate bracing system.

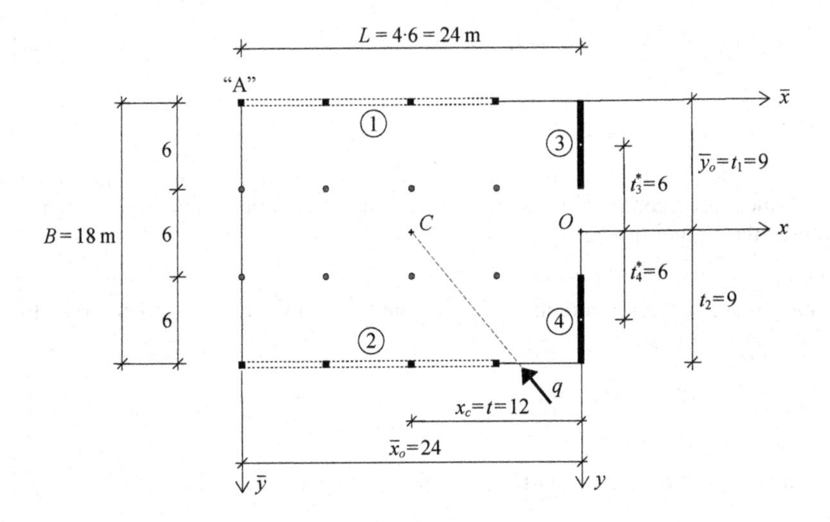

Figure 6.4 Case 2: a more balanced bracing system arrangement.

6.3.3.1 Stability analysis

The three basic (sway in directions x and y and pure torsional) critical loads will be calculated, then the question of the coupling of the basic modes will be addressed.

Direction x

Buckling in direction x is resisted by the two frames with some contribution from the two shear walls.

The effective shear stiffness of the system (with the two frames) is

$$K_e = \sum_{i=1}^{f} K_i s_i = 2 \cdot 166.0 \cdot 0.9801 = 325.4 \text{ MN} \qquad \{5.10\}$$

The effectiveness factor for the whole system is

$$s = \frac{K_e}{K} = \frac{325.4}{2 \cdot 166.0} = 0.9801 \qquad \{5.11\}$$

The total local bending stiffness of the system is obtained by adding up the bending stiffnesses of the vertical structural units:

$$EI = 2 \cdot 25 \cdot 10^3 (0.002477 + 0.0135) = 798.9 \text{ MNm}^2 \qquad \{5.12\}$$

The local bending critical load of the system is

$$N_l = \frac{7.837 r_s EI}{H^2} = \frac{7.837 \cdot 0.834 \cdot 798.9}{24^2} = 9.07 \text{ MN} \qquad \{5.13\}$$

With part critical load ratio

$$\beta = \frac{K_e}{N_l} = \frac{325.4}{9.07} = 35.9 \qquad \{5.17\}$$

the critical load parameter is

$$\alpha = 52.3 \qquad \{\text{Table 5.2}\}$$

The critical load in direction y is now obtained as

$$N_{cr,x} = N_l + K_e + (\alpha - \beta - 1) s N_l \qquad \{5.18\}$$

$$= 9.07 + 325.4 + (52.3 - 35.9 - 1) 0.9801 \cdot 9.07 = 471.4 \text{ MN}$$

Direction y

Buckling in direction y is resisted by the two shear walls. The corresponding critical load is

$$N_{cr,y} = 2 \cdot 1532 = 3064 \text{ MN} \qquad \{2.76 \text{ in Section 6.3.1}\}$$

Pure torsional buckling

Because of the symmetric arrangement of the two shear walls and of the two frames, the location of the shear centre is readily available:

$$\bar{x}_o = 24.0 \text{ m}, \qquad \bar{y}_o = 9.0 \text{ m}$$

The distances of the bracing units from the shear centre (Figure 6.4) are

$t_1 = t_2 = 9.0$ m, $\quad t_3 = t_4 = 0 \quad$ and $\quad t_3^* = t_4^* = 6$ m

With the distance between the shear centre and the centroid of the layout

$$t = x_c = \bar{x}_o - \frac{L}{2} = 24 - \frac{24}{2} = 12.0 \text{ m} \qquad \{5.33\}$$

the radius of gyration is

$$i_p = \sqrt{\frac{L^2 + B^2}{12} + t^2} = \sqrt{\frac{24^2 + 18^2}{12} + 12^2} = \sqrt{219} = 14.8 \text{ m} \qquad \{5.32\}$$

The "original" Saint-Venant stiffness of the system is

$$(GJ) = \sum_{k=1}^{m} GJ_k + \sum_{i=1}^{f} \left((K_i)_x y_i^2 + (K_i)_y x_i^2 \right) \qquad \{5.38\}$$

$$= 2\left(10.4 \cdot 10^3 (0.054 + 0.03413) + 166 \cdot 9^2\right) = 28725 \text{ MNm}^2$$

With the effectiveness factor [{5.9} in Section 6.3.1], the effective Saint-Venant stiffness is

$$(GJ)_e = \sum_{k=1}^{m} GJ_k + \sum_{k=1}^{f} \left((K_{e,i})_x y_i^2 + (K_{e,i})_y x_i^2 \right) \qquad \{5.29\}$$

$$= 2\left(10.4 \cdot 10^3 (0.054 + 0.03413) + 166 \cdot 0.9801 \cdot 9^2\right) = 28190 \text{ MNm}^2$$

The Saint-Venant torsional critical load that is associated with this stiffness is

$$N_t = \frac{(GJ)_e}{i_p^2} = \frac{28190}{219} = 128.7 \text{ MN} \qquad \{5.36\}$$

The effectiveness factor is

$$s_\varphi = \frac{(GJ)_e}{(GJ)} = \frac{28190}{28725} = 0.9814 \qquad \{5.37\}$$

The warping stiffness comes from three sources {5.30} but both the contribution of the columns of the frames (in directions x and y) and that of the shear walls are relatively small:

$$EI_\omega = E_w \sum_{k=1}^{m} \left(I_{\omega,k} + (I_{w,k})_x y_k^2 + (I_{w,k})_y x_k^2 \right) + E_c \sum_{i=1}^{f} \left((I_{c,i} r_i)_x y_i^2 + (I_{c,i} r_i)_y x_i^2 \right)$$

$$= 25000\left(0.002477 \cdot 9^2 + 0.0021\dot{3}(6^2 + 12^2 + 18^2 + 24^2) + 0.0135 \cdot 6^2\right)2 = 149530 \text{ MNm}^4$$

The warping torsional critical load of the system is

$$N_\omega = \frac{7.837 r_s EI_\omega}{i_p^2 H^2} = \frac{7.837 \cdot 0.834 \cdot 149530}{219 \cdot 24^2} = 7.75 \text{ MN} \qquad \{5.35\}$$

With part critical load ratio

$$\beta_\varphi = \frac{N_t}{N_\omega} = \frac{128.7}{7.75} = 16.61 \qquad \{5.39\}$$

the critical load parameter is

$$\alpha_\varphi = 18.118 + \frac{31.82 - 18.118}{20 - 10}(16.61 - 10) = 27.18 \qquad \{\text{Table } 5.2\}$$

Using the part torsional critical loads and the critical load parameter, the critical load for pure torsional buckling is obtained as

$$N_{cr,\varphi} = N_\omega + N_t + (\alpha_\varphi - \beta_\varphi - 1)s_\varphi N_\omega \qquad \{5.34\}$$

$$= 7.75 + 128.7 + (27.18 - 16.61 - 1)0.9814 \cdot 7.75 = 209.2 \text{ MN}$$

With the three basic modes available, their coupling must be considered.

Mode coupling

As in the previous case, the arrangement of the bracing system is monosymmetric and the centroid of the vertical load of the building lies on axis x. Two things may happen. Sway buckling may develop in direction x (defined by $N_{cr,x}$) or buckling in direction y ($N_{cr,y}$) couples with pure torsional buckling ($N_{cr,\varphi}$). The critical load of this coupled buckling is obtained approximately as

$$N_{y\varphi} = \left(\frac{1}{N_{cr,y}} + \frac{1}{N_{cr,\varphi}} \right)^{-1} = \left(\frac{1}{3064} + \frac{1}{209.2} \right)^{-1} = 196 \text{ MN} \qquad \{5.45\}$$

As basic critical load $N_{cr,x}$ is greater than this combined critical load, the above value is the global critical load of the building:

$$N_{cr} = \min(N_{cr,x}, N_{y\varphi}) = N_{y\varphi} = 196 \text{ MN} \qquad \{5.46\}$$

With the total load on the building, the global critical load ratio is

$$\lambda = \frac{N_{cr}}{N} = \frac{196}{34.56} = 5.7 \qquad \{6.3\}$$

indicating a theoretically stable building and a big improvement on Case 1, as far as stability is concerned. The above values of the basic and global critical loads indicate that the stability of the system is limited by the relatively small value of pure torsional buckling.

The situation regarding the fundamental frequency and the maximum deflection of the building will be looked at in the next two sections.

6.3.3.2 Frequency analysis

The three basic (lateral in directions x and y and pure torsional) frequencies will be calculated first, then the coupling of the modes will be considered. The behaviour of the building is very similar to that with the stability analysis.

Direction x

The participating units in direction x are the two frames and the two shear walls (utilizing their I_{min}).

The lateral frequency which is associated with shear deformation is determined using the effective shear stiffness of the two frames:

$$f_s^2 = \frac{1}{(4H)^2}\frac{r_f^2 K_e}{m} = \frac{0.892^2 \cdot 2 \cdot 184353}{(4 \cdot 24)^2 \cdot 132.1} = 0.2410 \text{ Hz}^2 \qquad \{4.13\}$$

With the bending stiffness of the two shear walls and the columns of the two frames, the lateral frequency of the system in bending is obtained from Equation (4.15) as

$$f_b^2 = \frac{0.313 r_f^2 EI}{H^4 m} = \frac{0.313 \cdot 0.892^2 \cdot 25 \cdot 10^6 \cdot 2(0.0135 + 0.002834)}{24^4 \cdot 132.1} = 0.00464 \text{ Hz}^2$$

With the non-dimensional parameter

$$k = H\sqrt{\frac{K_e}{EI}} = 24\sqrt{\frac{2 \cdot 184353}{25 \cdot 10^6 \cdot 2(0.0135 + 0.002834)}} = 16.1 \qquad \{4.17\}$$

the frequency parameter is obtained as

$$\eta = 4.284 + \frac{4.408 - 4.284}{16.5 - 16}(16.1 - 16) = 4.309 \qquad \{\text{Table 4.2}\}$$

and the lateral frequency in direction x is

$$f_x = \sqrt{f_b^2 + f_s^2 + \left(\frac{\eta^2}{0.313} - \frac{k^2}{5} - 1\right) s_f f_b^2} \qquad \{4.18\}$$

$$= \sqrt{0.00464 + 0.241 + \left(\frac{4.309^2}{0.313} - \frac{16.1^2}{5} - 1\right) 0.9852 \cdot 0.00464} = 0.525\ \text{Hz}$$

Direction y
Vibration in direction y is resisted by the two shear walls. The corresponding lateral frequency is

$$f_y = \frac{0.56 r_f}{H^2}\sqrt{\frac{EI_x}{m}} = \frac{0.56 \cdot 0.892}{24^2}\sqrt{\frac{25 \cdot 10^6 \cdot 2 \cdot 5.4}{132.1}} = 1.24\ \text{Hz} \qquad \{2.75\}$$

Pure torsional vibration
The location of the shear centre, the radius of gyration and the distances of the bracing units are the same as with the stability analysis in Section 6.3.3.1.

The "original" Saint-Venant stiffness is

$$(GJ) = \sum_{k=1}^{m} GJ_k + \sum_{i=1}^{f}\left((K_i)_x y_i^2 + (K_i)_y x_i^2\right) \qquad \{4.30\}$$

$$= 2\left(10.4 \cdot 10^6 (0.054 + 0.03413) + 189937 \cdot 9^2\right) = 32.60 \cdot 10^6\ \text{kNm}^2$$

and the effective Saint-Venant stiffness is

$$(GJ)_e = \sum_{k=1}^{m} GJ_k + \sum_{i=1}^{f}\left((K_{e,i})_x y_i^2 + (K_{e,i})_y x_i^2\right) \qquad \{4.21\}$$

$$= 2\left(10.4 \cdot 10^6 (0.054 + 0.03413) + 184353 \cdot 9^2\right) = 31.70 \cdot 10^6\ \text{kNm}^2$$

The Saint-Venant torsional frequency that is associated with this stiffness is

$$f_t^2 = \frac{r_f^2 (GJ)_e}{16 i_p^2 H^2 m} = \frac{0.892^2 \cdot 31.70 \cdot 10^6}{16 \cdot 219 \cdot 24^2 \cdot 132.1} = 0.0946\ \text{Hz}^2 \qquad \{4.28\}$$

The effectiveness of the Saint-Venant stiffness is expressed by the factor

$$s_\varphi = \sqrt{\frac{(GJ)_e}{(GJ)}} = \sqrt{\frac{31.70}{32.60}} = 0.9861 \qquad \{4.29\}$$

The warping stiffness of the system {4.22} originates from the contribution of two shear walls and the columns of the frames (in directions x and y):

$$EI_\omega = E_W \sum_{k=1}^{m} \left(I_{\omega,k} + (I_{w,k})_x y_k^2 + (I_{w,k})_y x_k^2 \right) + E_c \sum_{i=1}^{f} \left((I_{c,i} r_i)_x y_i^2 + (I_{c,i} r_i)_y x_i^2 \right)$$

$$= 25 \cdot 10^6 \left(0.0135 \cdot 6^2 + 0.002835 \cdot 9^2 + 0.0021\dot{3}(6^2 + 12^2 + 18^2 + 24^2) \right) 2 = 151 \cdot 10^6 \text{kNm}^4$$

The warping torsional frequency of the system is

$$f_\omega^2 = \frac{0.313 r_f^2 EI_\omega}{i_p^2 H^4 m} = \frac{0.313 \cdot 0.892^2 \cdot 151 \cdot 10^6}{219 \cdot 24^4 \cdot 132.1} = 0.00392 \text{ Hz}^2 \qquad \{4.27\}$$

With torsion parameter

$$k_\varphi = H \sqrt{\frac{(GJ)_e}{EI_\omega}} = 24 \sqrt{\frac{31.70 \cdot 10^6}{151 \cdot 10^6}} = 11.0 \qquad \{4.31\}$$

the frequency parameter can now be obtained using Table 4.2 as $\eta_\varphi = 3.049$.

The frequency for pure torsional vibration is

$$f_\varphi = \sqrt{f_\omega^2 + f_t^2 + \left(\frac{\eta_\varphi^2}{0.313} - \frac{k_\varphi^2}{5} - 1 \right) s_\varphi f_\omega^2} \qquad \{4.26\}$$

$$= \sqrt{0.00392 + 0.0946 + \left(\frac{3.049^2}{0.313} - \frac{11.0^2}{5} - 1 \right) 0.9861 \cdot 0.00392} = 0.340 \text{ Hz}$$

The coupling of the basic frequencies is considered next.

Coupling of the vibration modes

Pure torsional vibration and lateral vibration in direction y combine. The value of the combined frequency can be approximated as

$$f_{y\varphi} = \left(\frac{1}{f_y^2} + \frac{1}{f_\varphi^2} \right)^{-\frac{1}{2}} = \left(\frac{1}{1.24^2} + \frac{1}{0.34^2} \right)^{-\frac{1}{2}} = 0.328 \text{ Hz} \qquad \{4.37\}$$

As the frequency of lateral vibration in direction x is greater, the fundamental frequency of the building is this smaller value:

$$f = \min(f_x, f_{y\varphi}) = f_{y\varphi} = 0.328 \text{ Hz} \qquad \{4.38\}$$

It is clear from the above values that resistance to torsion is relatively small.

6.3.3.3 Maximum deflection

The two components of the wind load will be treated separately.

Wind load in direction x is considered first. The arrangement is symmetric in this case and the wind is resisted by the two identical frames and the two shear walls (utilizing their I_{min}!). No rotation occurs. It is enough to work with half of the structure (one frame and one shear wall) and half of the load. As the number of frames in this system is one, it is possible to “push” the shear wall into the frame, in other words, to add the second moment of area of the shear wall to the second moment of area of the columns of the frame.

The local second moment of area is

$$I = 0.002835 + 0.0135 = 0.016335 \text{ m}^4$$

and the total second moment of area is

$$I_f = I + I_g = 0.016335 + 28.8 = 28.816335 \text{ m}^4 \qquad \{2.23; 3.44\}$$

With auxiliary quantities

$$s = 1 + \frac{I}{I_g} = 1 + \frac{0.016335}{28.8} = 1.000567 \qquad \{2.14\}$$

$$\kappa = \sqrt{\frac{Ks}{EI}} = \sqrt{\frac{189937 \cdot 1.000567}{25 \cdot 10^6 \cdot 0.016335}} = 0.6822, \qquad \kappa H = 16.37$$

and wind load $w_x = 0.5 \cdot 19.28 = 9.64$ kN/m in direction x, the maximum deflection is

$$u_{\max} = \frac{9.64 \cdot 24^4}{8 \cdot 25 \cdot 10^6 \cdot 28.82} + \frac{9.64 \cdot 24^2}{2 \cdot 189937 \cdot 1.000567^2} - \qquad \{2.24\}$$

$$- \frac{9.64 \cdot 25 \cdot 10^6 \cdot 0.016335}{189937^2 \cdot 1.000567^3}\left(\frac{1 + 16.37 \sinh 16.37}{\cosh 16.37} - 1\right) = 0.0135 \text{ m}$$

When the component of the wind in direction y is considered

($w_y = 22.98$ kN/m), the maximum deflection is obtained in two steps. First, the deflection of the shear centre axis has to be determined, caused by the wind load acting through the shear centre, then the additional deflection due to the rotation of the building around the shear centre has to be added.

In the first case, the load is resisted by the two shear walls and the maximum deflection is

$$v_o = \frac{22.98 \cdot 24^4}{8 \cdot 25 \cdot 10^6 \cdot 2 \cdot 5.4} = 0.00353\,\text{m} \qquad \{2.74\}$$

In addition to this uniform deflection, the torsional moment

$$m_z = w_y x_c = 22.98 \cdot 12 = 275.76\ \text{kNm/m} \qquad \{3.53\}$$

develops rotation around the shear centre, which leads to additional deflection. The torsional moment is mainly balanced by the torsional resistance of the two frames. The two shear walls also have some small contribution but they can only utilize their $I_{\min}$ with their torsion arm of $t_3^* = t_4^* = 6$ metres as the torsion arm that belongs to $I_{\max}$ is zero. There is also some contributions from the columns of the frames acting perpendicularly to the plane of the frames. However, this contribution is small and is therefore neglected in the calculation below.

The torsional stiffnesses are

$$S_{\omega,1} = S_{\omega,2} = \frac{t_1^2}{y_1} = \frac{9.0^2}{0.0344} = 2355\,\text{m}, \quad S_{\omega,3} = S_{\omega,4} = \frac{t_3^{*2}}{y_3} = \frac{6.0^2}{2.824} = 12.75\,\text{m} \qquad \{3.54\}$$

With the sum of the torsional stiffnesses

$$S_\omega = \sum_{i=1}^{f+m} S_{\omega,i} = 2(2355 + 12.75) = 4735.5\,\text{m} \qquad \{3.55\}$$

the torsional apportioners can now be determined. Choosing, say, Bracing Unit 3, its torsional apportioner is

$$q_{\omega,3} = \frac{S_{\omega,3}}{S_\omega} = \frac{12.75}{4735.5} = 0.002692 \qquad \{3.56\}$$

The warping constant that belongs to Bracing Unit 3 is

$$I_{\omega,3} = I_{3\min} t_3^{*2} = 0.0135 \cdot 6^2 = 0.486\,\text{m}^6$$

and its moment share is

$$m_{z,3} = q_{\omega,3} m_z = 0.002692 \cdot 275.76 = 0.7423 \text{ kNm/m} \qquad \{3.57\}$$

The maximum rotation of the building can now be determined from Equation (3.59) and using Bracing Unit 3. As Bracing Unit 3 is a shear wall, the equation reduces to its first term (and $I_{f\omega}$ reduces to I_ω):

$$\varphi_{\max} = \varphi_3(H) = \frac{m_{z,3} H^4}{8EI_{\omega,3}} = \frac{0.7423 \cdot 24^4}{8 \cdot 25 \cdot 10^6 \cdot 0.486} = 0.002534 \text{ rad} \qquad \{3.59\}$$

The maximum deflection of the building develops at the top floor at the left-hand side corner of the plan of the building (point "*A*" in Figure 6.4) and, making use of the angle of rotation, is obtained as

$$v_{\max} = v(H) = v_o + \varphi x_{\max} = 0.00353 + 0.002534 \cdot 24 = 0.0643 \text{ m} \qquad \{3.66\}$$

where $x_{\max}$ is the distance of the corner point (where maximum deflection occurs) from the shear centre. It is clear from Equation {3.66} that the deflection of the building is dominated by the effect of rotation—see second term.

The maximum deflection of the building is

$$d_{\max} = \sqrt{0.0135^2 + 0.0643^2} = 0.066 \text{ m}$$

This value is much smaller than the maximum deflection of "Case 1", but it is still greater than the recommended maximum ($d_{ASCE} = 0.048$ m).

As the increase in the value of the critical load ratio indicated, the performance of "Case 2" improved drastically, compared to "Case 1": the critical load and the fundamental frequency increased considerably and the maximum deflection decreased. However, as the next section will demonstrate, further improvements are possible.

6.3.4 Case 3: an effective bracing system arrangement

As the results in the previous section show, the efficiency of "Case 2" was limited mainly because of its relatively poor performance in torsion. It is therefore the torsional resistance of the system that is improved in this section by rearranging the bracing units again in such a way that Bracing Units 1 and 3 are exchanged. The resulting bracing system is doubly symmetric (Figure 6.5), still has considerable lateral stiffness in both principal directions and its torsional resistance is increased by the fact that all four bracing units now have effective "torsion arms" (i.e. perpendicular distances from the shear centre).

6.3.4.1 Stability analysis

The three basic (sway in directions x and y and pure torsional) critical loads will be calculated first and then the coupling of the modes will be looked at.

Direction x
Buckling in direction x is resisted by the two frames. The corresponding critical load is identical to that of "Case 2":

$$N_{cr,x} = 471.4 \text{ MN} \qquad \{5.18 \text{ in Section 6.3.3.1}\}$$

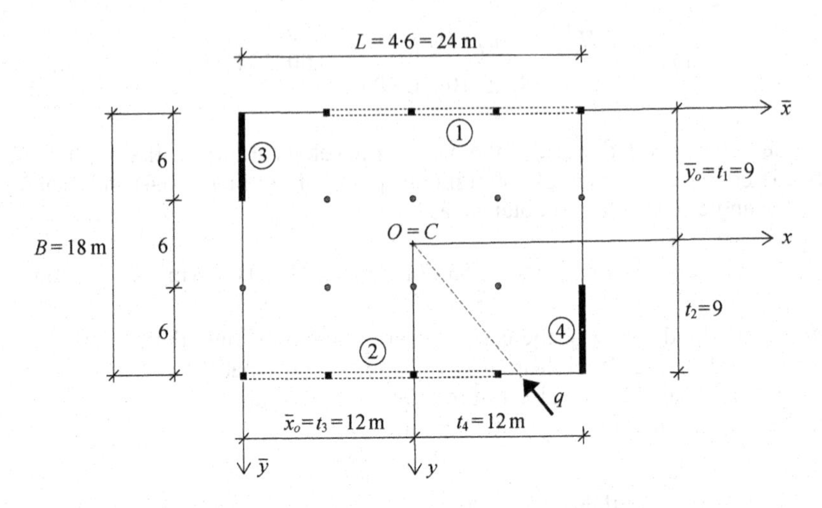

Figure 6.5 Case 3: a doubly symmetric bracing system arrangement.

Direction y
Buckling in direction y is resisted by the two shear walls. The corresponding critical load is again identical to that of "Case 2":

$$N_{cr,y} = 2 \cdot 1532 = 3064 \text{ MN} \qquad \{2.76 \text{ in Section 6.3.3.1}\}$$

Pure torsional buckling
Because of the doubly symmetric arrangement of the two shear walls and of the two frames, the location of the shear centre is readily available:

$$\bar{x}_o = 12 \text{ m} \qquad \text{and} \qquad \bar{y}_o = 9 \text{ m}$$

The radius of gyration is

$$i_p = \sqrt{\frac{L^2 + B^2}{12} + t^2} = \sqrt{\frac{24^2 + 18^2}{12}} = \sqrt{75} = 8.66 \text{ m} \qquad \{5.32\}$$

The "original" and effective Saint-Venant stiffnesses of the two frames are identical to that of "Case 2" (Section 6.3.3):

$$(GJ) = 2\left(10.4 \cdot 10^3 (0.054 + 0.03413) + 166 \cdot 9^2\right) = 28725 \text{ MNm}^2 \qquad \{5.38\}$$

$$(GJ)_e = 2\left(10.4\cdot 10^3(0.054+0.03413)+166\cdot 0.9801\cdot 9^2\right) = 28190 \text{ MNm}^2 \quad \{5.29\}$$

The effectiveness of the Saint-Venant stiffness is expressed by the factor

$$s_\varphi = \frac{(GJ)_e}{(GJ)} = \frac{28190}{28725} = 0.9814 \quad \{5.37\}$$

The effect of the columns of the frames on the warping stiffness is negligible and the warping stiffness is dominated by the contribution of the two shear walls

$$EI_\omega = E_w \sum_{k=1}^{m}\left(I_{\omega,k} + (I_{w,k})_x y_k^2 + (I_{w,k})_y x_k^2\right) \quad \{5.30\}$$

$$= 25\cdot 10^3(5.4\cdot 12^2 + 0.0135\cdot 6^2)2 = 38.90\cdot 10^6 \text{ MNm}^4$$

With the part critical loads

$$N_\omega = \frac{7.837 r_s EI_\omega}{i_p^2 H^2} = \frac{7.837\cdot 0.834\cdot 38.90\cdot 10^6}{75\cdot 24^2} = 5885 \text{ MN} \quad \{5.35\}$$

$$N_t = \frac{(GJ)_e}{i_p^2} = \frac{28190}{75} = 375.9 \text{ MN} \quad \{5.36\}$$

the critical load parameter can be obtained as a function of

$$\beta_\varphi = \frac{N_t}{N_\omega} = \frac{375.9}{5885} = 0.0639 \quad \{5.39\}$$

from Table 5.2 as

$$\alpha_\varphi = 1.1782 + \frac{1.2075 - 1.1782}{0.07 - 0.06}(0.0639 - 0.06) = 1.190 \quad \{\text{Table 5.2}\}$$

The critical load for pure torsional buckling is

$$N_{cr,\varphi} = N_\omega + N_t + (\alpha_\varphi - \beta_\varphi - 1)s_\varphi N_\omega \quad \{5.34\}$$

$$= 5885 + 375.9 + (1.19 - 0.0639 - 1)0.9814\cdot 5885 = 6989 \text{ MN}$$

As the arrangement of the bracing system is doubly symmetric and the centroid of the vertical load of the building coincides with the shear centre of the bracing system, the critical load of the building is the smallest one of the three basic critical loads:

$$N_{cr} = \min(N_{cr,x}, N_{cr,y}, N_{cr,\varphi}) = N_{cr,x} = 471.4\,\text{MN} \qquad \{5.47\}$$

With the total vertical load on the building ($N = 34.56$MN), the global critical load ratio is now

$$\lambda = \frac{N_{cr}}{N} = \frac{471.4}{34.54} = 13.6 > 10 \qquad \{6.3\} \text{ and } \{6.5\}$$

indicating a stable building and a big improvement on "Case 2". Its value also exceeds the recommended minimum value. The situation regarding the fundamental frequency and the maximum deflection of the building will be looked at in the next two sections.

6.3.4.2 Frequency analysis

The three basic frequencies will be determined first and then their coupling will be considered.

Lateral vibration

The situation is similar to that with the stability analysis in that the lateral frequencies are unchanged (i.e. identical to those of "Case 2"):

$$f_x = 0.525\ \text{Hz} \qquad \{\text{Section } 6.3.3.2\}$$

$$f_y = 1.24\,\text{Hz} \qquad \{\text{Section } 6.3.3.2\}$$

Pure torsional vibration

The "original" Saint-Venant stiffness, the effective Saint-Venant stiffness and the effectiveness factor are unchanged and can be taken from "Case 2" (Section 6.3.3.2) as

$$(GJ) = 2\left(10.4 \cdot 10^6 (0.054 + 0.03413) + 189937 \cdot 9^2\right) = 32.60 \cdot 10^6\ \text{kNm}^2 \qquad \{4.30\}$$

$$(GJ)_e = 2\left(10.4 \cdot 10^6 (0.054 + 0.03413) + 184353 \cdot 9^2\right) = 31.70 \cdot 10^6\,\text{kNm}^2 \qquad \{4.21\}$$

$$s_\varphi = \sqrt{\frac{(GJ)_e}{(GJ)}} = \sqrt{\frac{31.70}{32.60}} = 0.9861 \qquad \{4.29\}$$

The warping stiffness is dominated by the contribution of the two shear walls and can be taken from the stability analysis above:

$$EI_\omega = 25 \cdot 10^3 (5.4 \cdot 12^2 + 0.0135 \cdot 6^2) 2 = 38.90 \cdot 10^6 \text{ MNm}^4 \qquad \{4.22;\ 5.30\}$$

The two contributors to the pure torsional frequency are:

$$f_\omega^2 = \frac{0.313 r_f^2 EI_\omega}{i_p^2 H^4 m} = \frac{0.313 \cdot 0.892^2 \cdot 38.90 \cdot 10^9}{75 \cdot 24^4 \cdot 132.1} = 2.947 \text{ Hz}^2 \qquad \{4.27\}$$

$$f_t^2 = \frac{r_f^2 (GJ)_e}{16 i_p^2 H^2 m} = \frac{0.892^2 \cdot 31.7 \cdot 10^6}{16 \cdot 75 \cdot 24^2 \cdot 132.1} = 0.276 \text{ Hz}^2 \qquad \{4.28\}$$

With torsion parameter

$$k_\varphi = H \sqrt{\frac{(GJ)_e}{EI_\omega}} = 24 \sqrt{\frac{31.7 \cdot 10^6}{38.90 \cdot 10^9}} = 0.685 \qquad \{4.31\}$$

the frequency parameter is obtained using Table 4.2:

$$\eta_\varphi = 0.5959 + \frac{0.6084 - 0.5959}{0.7 - 0.6}(0.685 - 0.6) = 0.6065 \qquad \{\text{Table 4.2}\}$$

The frequency for pure torsional vibration can now be determined:

$$f_\varphi = \sqrt{f_\omega^2 + f_t^2 + \left(\frac{\eta_\varphi^2}{0.313} - \frac{k_\varphi^2}{5} - 1\right) s_\varphi f_\omega^2} \qquad \{4.26\}$$

$$= \sqrt{2.947 + 0.276 + \left(\frac{0.6065^2}{0.313} - \frac{0.685^2}{5} - 1\right) 0.9861 \cdot 2.947} = 1.860 \text{ Hz}$$

As the arrangement of the bracing system is doubly symmetric and the centroid of the mass of the building coincides with the shear centre of the bracing system, no coupling occurs and the fundamental frequency of the building is the smallest one of f_x, f_y and f_φ, i.e.:

$$f = \min(f_x,\ f_y,\ f_\varphi) = f_x = 0.525 \text{ Hz} \qquad \{4.39\}$$

6.3.4.3 Maximum deflection

When wind load in direction x is considered, the arrangement is symmetric and no rotation occurs. The wind is resisted by the two identical frames. The situation is identical to that discussed in detail in Section 6.3.3 with "Case 2" and the corresponding deflection is

$$u_{max} = 0.0135\ \text{m} \qquad \{2.24 \text{ in Section } 6.3.3.3\}$$

The wind load in direction y is resisted by the two shear walls in a symmetric arrangement and the corresponding deflection is

$$v_{max} = \frac{q_y H^4}{8EI_x} = \frac{22.98 \cdot 24^4}{8 \cdot 25 \cdot 10^6 \cdot 2 \cdot 5.4} = 0.0035\ \text{m} \qquad \{2.74\}$$

The maximum deflection of the building is

$$d_{max} = \sqrt{0.0135^2 + 0.0035^2} = 0.014\ \text{m}$$

which is smaller than the recommended maximum ($d_{ASCE} = H/500 = 0.048$ m).

The two weaknesses of the original bracing arrangement have been eliminated in two steps. First, the problem of the practically non-existent lateral stiffness of "Case 1" was addressed with "Case 2" and then the still relatively poor torsional behaviour was improved with "Case 3". The change in the value of the global critical load ratio spectacularly shows its usefulness in monitoring the efficiency of the bracing system.

6.4 PRACTICAL APPLICATION NO. 2: KOLLÁR'S CLASSIC FIVE-STOREY BUILDING

Kollár's (1977) classic five-storey building was originally used to show how a single core is best used to be effective against torsional buckling. The worked example here will demonstrate that, in addition to stability, the global critical load ratio also "handles" frequencies, rotations and deflections and identifies efficient and inefficient bracing system arrangements.

The vertical load of the building is carried by columns but, because of their pinned joints, their lateral and torsional stiffnesses cannot be taken into account in the calculations. It is assumed that the lateral and torsional stiffness of the building is provided by a U-core (Figure 6.6/b). The basic geometrical, stiffness and loading characteristics are as follows:

Size of ground plan: $L = 26$ m and $B = 13$ m
Storey height: $h = 3.0$ m, number of storeys: 5
Height of building: $H = 15$ m
Modulus of elasticity: $E = 23000$ MN/m^2
Modulus of elasticity in shear: $G = 9580$ MN/m^2

Floor load (for the stability analysis): $Q = 10$ kN/m^2
Wind load (for the deflection analysis): $w = 1.0$ kN/m^2 in direction y
Weight per unit volume of the building (for frequency analysis): $\gamma = 3$ kN/m^3

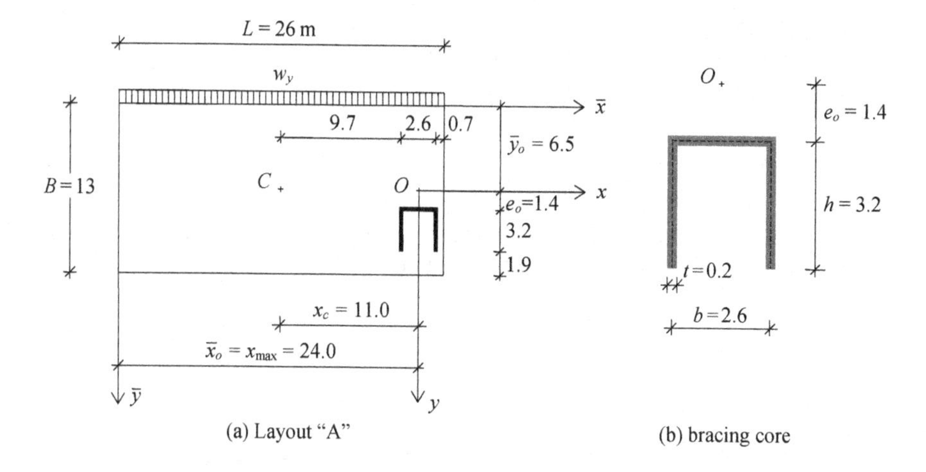

Figure 6.6 Kollár's building. Layout "A" with bracing U-core.

The load distribution factor for the stability analysis is

$$r_s = 0.759 \qquad \{\text{Table 5.1}\}$$

and the mass distribution factor for the frequency analysis is

$$r_f = 0.842 \qquad \{\text{Table 4.1}\}$$

As the size of the building and of the sole bracing unit is given, the structural performance of the building is governed by the location of the core. Two locations will be considered with and without taking into consideration the possibility that the open U-core can partially be closed. This results in four cases. In the four cases the maximum rotation and the maximum deflection, the fundamental frequency, the global critical load and the global critical load ratio of the building will be determined.

The equations related to U-cores and given in Tables 2.6 and 2.7 are used for determining the stiffness characteristics of the U-core.

6.4.1 Layout "A": an open core on the right side of the layout

The bracing core is first placed on the lower right-hand section of the layout in such a way that the shear centre of the core lies in the middle of size B of the building (Figure 6.6/a). The geometrical and stiffness characteristics of this case are collected in the first row in Table 6.3.

Table 6.3 Geometrical and stiffness characteristics for Kollár's building.

Layout	I_x	I_y	J	I_ω	$\overline{x}_o$	$\overline{y}_o$	x_c	y_c	i_p
"A"	2.045	2.466	0.024	2.51	24.0	6.50	11.0	0.00	13.84
"B"	2.045	2.466	0.024	2.51	13.0	6.50	0.0	0.00	8.39
"C"	2.479	2.599	1.450	0.00	24.0	8.20	11.0	1.70	13.94
"D"	2.479	2.599	1.450	0.00	13.0	6.50	0.0	0.00	8.39

6.4.1.1 Maximum rotation and deflection

The $w = 1$ kN/m^2 wind load intensity represents a $w_y = Lw = 26.0$ kN/m wind load per unit height in direction y. The total torsional moment is

$$m_z = w_y x_c = 26 \cdot 11 = 286 \text{ kNm/m} \qquad \{3.53\}$$

With torsion parameter

$$k = H\sqrt{\frac{GJ}{EI_\omega}} = 15\sqrt{\frac{9580 \cdot 0.024}{23000 \cdot 2.51}} = 0.947 \qquad \{2.88\}$$

the maximum rotation of the building can be calculated as

$$\varphi_{\max} = \varphi(H) = \frac{m_z H^2}{GJ}\left(\frac{\cosh k - 1}{k^2 \cosh k} - \frac{\tanh k}{k} + \frac{1}{2}\right) \qquad \{2.89\}$$

$$= \frac{286 \cdot 15^2}{9.58 \cdot 10^6 \cdot 0.024}\left(\frac{\cosh(0.947) - 1}{0.947^2 \cosh(0.947)} - \frac{\tanh(0.947)}{0.947} + \frac{1}{2}\right) = 0.02335 \text{ rad}$$

Maximum deflection develops at the left-hand side of the building at $x_{\max} = 24.0$. It consists of two parts. The top layout of the building undergoes a uniform translation of

$$v_o(H) = \frac{w_y H^4}{8EI_x} = \frac{26 \cdot 15^4}{8 \cdot 23 \cdot 10^6 \cdot 2.045} = 0.0035 \text{ m} \qquad \{2.85\}$$

The rotation around the shear centre causes additional deflection and the total maximum deflection is calculated as

$$v_{\max} = v_o(H) + \varphi(H)\, x_{\max} = 0.0035 + 0.02335 \cdot 24 = 0.5639 \text{ m} \qquad \{3.66\}$$

As the recommended maximum deflection is

$$v_{ASCE} = \frac{H}{500} = \frac{15}{500} = 0.030\,\text{m} \qquad \{3.67\}$$

the bracing of the building is clearly unacceptable.

6.4.1.2 Fundamental frequency

The mass density per unit height for the building is

$$m = \rho A = \frac{\gamma}{g}A = \frac{3}{9.81}26 \cdot 13 = 103.4\,\text{kg/m} \qquad \{4.1\}$$

The lateral frequencies in directions x and y are calculated using the formulae given for cores but with the above mass density (which relates to the whole building):

$$f_x = \frac{0.56 r_f}{H^2}\sqrt{\frac{EI_y}{m}} = \frac{0.56 \cdot 0.842}{15^2}\sqrt{\frac{23 \cdot 10^6 \cdot 2.466}{103.4}} = 1.55\,\text{Hz} \qquad \{2.92\}$$

$$f_y = \frac{0.56 r_f}{H^2}\sqrt{\frac{EI_x}{m}} = \frac{0.56 \cdot 0.842}{15^2}\sqrt{\frac{23 \cdot 10^6 \cdot 2.045}{103.4}} = 1.41\,\text{Hz} \qquad \{2.92\}$$

The frequency of pure torsional vibration of the building is obtained using the formula given for a single core but with the radius gyration that relates to the whole layout area as the mass is distributed over the whole floor area of the building:

$$i_p = \sqrt{\frac{L^2 + B^2}{12} + t^2} = \sqrt{\frac{L^2 + B^2}{12} + x_c^2} = \sqrt{\frac{26^2 + 13^2}{12} + 11^2} = 13.84\,\text{m} \qquad \{4.24\}$$

With using the torsion parameter calculated earlier ($k = 0.947$), the frequency parameter for pure torsional vibration is obtained from Table 4.2:

$$\eta_\varphi = 0.6376 + \frac{0.6542 - 0.6376}{1.0 - 0.9}(0.947 - 0.9) = 0.645 \qquad \{\text{Table 4.2}\}$$

and the pure torsional frequency is

$$f_\varphi = \frac{\eta_\varphi r_f}{i_p H^2}\sqrt{\frac{EI_\omega}{m}} = \frac{0.645 \cdot 0.842}{13.84 \cdot 15^2}\sqrt{\frac{23 \cdot 10^6 \cdot 2.51}{103.4}} = 0.130\,\text{Hz} \qquad \{2.94\}$$

The centroid of the layout lies on axis x of the coordinate system whose origin is the shear centre of the core—which now is the whole bracing system—and therefore there is a coupling of lateral vibration in direction y and pure torsional vibration. One of the frequencies (f_φ) is much smaller than the other. In such cases, it is favourable to use the approximate formulae for taking into account the effect of coupling as they offer a very simple solution with good accuracy. Hence

$$f_{y\varphi} = \left(\frac{1}{f_y^2} + \frac{1}{f_\varphi^2}\right)^{-\frac{1}{2}} = \left(\frac{1}{1.41^2} + \frac{1}{0.13^2}\right)^{-\frac{1}{2}} = 0.129 \text{ Hz} \qquad \{4.37\}$$

This frequency is smaller than that of lateral vibration in direction x, so the fundamental frequency of the building is

$$f = \min(f_x, f_{y\varphi}) = f_{y\varphi} = 0.129 \text{ Hz} \qquad \{4.38\}$$

This is a very small value considering other buildings of similar size and mass.

6.4.1.3 Global critical load and critical load ratio

The critical loads for sway buckling in directions x and y are calculated using the relevant second moment of area of the core:

$$N_{cr,x} = \frac{7.837 EI_y r_s}{H^2} = \frac{7.837 \cdot 23 \cdot 10^3 \cdot 2.466 \cdot 0.759}{15^2} = 1499 \text{ MN} \qquad \{2.96\}$$

$$N_{cr,y} = \frac{7.837 EI_x r_s}{H^2} = \frac{7.837 \cdot 23 \cdot 10^3 \cdot 2.045 \cdot 0.759}{15^2} = 1244 \text{ MN} \qquad \{2.96\}$$

The critical load of pure torsional buckling is obtained using the formula given for a single core but with the radius gyration that relates to the whole layout area as the load is distributed over the whole floor area of the building. The radius of gyration is

$$i_p = \sqrt{\frac{L^2 + B^2}{12} + t^2} = \sqrt{\frac{L^2 + B^2}{12} + x_c^2} = \sqrt{\frac{26^2 + 13^2}{12} + 11^2} = 13.84 \text{ m} \qquad \{5.32\}$$

The non-dimensional parameter

$$k_s = H\sqrt{\frac{GJ}{r_s EI_\omega}} = 15\sqrt{\frac{9580 \cdot 0.024}{0.759 \cdot 23000 \cdot 2.51}} = 1.087 \qquad \{2.98\}$$

is also needed as the critical load parameter α is obtained as a function of k_s:

$$\alpha = 10.77 + \frac{11.37 - 10.77}{1.1 - 1.0}(1.087 - 1.0) = 11.29 \qquad \{\text{Table 2.9}\}$$

The critical load for pure torsional buckling is now obtained as

$$N_{cr,\varphi} = \frac{\alpha r_s EI_\omega}{i_p^2 H^2} = \frac{11.29 \cdot 0.759 \cdot 23 \cdot 10^3 \cdot 2.51}{13.84^2 \cdot 15^2} = 11.5 \text{ MN} \qquad \{2.97\}$$

When the coupling of the basic critical loads is considered, the situation is similar to that of vibration. There is a coupling of sway buckling in direction y and pure torsional buckling. One of the critical loads ($N_{cr,\varphi}$) is much smaller than the other one and the relevant summation formula is used:

$$N_{y\varphi} = \left(\frac{1}{N_{cr,y}} + \frac{1}{N_{cr,\varphi}} \right)^{-1} = \left(\frac{1}{1244} + \frac{1}{11.5} \right)^{-1} = 11.4 \text{ MN} \qquad \{5.45\}$$

This critical load is smaller than that of sway buckling in direction x, so the critical load of the building is

$$N_{cr} = \min(N_{cr,x}, N_{y\varphi}) = N_{y\varphi} = 11.4 \text{ MN} \qquad \{5.46\}$$

The total vertical load on the five floors is

$$N = LBQn = 26 \cdot 13 \cdot 0.01 \cdot 5 = 16.9 \text{ MN} \qquad \{6.2\}$$

and the global critical load ratio

$$\lambda = \frac{N_{cr}}{N} = \frac{11.4}{16.9} = 0.67 \qquad \{6.3\}$$

reveals an unstable building.

The bracing of the building is totally unacceptable and the examination of the relevant figures related to the top deflection, fundamental frequency and stability points to a weak torsional performance. The torsional resistance of the building is small, for two reasons. One, the torsional stiffness of the core is small and two, the distance between the centroid of the layout and the shear centre is great.

In the following section an attempt is made to remedy the situation by moving the core to a more favourable position.

6.4.2 Layout "B": an open core in the centre of the layout

The bracing core is moved to the centre in such a way that its shear centre and the centroid of the layout coincide (Figure 6.7).

The geometrical and stiffness characteristics of this case are collected in the second row in Table 6.3 in the previous section.

6.4.2.1 Maximum rotation and deflection

As the resultant of the wind load passes through the shear centre, there is no rotation around the shear centre:

$$\varphi = 0 \qquad \{2.89\}$$

It also follows that the deflection of the building is entirely made up from the uniform part of the deflection. This was calculated in the previous case, so the top deflection of the building is readily available as

$$v_{max} = v_o + \varphi x_{max} = 0.0035 + 0 = 0.0035 \text{ m} \qquad \{3.66\}$$

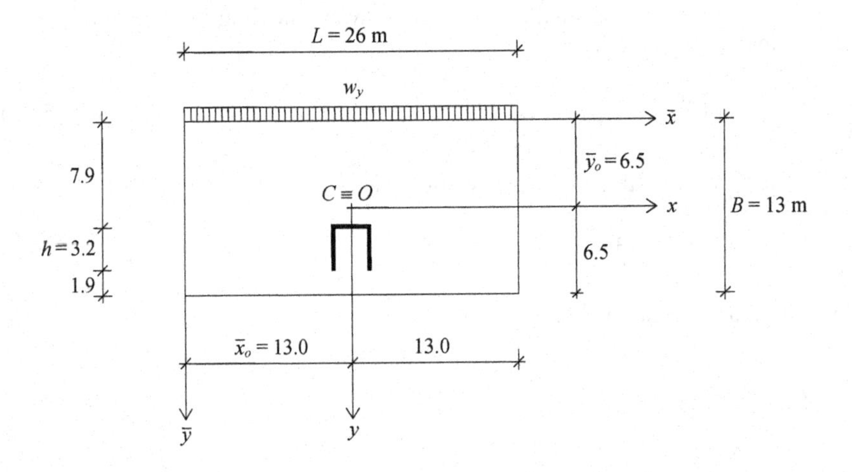

Figure 6.7 Kollár's building. Layout "B": bracing core in the centre.

6.4.2.2 Fundamental frequency

By moving the core to the centre, the values of the lateral vibration in the principal directions do not change and the results obtained in the previous section hold:

$$f_x = \frac{0.56 r_f}{H^2}\sqrt{\frac{EI_y}{m}} = 1.55 \text{ Hz}, \qquad f_y = \frac{0.56 r_f}{H^2}\sqrt{\frac{EI_x}{m}} = 1.41 \text{ Hz} \qquad \{2.92\}$$

The situation is different when the torsional behaviour is considered. The distance between the shear centre and the centroid of the mass is now reduced to zero and this fact alters the value of the radius of gyration:

$$i_p = \sqrt{\frac{L^2 + B^2}{12} + t^2} = \sqrt{\frac{26^2 + 13^2}{12} + 0} = 8.39 \text{ m} \qquad \{4.24\}$$

The frequency parameter is unchanged at $\eta_\varphi = 0.645$ and the pure torsional frequency is

$$f_\varphi = \frac{\eta_\varphi r_f}{i_p H^2}\sqrt{\frac{EI_\omega}{m}} = \frac{0.645 \cdot 0.842}{8.39 \cdot 15^2}\sqrt{\frac{23 \cdot 10^6 \cdot 2.51}{103.4}} = 0.215 \text{ Hz} \qquad \{2.94\}$$

As the centroid and the shear centre coincide, there is no coupling among the two lateral and pure torsional vibrations and the fundamental frequency is the smallest one of the three:

$$f = \min(f_x, f_y, f_\varphi) = f_\varphi = 0.215 \text{ Hz} \qquad \{4.39\}$$

6.4.2.3 Global critical load and critical load ratio

The situation concerning stability is very similar to that of vibration. The sway critical loads are unchanged from the previous case at

$$N_{cr,x} = \frac{7.837 EI_y r_s}{H^2} = 1499 \text{ MN} \qquad \{2.96\}$$

$$N_{cr,y} = \frac{7.837 EI_x r_s}{H^2} = 1244 \text{ MN} \qquad \{2.96\}$$

However, due to the change in the value of the radius of gyration, the value of pure torsional buckling changes. With $\alpha = 11.29$, determined in the previous case, the pure torsional critical load is

$$N_{cr,\varphi} = \frac{\alpha r_s EI_\omega}{i_p^2 H^2} = \frac{11.29 \cdot 0.759 \cdot 23 \cdot 10^3 \cdot 2.51}{8.39^2 \cdot 15^2} = 31.2 \text{ MN} \qquad \{2.97\}$$

As there is no coupling, this is also the global critical load of the building:

$$N_{cr} = \min(N_{cr,x}, N_{cr,y}, N_{cr,\varphi}) = N_{cr,\varphi} = 31.2 \text{ MN} \qquad \{5.47\}$$

To sum it up, everything has improved compared to the previous case: the maximum deflection decreased enormously, the fundamental frequency increased and the critical load also increased nearly three-fold. The global critical load ratio reflects these favourable changes:

$$\lambda = \frac{N_{cr}}{N} = \frac{31.2}{16.9} = 1.85 \qquad \{6.3\}$$

However, although the building is now theoretically stable, the value of the global critical load ratio is far from the recommended value of $\lambda = 10$ [Equation (6.5)].

As mentioned above in connection with Layout "A", the poor torsional performance of the building was caused by the small torsional stiffness of the core and the relatively great distance between the centroid of the layout and the shear centre.

The situation has improved by moving the core to the centre but the improvement is not big enough. In the following section another attempt is made to remedy the original situation by increasing the torsional stiffness of the core (while leaving the core at its original position).

6.4.3 Layout "C": a partially closed core on the right side of the layout

In practical situations it is normally possible and feasible to close the U-core partially. It is done in this case by adding small lip-sections and connecting the wall sections at the opening by beams at floor levels (Figure 6.8/b). The core remains in its original position (Figure 6.8/a). The thickness and the depth of the connecting beams are $t_b = 0.20$ m and $d = 0.65$ m. The distance of the connecting beams is equal to the storey height at $s = 3$ m.

Due to this alteration, the value of the second moments of area of the core slightly changes—see the third row in Table 6.3 and the downloadable files 643Kol_C.xmcd and 643Kol_C.xls for details. The value of the warping constant dramatically decreases and, as a conservative estimate, it is ignored in the following calculation. The lips add a little to the value of the original Saint-Venant torsional constant:

$$J = \frac{1}{3}\sum_{i=1}^{m} h_i t_i^3 = \frac{0.2^3}{3}(2\cdot 3.1 + 2.6 + 2\cdot 0.4) = 0.026\,\text{m}^4 \qquad \{2.77\}$$

Because of the partial closure, however, the value of the Saint-Venant constant drastically increases. According to Vlasov (1961), this increase is calculated as

$$\bar{J} = \frac{4A_o^2}{\dfrac{l^3 sG}{12EI_b} + \dfrac{1.2ls}{A_b}} = \frac{4\cdot 3.1^2\cdot 2.6^2}{\dfrac{1.8^3\cdot 3\cdot 9580}{12\cdot 23000\cdot 0.004577} + \dfrac{1.2\cdot 1.8\cdot 3}{0.13}} = 1.424\,\text{m}^4 \qquad \{2.81\}$$

where

$$A_b = t_b d = 0.2 \cdot 0.65 = 0.13\ \text{m}^2 \quad \text{and} \quad I_b = \frac{t_b d^3}{12} = \frac{0.2 \cdot 0.65^3}{12} = 0.004577\ \text{m}^4$$

are the cross-sectional area and the second moment of area of the connecting beams.

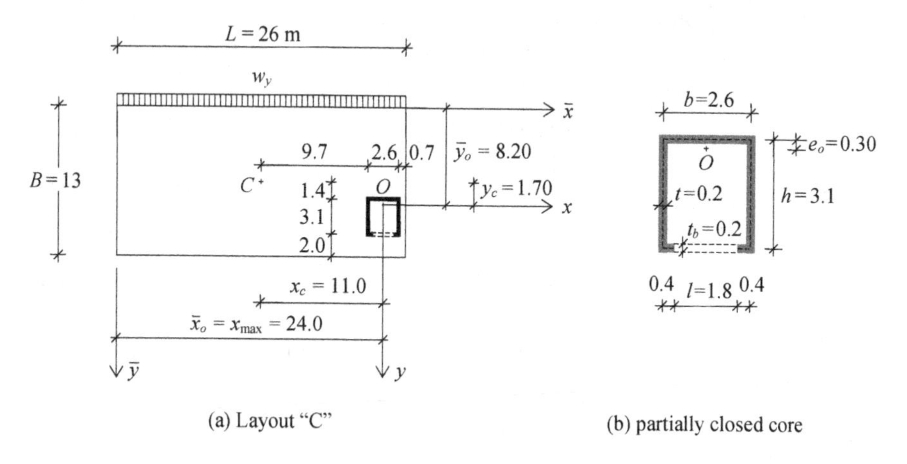

(a) Layout "C" (b) partially closed core

Figure 6.8 Kollár's building. Layout "C".

The value of the total Saint-Venant constant is

$$J + \bar{J} = 0.026 + 1.424 = 1.45\ \text{m}^4$$

Because of the partial closure, the location of the shear centre of the core also changes. It is a move "backwards", definitely towards the centre of the core, but its exact value is very difficult to establish. Using an equivalent thickness of

$$t_w^* = \frac{d}{s} t_b = \frac{0.65}{3.0} 0.2 \cong 0.04\ \text{m} \qquad \{2.83\}$$

the computer program PROSEC (1994) gives

$$e_o = 0.30\ \text{m}$$

for the location of the shear centre (Figure 6.8/b).

The geometrical and stiffness characteristics related to Layout "C" are collected in the third row of Table 6.3 in Section 6.4.1.

6.4.3.1 Maximum rotation and deflection

The situation is similar to the one with Layout "A". Maximum deflection develops at the left-hand side of the building at $x_{max} = 24.0$. It consists of two parts. The building undergoes a uniform deflection of

$$v_o = v(H) = \frac{w_y H^4}{8EI_x} = \frac{26 \cdot 15^4}{8 \cdot 23 \cdot 10^6 \cdot 2.479} = 0.0029 \text{ m} \qquad \{2.85\}$$

The rotation around the shear centre is

$$\varphi_{\max} = \varphi(H) = \frac{m_z H^2}{2GJ} = \frac{286 \cdot 15^2}{2 \cdot 9.58 \cdot 10^6 \cdot 1.45} = 0.00232 \qquad \{2.91\}$$

and this rotation—with a "torsion arm" of $x_{\max} = 24.0$ m—leads to additional deflection. The total deflection is

$$v_{\max} = v_o + \varphi x_{\max} = 0.0029 + 0.00232 \cdot 24 = 0.0586 \text{ m} \qquad \{3.66\}$$

This value is greater than the recommended maximum deflection of 0.030 m.

6.4.3.2 Fundamental frequency

The three basic frequencies are needed first.

The lateral frequencies in directions x and y are calculated using the formulae given for cores but with the mass density which relates to the whole building:

$$f_x = \frac{0.56 r_f}{H^2}\sqrt{\frac{EI_y}{m}} = \frac{0.56 \cdot 0.842}{15^2}\sqrt{\frac{23 \cdot 10^6 \cdot 2.599}{103.4}} = 1.593 \text{ Hz} \qquad \{2.92\}$$

$$f_y = \frac{0.56 r_f}{H^2}\sqrt{\frac{EI_x}{m}} = \frac{0.56 \cdot 0.842}{15^2}\sqrt{\frac{23 \cdot 10^6 \cdot 2.479}{103.4}} = 1.556 \text{ Hz} \qquad \{2.92\}$$

The radius of gyration is

$$i_p = \sqrt{\frac{L^2 + B^2}{12} + t^2} = \sqrt{\frac{26^2 + 13^2}{12} + 11^2 + 1.70^2} = 13.94 \text{ m} \qquad \{4.24\}$$

The frequency of pure torsional vibration of the building is obtained using the formula given for a single core (with GJ only) but with the radius gyration that relates to the whole layout area as the mass is distributed over the whole floor area of the building:

$$f_\varphi = \frac{r_f}{4Hi_p}\sqrt{\frac{GJ}{m}} = \frac{0.842}{4 \cdot 15 \cdot 13.94}\sqrt{\frac{9.58 \cdot 10^6 \cdot 1.45}{103.4}} = 0.369 \text{ Hz} \qquad \{2.95\}$$

The arrangement is asymmetric and triple coupling occurs. As one of the basic frequencies is much smaller than the other two, the effect of coupling can be approximated with good accuracy using the Föppl-Papkovich formula:

$$f=\left(\frac{1}{f_x^2}+\frac{1}{f_y^2}+\frac{1}{f_\varphi^2}\right)^{-\frac{1}{2}}=\left(\frac{1}{1.593^2}+\frac{1}{1.556^2}+\frac{1}{0.369^2}\right)^{-\frac{1}{2}}=0.350\text{ Hz} \quad \{4.36\}$$

6.4.3.3 Global critical load and critical load ratio

The critical loads for sway buckling are calculated using the relevant second moment of area of the core:

$$N_{cr,x}=\frac{7.837EI_y r_s}{H^2}=\frac{7.837\cdot 23\cdot 10^3\cdot 2.599\cdot 0.759}{15^2}=1580\text{ MN} \quad \{2.96\}$$

$$N_{cr,y}=\frac{7.837EI_x r_s}{H^2}=\frac{7.837\cdot 23\cdot 10^3\cdot 2.479\cdot 0.759}{15^2}=1507\text{ MN} \quad \{2.96\}$$

Torsion is resisted by the Saint-Venant stiffness and the critical load of pure torsional buckling is

$$N_{cr,\varphi}=\frac{GJ}{i_p^2}=\frac{9580\cdot 1.45}{13.94^2}=71.5\text{ MN} \quad \{2.99\}$$

Because of the triple coupling, this critical load is reduced and the global critical load of the building is

$$N_{cr}=\left(\frac{1}{N_{cr,x}}+\frac{1}{N_{cr,y}}+\frac{1}{N_{cr,\varphi}}\right)^{-1}=\left(\frac{1}{1580}+\frac{1}{1507}+\frac{1}{71.5}\right)^{-1}=65.4\text{ MN} \quad \{5.44\}$$

The global critical load ratio

$$\lambda=\frac{N_{cr}}{N}=\frac{65.4}{16.9}=3.87 \quad \{6.3\}$$

shows a stable structure but the recommended margin is not yet achieved. (The maximum deflection also exceeds the recommended value.) However, the situation can further be improved.

6.4.4 Layout "D": a partially closed core in the centre of the layout

In combining the previous two actions, the partially closed core is now moved to the centre in such a way that its shear centre and the centroid of the layout coincide (Figure 6.9). The geometrical and stiffness characteristics of this case are collected in the fourth row in Table 6.3 in Section 6.4.1.

6.4.4.1 Maximum rotation and deflection

As the resultant of the wind load passes through the shear centre, there is no rotation around the shear centre:

$$\varphi = 0$$

It also follows that the deflection of the building is entirely made up from the uniform part of the deflection. This was calculated in the previous case so the top deflection of the building is readily available as

$$v_{max} = v_o + \varphi x_{max} = 0.0029 + 0.0 = 0.0029\,\text{m} \qquad \{3.66\}$$

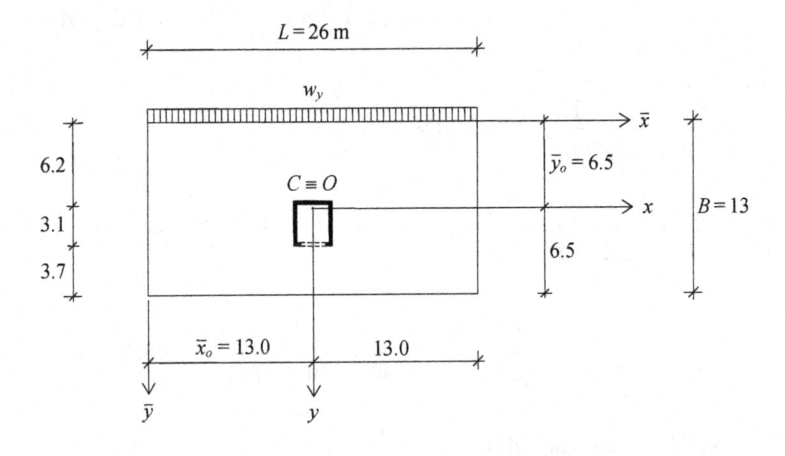

Figure 6.9 Kollár's building. Layout "D": partially closed core in the centre.

6.4.4.2 Fundamental frequency

By moving the core to the centre, the values of the lateral vibration do not change and the results obtained in the previous section hold:

$$f_x = \frac{0.56 r_f}{H^2}\sqrt{\frac{EI_y}{m}} = 1.593\,\text{Hz}, \qquad f_y = \frac{0.56 r_f}{H^2}\sqrt{\frac{EI_x}{m}} = 1.556\,\text{Hz} \qquad \{2.92\}$$

The situation is different when the torsional behaviour is considered. The distance between the shear centre and the centroid of the mass is now reduced to zero and this (favourable) fact alters the value of the radius of gyration:

$$i_p = \sqrt{\frac{L^2 + B^2}{12} + t^2} = \sqrt{\frac{26^2 + 13^2}{12} + 0} = 8.39 \text{ m} \qquad \{4.24\}$$

The pure torsional frequency is

$$f_\varphi = \frac{r_f}{4Hi_p}\sqrt{\frac{GJ}{m}} = \frac{0.842}{4 \cdot 15 \cdot 8.39}\sqrt{\frac{9.58 \cdot 10^6 \cdot 1.45}{103.4}} = 0.613 \text{ Hz} \qquad \{2.95\}$$

As the centroid and the shear centre coincide, there is no coupling among the two lateral and pure torsional modes and the fundamental frequency is the smallest one of the three basic frequencies:

$$f = \min(f_x, f_y, f_\varphi) = f_\varphi = 0.613 \text{ Hz} \qquad \{4.39\}$$

6.4.4.3 Global critical load and critical load ratio

The situation concerning stability is very similar to that of vibration. The sway critical loads are unchanged from the previous case:

$$N_{cr,x} = \frac{7.837 EI_y r_s}{H^2} = 1580 \text{ MN} \qquad \{2.96\}$$

$$N_{cr,y} = \frac{7.837 EI_x r_s}{H^2} = 1507 \text{ MN} \qquad \{2.96\}$$

However, due to the change in the value of the radius of gyration, the value of pure torsional buckling changes:

$$N_{cr,\varphi} = \frac{GJ}{i_p^2} = \frac{9580 \cdot 1.45}{8.39^2} = 197 \text{ MN} \qquad \{2.99\}$$

As there is no coupling, this, being the smallest one of the three basic critical loads, is also the global critical load of the building:

$$N_{cr} = \min(N_{cr,x}, N_{cr,y}, N_{cr,\varphi}) = N_{cr,\varphi} = 197 \text{ MN} \qquad \{5.47\}$$

To sum it up, everything has improved compared to the previous case: the maximum deflection decreased enormously, the fundamental frequency increased and the critical load increased three-fold. The global critical load ratio reflects these favourable changes:

$$\lambda = \frac{N_{cr}}{N} = \frac{197}{16.9} = 11.7 > 10 \qquad \{6.3\}$$

The results of the four arrangements are summarized in Table 6.4.

Table 6.4 Kollár's building: a summary.

Layout	Maximum rotation [°]	Maximum deflection [mm]	Fundamental frequency [Hz]	Global critical load [MN]	Global critical load ratio [-]
"A"	1.3	564	0.129	11.4	0.67
"B"	0	3.5	0.215	31.2	1.85
"C"	0.13	58.6	0.350	65.4	3.87
"D"	0	2.9	0.613	197	11.7

6.5 PRACTICAL APPLICATION NO. 3: TEN-STOREY ASYMMETRIC BUILDING

The structural engineer is normally responsible for the building as a whole, not only for certain units or individual aspects of its behaviour. Before the building is constructed, all areas relating to structural behaviour have to be looked at. This practical example shows how such global analysis is carried out using a real building. A comprehensive global structural analysis may start with the stability analysis, as through the determination of the critical load ratio the structural engineer gets a clear picture of the three-dimensional behaviour of the building and of the adequacy of the bracing system. The investigation then moves on to determine the maximum deflection of the building under horizontal load. When the building location is in a seismic zone, the fundamental frequency of the building may also be needed. Many stiffness characteristics needed for the individual investigations are identical, and they only have to be established once and then can be reused. Hence the resulting global analysis covering the three different areas requires much less work than three individual analyses separately.

The subject of this section's worked example is a ten-storey building braced by five bracing units: one U-core, two shear walls (W), frame F1 and frame F7 (Figure 6.10 and Figure 3.1). The U-core has a wall-thickness of 0.20 m with size $b = 6$ m and $h = 3$ m. The wall-thickness of the shear walls is 0.20 m with size $h = 3$ m. It is assumed for the analysis that the U-core and the shear walls only develop bending deformation. The cross-sections of both the beams and columns of both frames F1 and F7 are 0.4 m/0.4 m. Both frames are part of the collection used for a comprehensive accuracy analysis in Chapter 7 (see Figure 7.1).

The cross-sectional characteristics of the U-core and shear wall W are collected in Table 6.5. Three individual concrete columns are also part of the vertical load carrying system but their contribution to the lateral and torsional stiffness is small compared to that of the bracing units and is therefore ignored in the calculations.

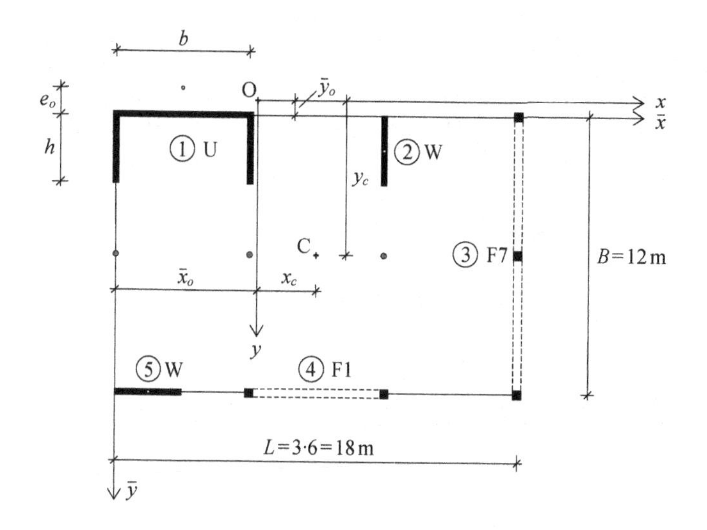

Figure 6.10 Layout of ten-storey asymmetric building.

The basic data of the building is given below.

Size of ground plan: $L = 18.0$ m and $B = 12.0$ m.
Storey height: $h = 3$ m. Number of storeys: 10. Height of structure: $H = 30$ m.
Modulus of elasticity: $E = 25000$ MN/m^2. Shear modulus: $G = 10400$ MN/m^2.

Table 6.5 Cross-sectional characteristics of bracing units U and W.

Bracing unit	I_{min} [m^4]	I_{max} [m^4]	J [m^4]	I_ω [m^6]	e_o [m]
U	2.257	14.42	0.032	14.18	1.125
W	0.002	0.45	0.008	0	–

The weight per unit volume of the building (for the dynamic analysis) is assumed to be $\gamma = 4.0$ kN/m^3. A vertical load of $Q = 12$ kN/m^2 is used for the stability analysis and for the determination of the global critical load ratio. When the structure is subjected to lateral load and the top rotation and deflection are calculated, a uniformly distributed horizontal load of intensity 1.3 kN/m^2 is taken into account in direction y, which results in a wind load of $w_y = 23.4$ kN/m.

The numbers of the equations used for the calculations will be given on the right-hand side in curly brackets. The investigation starts with the stability analysis.

6.5.1 Stability analysis

The load distribution factor is obtained from Table 5.1 as $r_s = 0.863$.

6.5.1.1 Individual bracing units

Before the actual stability analysis is carried out, the basic characteristics of the individual bracing units are given bellow. Some of these data will also be used with the frequency and deflection analyses.

Bracing Unit 1 (U-core)
The critical loads in the principal directions are

$$N_{cr,1x} = \frac{7.837EIr_s}{H^2} = \frac{7.837 \cdot 25000 \cdot 14.42 \cdot 0.863}{30^2} = 2709 \text{ MN} \qquad \{2.76\}$$

$$N_{cr,1y} = \frac{7.837EIr_s}{H^2} = \frac{7.837 \cdot 25000 \cdot 2.257 \cdot 0.863}{30^2} = 424 \text{ MN} \qquad \{2.76\}$$

Bracing Units 2 and 5 (shear wall W)
Only the in-plane critical load will be taken into account:

$$N_{cr,2} = \frac{7.837EIr_s}{H^2} = \frac{7.837 \cdot 25000 \cdot 0.45 \cdot 0.863}{30^2} = 84.54 \text{ MN} \qquad \{2.76\}$$

Bracing Unit 3 (frame F7)
The part of the shear stiffness which is associated with the beams is

$$K_{b,3} = \sum_{j=1}^{n-1} \frac{12EI_{b,j}}{l_j h} = 2\frac{12 \cdot 25000 \cdot 0.4^4}{12 \cdot 6 \cdot 3} = 71.11 \text{ MN} \qquad \{2.53\}$$

The part of the shear stiffness which is associated with the columns is

$$K_{c,3} = \sum_{j=1}^{n} \frac{\pi^2 EI_{c,j}}{h^2} = 3\frac{\pi^2 \cdot 25000 \cdot 0.4^4}{12 \cdot 3^2} = 175.46 \text{ MN} \qquad \{2.54\}$$

The above two part stiffnesses define reduction factor r as

$$r_3 = \frac{K_{c,3}}{K_{b,3} + K_{c,3}} = \frac{175.46}{71.11 + 175.46} = 0.7116 \qquad \{2.56\}$$

The shear stiffness of the frame can now be determined:

$$K_3 = K_{b,3} r_3 = K_{b,3} \frac{K_{c,3}}{K_{b,3} + K_{c,3}} = 71.11 \cdot 0.7116 = 50.60 \text{ MN} \qquad \{2.55\}$$

For the local bending stiffness ($EI = EI_c r$), the sum of the second moments of area of the columns should be produced:

$$I_3 = r_3 \sum_{j=1}^{n} I_{c,j} = 0.7116 \cdot 3 \cdot \frac{0.4^4}{12} = 0.004554 \text{ m}^4 \qquad \{2.58\}$$

The global second moment of area is

$$I_{g,3} = \sum_{j=1}^{n} A_{c,j} t_j^2 = 0.4 \cdot 0.4 (6^2 + 6^2) = 11.52 \text{ m}^4 \qquad \{2.57\}$$

The global bending critical load is

$$N_{g,3} = \frac{7.837 EI_g r_s}{H^2} = \frac{7.837 \cdot 25000 \cdot 11.52 \cdot 0.863}{30^2} = 2164 \text{ MN} \qquad \{2.65\}$$

With stiffness ratios

$$\frac{I}{I_f} = \frac{0.004554}{11.5245} = 0.000395 \quad \text{and} \quad \frac{KH^2}{EI_f r_s} = \frac{50.6 \cdot 30^2}{25000 \cdot 11.5245 \cdot 0.863} = 0.183 \qquad \{2.62\}$$

the stability parameter is obtained from Table 2.1 as $c_1 = 0.222$.

Using this stability parameter, the critical load of Bracing Unit 3 (frame F7) is

$$N_{cr,3} = c_1 \frac{EI_f r_s}{H^2} = 0.222 \frac{25000 \cdot 11.5245 \cdot 0.863}{30^2} = 61.33 \text{ MN} \qquad \{2.60\}$$

The effectiveness factor and the effective shear stiffness will also be needed later on

$$s_3 = \frac{N_{g,3}}{K_3 + N_{g,3}} = \frac{2164}{50.6 + 2164} = 0.9772 \qquad \{5.9\}$$

$$K_{e3} = s_3 K_3 = 0.9772 \cdot 50.6 = 49.45 \text{ MN} \qquad \{5.10\}$$

Bracing Unit 4 (frame F1)

The part of the shear stiffness which is associated with the beams is

$$K_{b,4} = \sum_{j=1}^{n-1} \frac{12EI_{b,j}}{l_j h} = \frac{12 \cdot 25000 \cdot 0.4^4}{12 \cdot 6 \cdot 3} = 35.56 \text{ MN} \qquad \{2.53\}$$

The part of the shear stiffness which is associated with the columns is

$$K_{c,4} = \sum_{j=1}^{n} \frac{\pi^2 EI_{c,j}}{h^2} = 2\frac{\pi^2 \cdot 25000 \cdot 0.4^4}{12 \cdot 3^2} = 117.0 \text{ MN} \qquad \{2.54\}$$

The above two part stiffnesses define reduction factor r as

$$r_4 = \frac{K_{c,4}}{K_{b,4} + K_{c,4}} = \frac{117}{35.56 + 117} = 0.7669 \qquad \{2.56\}$$

The shear stiffness of the frame can now be determined:

$$K_4 = K_{b,4} r_4 = K_{b,4} \frac{K_{c,4}}{K_{b,4} + K_{c,4}} = 35.56 \cdot 0.7669 = 27.27 \text{ MN} \qquad \{2.55\}$$

For the local bending stiffness ($EI = EI_c r$), the sum of the second moments of area of the columns should be produced:

$$I_4 = r_4 \sum_{j=1}^{n} I_{c,j} = 0.7669 \cdot 2 \cdot \frac{0.4^4}{12} = 0.003272 \text{ m}^4 \qquad \{2.58\}$$

The global second moment of area is

$$I_{g,4} = \sum_{j=1}^{n} A_{c,j} t_j^2 = 0.4 \cdot 0.4 \cdot 3^2 \cdot 2 = 2.88 \text{ m}^4 \qquad \{2.57\}$$

The global bending critical load is

$$N_{g,4} = \frac{7.837 EI_g r_s}{H^2} = \frac{7.837 \cdot 25000 \cdot 2.88 \cdot 0.863}{30^2} = 541 \text{ MN} \qquad \{2.65\}$$

As a function of stiffness ratios

$$\frac{I}{I_f} = \frac{0.003272}{2.883272} = 0.001135 \quad \text{and} \quad \frac{KH^2}{EI_f r_s} = \frac{27.27 \cdot 30^2}{25000 \cdot 2.883272 \cdot 0.863} = 0.394 \qquad \{2.62\}$$

the stability parameter is obtained from Table 2.1 as $c_1 = 0.543$.

Using this stability parameter, the critical load of Bracing Unit 4 (frame F1) is

$$N_{cr,4} = c_1 \frac{EI_f r_s}{H^2} = 0.543 \frac{25000 \cdot 2.883272 \cdot 0.863}{30^2} = 37.53 \text{ MN} \qquad \{2.60\}$$

The effectiveness factor and the effective shear stiffness will also be needed:

$$s_4 = \frac{N_{g,4}}{K_4 + N_{g,4}} = \frac{541}{27.27 + 541} = 0.9520 \qquad \{5.9\}$$

$$K_{e4} = s_4 K_4 = 0.952 \cdot 27.27 = 25.96 \text{ MN} \qquad \{5.10\}$$

6.5.1.2 Sway buckling in direction x

The participating bracing units in direction x are Units 1, 4 and 5.

The bending stiffness originates from the U-core and shear wall W:

$$EI = E_c I_c + E_w I_w = 25000(14.42 + 0.45) = 371.8 \cdot 10^3 \text{ MNm}^2 \qquad \{5.12\}$$

where the contribution of the columns of frame F1—being negligible—is ignored.

The local bending critical load can now be calculated:

$$N_l = \frac{7.837 r_s EI}{H^2} = \frac{7.837 \cdot 0.863 \cdot 371.8 \cdot 10^3}{30^2} = 2794 \text{ MN} \qquad \{5.13\}$$

With the effective shear stiffness (from the previous section), as a function of the part critical load ratio

$$\beta = \frac{K_e}{N_l} = \frac{25.96}{2794} = 0.00929 \qquad \{5.17\}$$

the critical load parameter is obtained from Table 5.2 as

$$\alpha = 1.027 + \frac{1.03 - 1.027}{0.010 - 0.009}(0.00929 - 0.009) = 1.028 \qquad \{\text{Table 5.2}\}$$

The sway critical load in direction x can now be determined:

$$N_{cr,x} = N_l + K_e + (\alpha - \beta - 1)sN_l \qquad \{5.18\}$$

$$= 2794 + 25.96 + (1.028 - 0.00929 - 1)0.952 \cdot 2794 = 2870 \text{ MN}$$

6.5.1.3 *Sway buckling in direction y*

The participating bracing units in direction y are Units 1, 2 and 3.

The bending stiffness originates from the U-core and the shear wall:

$$EI = E_c I_c + E_w I_w = 25000(2.257 + 0.45) = 67.68 \cdot 10^3 \text{ MNm}^2 \qquad \{5.12\}$$

where the contribution of the columns of frame F7—being negligible—is ignored.

The local bending critical load can now be calculated:

$$N_l = \frac{7.837 r_s EI}{H^2} = \frac{7.837 \cdot 0.863 \cdot 67.68 \cdot 10^3}{30^2} = 509 \text{ MN} \qquad \{5.13\}$$

With the effective shear stiffness (from the previous section), as a function of the part critical load ratio

$$\beta = \frac{K_e}{N_l} = \frac{49.45}{509} = 0.0972 \qquad \{5.17\}$$

the critical load parameter is obtained from Table 5.2 as

$$\alpha = 1.2659 + \frac{1.2949 - 1.2659}{0.10 - 0.09}(0.0972 - 0.09) = 1.287 \qquad \{\text{Table 5.2}\}$$

and the sway critical load in direction y is:

$$N_{cr,y} = N_l + K_e + (\alpha - \beta - 1)sN_l \qquad \{5.18\}$$

$$= 509 + 49.45 + (1.287 - 0.0972 - 1)0.9772 \cdot 509 = 653 \text{ MN}$$

6.5.1.4 *Pure torsional buckling*

The shear centre is located using the critical loads of the bracing units:

$$\bar{x}_o = \frac{\sum_1^{f+m} N_{y,i}\bar{x}_i}{\sum_1^{f+m} N_{y,i}} = \frac{3 \cdot 424 + 12 \cdot 84.54 + 18 \cdot 61.33}{424 + 84.54 + 61.33} = 5.95 \text{ m} \qquad \{5.27\}$$

$$\bar{y}_o = \frac{\sum_1^{f+m} N_{x,i}\bar{y}_i}{\sum_1^{f+m} N_{x,i}} = \frac{-1.125 \cdot 2709 + 12(84.54 + 37.53)}{2709 + 84.54 + 37.53} = -0.559 \text{ m} \qquad \{5.27\}$$

A new *x-y* coordinate system is now established whose *O* origin is in the shear centre (Figure 6.10). The locations of the bracing units in this coordinate system are:

$$t_{1,x} = \bar{x}_o - 3 = 2.95 \text{ m}, \qquad t_{1,y} = e_o - |\bar{y}_o| = 0.566 \text{ m}, \qquad t_2 = 12 - \bar{x}_o = 6.05 \text{ m}$$

$$t_3 = 18 - \bar{x}_o = 12.05 \text{ m}, \qquad t_4 = t_5 = B + |\bar{y}_o| = 12.559 \text{ m}$$

With the coordinates of the centroid in the coordinate system *x-y*

$$x_c = \frac{L}{2} - \bar{x}_o = \frac{18}{2} - 5.95 = 3.05 \text{ m}, \quad y_c = \frac{B}{2} - \bar{y}_o = \frac{12}{2} + 0.559 = 6.559 \text{ m} \quad \{5.33\}$$

the distance between the shear centre and the centroid is

$$t = \sqrt{x_c^2 + y_c^2} = \sqrt{3.05^2 + 6.559^2} = 7.233 \text{ m}$$

With the above data, the radius of gyration is

$$i_p = \sqrt{\frac{L^2 + B^2}{12} + t^2} = \sqrt{\frac{18^2 + 12^2}{12} + 7.233^2} = \sqrt{91.32} = 9.556 \text{ m} \qquad \{5.32\}$$

The "original" Saint-Venant stiffness consists of two parts. In addition to the Saint-Venant stiffness of the individual bracing cores, those units that have shear stiffness (i.e. the frames) also contribute:

$$(GJ) = \sum_1^m GJ_k + \sum_1^f \left((K_i)_x y_i^2 + (K_i)_y x_i^2\right) \qquad \{5.38\}$$

$$= 10.4 \cdot 10^3 (0.032 + 2 \cdot 0.008) + 50.6 \cdot 12.05^2 + 27.27 \cdot 12.559^2 = 12148 \text{ MNm}^2$$

The real (effective) Saint-Venant stiffness is always smaller than the "original" one as the effect of the frames is limited by their effectiveness factor:

$$(GJ)_e = \sum_1^m GJ_k + \sum_1^f \left((K_{e,i})_x y_i^2 + (K_{e,i})_y x_i^2\right) \qquad \{5.29\}$$

$$= 10.4 \cdot 10^3 (0.032 + 2 \cdot 0.008) + 49.45 \cdot 12.05^2 + 25.96 \cdot 12.559^2 = 11774 \text{ MNm}^2$$

The effectiveness of the Saint-Venant stiffness for the whole system can now be determined:

$$s_\varphi = \frac{(GJ)_e}{(GJ)} = \frac{11774}{12148} = 0.9692 \qquad \{5.37\}$$

The warping stiffness of the system comes from three sources: the own warping stiffness of the core, the bending stiffness of the walls and the bending stiffness of the columns of the frames [Equation (5.30)]. When the first two items exist, the contribution of the third is normally negligible. This is the case now and the third source is neglected below. Hence, the warping stiffness is calculated as:

$$EI_\omega = E_w \sum_1^m \left(I_{\omega,k} + (I_{w,k})_x y_k^2 + (I_{w,k})_y x_k^2 \right) \qquad \{5.30\}$$

$$=25000[14.18+2.257 \cdot 2.95^2+14.42 \cdot 0.566^2+0.45(6.05^2+12.559^2)]=3147000 \text{ MNm}^4$$

With the above stiffnesses, the part torsional critical loads can now be determined. The warping and Saint-Venant torsional critical loads are

$$N_\omega = \frac{7.837 r_s EI_\omega}{i_p^2 H^2} = \frac{7.837 \cdot 0.863 \cdot 3147 \cdot 10^3}{91.32 \cdot 30^2} = 259 \text{ MN} \qquad \{5.35\}$$

$$N_t = \frac{(GJ)_e}{i_p^2} = \frac{11774}{91.32} = 129 \text{ MN} \qquad \{5.36\}$$

As a function of the ratio of the above part critical loads

$$\beta_\varphi = \frac{N_t}{N_\omega} = \frac{129}{259} = 0.5 \qquad \{5.39\}$$

the critical load parameter is obtained from Table 5.2 as

$$\alpha_\varphi = 2.3817 \qquad \{\text{Table } 5.2\}$$

Finally, the critical load of pure torsional buckling is

$$N_{cr,\varphi} = N_\omega + N_t + (\alpha_\varphi - \beta_\varphi - 1) s_\varphi N_\omega \qquad \{5.34\}$$

$$= 259 + 129 + (2.3817 - 0.5 - 1)0.9692 \cdot 259 = 609 \text{ MN}$$

The third term represents the effect of the interaction between the Saint-Venant and warping torsional modes. It amounts to a considerable 36% of the total torsional critical load.

6.5.1.5 Coupling of the basic critical loads: the global critical load of the building

The centroid of the layout does not coincide with the shear centre so there is a coupling of basic critical loads $N_{cr,x}$, $N_{cr,y}$ and $N_{cr,\varphi}$. As the centroid does not even lie on one of the principal axes, this coupling is a triple one. Any coupling reduces the value of the critical load, so the effect of coupling must be taken into account. Eccentricity parameters τ_x and τ_y are needed for the exact calculation of the coupling of the critical loads:

$$\tau_x = \frac{x_c}{i_p} = \frac{3.05}{9.556} = 0.319 \quad \text{and} \quad \tau_y = \frac{y_c}{i_p} = \frac{6.559}{9.556} = 0.686 \qquad \{5.43\}$$

With the eccentricity parameters, the smallest root of the cubic equation

$$N^3 + b_2 N^2 + b_1 N - b_0 = 0 \qquad \{5.41\}$$

is the critical load, where

$$b_0 = \frac{N_{cr,x} N_{cr,y} N_{cr,\varphi}}{1 - \tau_x^2 - \tau_y^2} = \frac{2870 \cdot 653 \cdot 609}{1 - 0.319^2 - 0.686^2} = 2.669 \cdot 10^9 \qquad \{5.42\}$$

$$b_1 = \frac{N_{cr,x} N_{cr,y} + N_{cr,\varphi} N_{cr,x} + N_{cr,\varphi} N_{cr,y}}{1 - \tau_x^2 - \tau_y^2} = 9.40 \cdot 10^6 \qquad \{5.42\}$$

$$b_2 = \frac{N_{cr,x}\tau_x^2 + N_{cr,y}\tau_y^2 - N_{cr,x} - N_{cr,y} - N_{cr,\varphi}}{1 - \tau_x^2 - \tau_y^2} = -8261 \qquad \{5.42\}$$

The critical load (the smallest root of the above cubic equation) is

$$N_{cr} = 461 \text{ MN}$$

6.5.1.6 The global critical load ratio

With the uniformly distributed vertical floor load of intensity $Q = 12$ kN/m², the total vertical load on the building is

$$N = LBQn = 18 \cdot 12 \cdot 0.012 \cdot 10 = 25.92 \text{ MN} \qquad \{6.2\}$$

The global critical load ratio is therefore

$$\lambda = \frac{N_{cr}}{N} = \frac{461}{25.92} = 17.8 \qquad \{6.3\}$$

indicating a stable, satisfactory bracing system.

The condition

$$\lambda \geq 10 \qquad \{6.5\}$$

is satisfied, so any vertical load bearing element can be considered as braced (by the bracing system) and the second-order effects (due to sway and torsion) can be neglected.

6.5.2 Maximum deflection

The building is subjected to a uniformly distributed horizontal load of intensity 1.3 kN/m^2 in direction y, which results in a wind load of $w_y = 23.4$ kN/m. Due to this load, the top of the building undergoes a uniform translation (defined by the translation of the shear centre axis) and an "uneven" translation (due to the rotation of the building around the shear centre axis). Accordingly, having determined the basic characteristics of the individual bracing units, the maximum translation is determined in two steps.

6.5.2.1 Individual bracing units

Before the deflection of the shear centre axis and the rotation are determined, the stiffness characteristics and the maximum deflection of the individual bracing units are needed.

Bracing Unit 1 (U-core)

The maximum deflection and the stiffness of the U-core in the principal directions are

$$u_1 = \frac{wH^4}{8EI_y} = \frac{23.4 \cdot 30^4}{8 \cdot 25 \cdot 10^6 \cdot 14.42} = 0.00657\,\text{m}$$

$$v_1 = \frac{wH^4}{8EI_x} = \frac{23.4 \cdot 30^4}{8 \cdot 25 \cdot 10^6 \cdot 2.257} = 0.0420\,\text{m} \qquad \{2.74\}$$

$$S_{1,x} = \frac{1}{u_1} = \frac{1}{0.00657} = 152.2\frac{1}{\text{m}} \quad \text{and} \quad S_{1,y} = \frac{1}{v_1} = \frac{1}{0.042} = 23.8\frac{1}{\text{m}} \qquad \{3.13\}$$

Bracing Units 2 and 5 (shear wall W)

The maximum deflection and the stiffness of shear wall W in its plane are

$$u_2 = \frac{wH^4}{8EI_y} = \frac{23.4 \cdot 30^4}{8 \cdot 25 \cdot 10^6 \cdot 0.45} = 0.2106\,\text{m} \qquad \{2.74\}$$

$$S_2 = \frac{1}{u_2} = \frac{1}{0.2106} = 4.748\frac{1}{\text{m}} \qquad \{3.13\}$$

Bracing Unit 3 (frame F7)
The part of the shear stiffness which is associated with the columns has to be recalculated:

$$K_{c,3} = \sum_{j=1}^{n} \frac{12EI_{c,j}}{h^2} = 3\frac{12 \cdot 25 \cdot 10^6 \cdot 0.4^4}{12 \cdot 3^2} = 213333\,\text{kN} \qquad \{2.29\}$$

The part of the shear stiffness which is associated with the beams is unchanged (but using kilonewton instead of meganewton):

$$K_{b,3} = \sum_{j=1}^{n-1} \frac{12EI_{b,j}}{l_j h} = 2\frac{12 \cdot 25 \cdot 10^6 \cdot 0.4^4}{12 \cdot 6 \cdot 3} = 71111\,\text{kN} \qquad \{2.28\}$$

The above two part stiffnesses define reduction factor r as

$$r_3 = \frac{K_{c,3}}{K_{b,3} + K_{c,3}} = \frac{213333}{71111 + 213333} = 0.75 \qquad \{2.30\}$$

The shear stiffness of the frame can now be determined:

$$K_3 = K_{b,3} r_3 = K_{b,3} \frac{K_{c,3}}{K_{b,3} + K_{c,3}} = 71111 \cdot 0.75 = 53333\ \text{kN} \qquad \{2.27\}$$

For the local bending stiffness ($EI = EI_c r$), the sum of the second moments of area of the columns should be produced (and multiplied by reduction factor r):

$$I_3 = r_3 \sum_{j=1}^{n} I_{c,j} = 0.75 \cdot 3 \cdot \frac{0.4^4}{12} = 0.0048\,\text{m}^4 \qquad \{2.31\}$$

The sum of the local and global second moments of area is

$$I_{f,3} = I_3 + I_{g,3} = 0.0048 + 11.52 = 11.5248\,\text{m}^4 \qquad \{2.23\}$$

Auxiliary parameters s, κ and κH are needed:

$$s = 1 + \frac{I_c r}{I_g} = 1 + \frac{0.0048}{11.52} = 1.000417 \qquad \{2.14\}$$

$$\kappa = \sqrt{\frac{Ks}{EI}} = \sqrt{\frac{53333 \cdot 1.000417}{25 \cdot 10^6 \cdot 0.0048}} = 0.6668 \frac{1}{\text{m}} \qquad \text{and} \qquad \kappa H = 20$$

With these auxiliary parameters, the maximum top deflection of the frame is

$$y_3 = \frac{wH^4}{8EI_f} + \frac{wH^2}{2Ks^2} - \frac{wEI}{K^2 s^3}\left(\frac{1 + \kappa H \sinh \kappa H}{\cosh \kappa H} - 1\right) \qquad \{2.24\}$$

$$= \frac{23.4 \cdot 30^4}{8 \cdot 25 \cdot 10^6 \cdot 11.5248} + \frac{23.4 \cdot 30^2}{2 \cdot 53333 \cdot 1.000417^2} -$$

$$- \frac{23.4 \cdot 25 \cdot 10^6 \cdot 0.0048}{53333^2 \cdot 1.000417^3}\left(\frac{1 + 20 \sinh 20}{\cosh 20} - 1\right) = 0.18676 \text{ m}$$

and its stiffness is

$$S_3 = \frac{1}{y_3} = \frac{1}{0.18676} = 5.355 \frac{1}{\text{m}} \qquad \{3.13\}$$

Bracing Unit 4 (frame F1)

The part of the shear stiffness which is associated with the columns has to be recalculated:

$$K_{c,4} = \sum_{j=1}^{n} \frac{12EI_{c,j}}{h^2} = 2\frac{12 \cdot 25 \cdot 10^6 \cdot 0.4^4}{12 \cdot 3^2} = 142222 \text{ kN} \qquad \{2.29\}$$

The part of the shear stiffness which is associated with the beams is unchanged (but using kilonewton instead of meganewton):

$$K_{b,4} = \sum_{j=1}^{n-1} \frac{12EI_{b,j}}{l_j h} = \frac{12 \cdot 25 \cdot 10^6 \cdot 0.4^4}{12 \cdot 6 \cdot 3} = 35556 \text{ kN} \qquad \{2.28\}$$

The above two part stiffnesses define reduction factor r as

$$r_4 = \frac{K_{c,4}}{K_{b,4} + K_{c,4}} = \frac{142222}{35556 + 142222} = 0.8 \qquad \{2.30\}$$

The shear stiffness of the frame can now be determined:

$$K_4 = K_{b,4} r_4 = K_{b,4} \frac{K_{c,4}}{K_{b,4} + K_{c,4}} = 35556 \cdot 0.8 = 28445 \text{ kN} \qquad \{2.27\}$$

For the local bending stiffness ($EI = EI_c r$), the sum of the second moments of area of the columns should be produced (and multiplied by reduction factor r):

$$I_4 = r_4 \sum_{j=1}^{n} I_{c,j} = 0.8 \cdot 2 \cdot \frac{0.4^4}{12} = 0.00341\dot{3} \text{ m}^4 \qquad \{2.31\}$$

The sum of the local and global second moments of area is

$$I_{f,4} = I_4 + I_{g,4} = 0.0034 + 2.88 = 2.8834 \text{ m}^4 \qquad \{2.23\}$$

With the auxiliary parameters s, κ and κH

$$s = 1 + \frac{I_c r}{I_g} = 1 + \frac{0.0034}{2.88} = 1.00118 \qquad \{2.14\}$$

$$\kappa = \sqrt{\frac{Ks}{EI}} = \sqrt{\frac{28445 \cdot 1.00118}{25 \cdot 10^6 \cdot 0.0034}} = 0.5788 \frac{1}{\text{m}} \qquad \text{and} \qquad \kappa H = 17.36$$

the maximum top deflection of the frame is

$$y_4 = \frac{wH^4}{8EI_f} + \frac{wH^2}{2Ks^2} - \frac{wEI}{K^2 s^3}\left(\frac{1 + \kappa H \sinh \kappa H}{\cosh \kappa H} - 1\right) \qquad \{2.24\}$$

$$= \frac{23.4 \cdot 30^4}{8 \cdot 25 \cdot 10^6 \cdot 2.8834} + \frac{23.4 \cdot 30^2}{2 \cdot 28444 \cdot 1.00118^2} -$$

$$- \frac{23.4 \cdot 25 \cdot 10^6 \cdot 0.0034}{28444^2 \cdot 1.00118^3}\left(\frac{1 + 17.36 \sinh 17.36}{\cosh 17.36} - 1\right) = 0.362 \text{ m}$$

and its stiffness is

$$S_4 = \frac{1}{y_4} = \frac{1}{0.362} = 2.762 \frac{1}{\text{m}} \qquad \{3.13\}$$

6.5.2.2 Deflection of the shear centre axis

The participating bracing units in direction y are Units 1, 2 and 3. As there is only one frame—F7— among the bracing units, the bending stiffness of the U-core and shear wall W can be added to the local bending stiffness of the frame:

$$I = 0.0048 + 2.257 + 0.45 = 2.712\ \text{m}^4 \qquad \{3.39\}$$

The rest of the stiffnesses are unchanged so

$$I_g = 11.52\ \text{m}^4, \qquad I_f = I + I_g = 14.23\ \text{m}^4 \qquad \{3.44\}$$

$$K = 53333\ \text{kN} \qquad \{2.27\}$$

From now on, the procedure for a single frame is followed.
With auxiliary parameters

$$s = 1 + \frac{I}{I_g} = 1 + \frac{2.712}{11.52} = 1.235\,, \qquad \kappa = \sqrt{\frac{Ks}{EI}} = \sqrt{\frac{53333 \cdot 1.235}{25 \cdot 10^6 \cdot 2.712}} = 0.0312$$

$$\kappa H = 0.0312 \cdot 30 = 0.935 \qquad \{2.14\}$$

the maximum deflection of the shear centre axis is

$$v_o = \frac{wH^4}{8EI_f} + \frac{wH^2}{2Ks^2} - \frac{wEI}{K^2 s^3}\left(\frac{1 + \kappa H \sinh \kappa H}{\cosh \kappa H} - 1\right) \qquad \{2.24\}$$

$$= \frac{23.4 \cdot 30^4}{8 \cdot 25 \cdot 10^6 \cdot 14.23} + \frac{23.4 \cdot 30^2}{2 \cdot 53333 \cdot 1.235^2} -$$

$$- \frac{23.4 \cdot 25 \cdot 10^6 \cdot 2.712}{53333^2 \cdot 1.235^3}\left(\frac{1 + 0.935 \sinh 0.935}{\cosh 0.935} - 1\right) = 0.0278\ \text{m}$$

6.5.2.3 Rotation around the shear centre axis

The participating bracing units are Units 1, 2, 3, 4 and 5. The location of the shear centre is needed first. Using the relevant stiffnesses and the locations of the bracing units in coordinate system $\bar{x} - \bar{y}$, the two coordinates are given by Equation (3.46) as

$$\bar{x}_o = \frac{\sum_{i=1}^{f+m} S_{y,i}\bar{x}_i}{\sum_{i=1}^{f+m} S_{y,i}} = \frac{3S_{1y} + 12S_2 + 18S_3}{S_{1y} + S_2 + S_3} = \frac{3 \cdot 23.8 + 12 \cdot 4.748 + 18 \cdot 5.355}{23.8 + 4.748 + 5.355} = 6.629\ \text{m}$$

$$\bar{y}_o = \frac{\sum_{i=1}^{f+m} S_{x,i}\bar{y}_i}{\sum_{i=1}^{f+m} S_{x,i}} = \frac{-e_o S_{1x} + 12S_4 + 12S_5}{S_{1x} + S_4 + S_5} = \frac{-1.125 \cdot 152.2 + 12(2.762 + 4.748)}{152.2 + 2.762 + 4.748} = -0.508\,\text{m}$$

A new *x*-*y* coordinate system is now established whose *O* origin is in the shear centre (Figure 6.10). The locations of the bracing units in this coordinate system are:

$$t_{1,x} = \bar{x}_o - 3 = 3.629\,\text{m}, \qquad t_{1,y} = e_o - |\bar{y}_o| = 0.617\,\text{m}, \qquad t_2 = 12 - \bar{x}_o = 5.371\,\text{m}$$

$$t_3 = 18 - \bar{x}_o = 11.371\,\text{m}, \qquad t_4 = t_5 = B + |\bar{y}_o| = 12.508\,\text{m}$$

The torsional moment causing rotation around the shear centre is

$$m_z = wx_c = w\left(\frac{L}{2} - \bar{x}_o\right) = 23.4\left(\frac{18}{2} - 6.629\right) = 55.48\,\text{kNm/m} \qquad \{3.53\}$$

The "governing" torsional stiffnesses of the bracing units are obtained using their perpendicular distance from the shear centre and their maximum deflection:

$$S_{\omega,1y} = \frac{t_{1x}^2}{v_1} = \frac{3.629^2}{0.042} = 313.6\,\text{m}, \qquad S_{\omega,1x} = \frac{t_{1y}^2}{u_1} = \frac{0.617^2}{0.00657} = 57.94\,\text{m} \qquad \{3.54\}$$

$$S_{\omega,2} = \frac{t_2^2}{u_2} = \frac{5.371^2}{0.2106} = 137.0\,\text{m}, \qquad S_{\omega,3} = \frac{t_3^2}{y_3} = \frac{11.371^2}{0.187} = 691.4\,\text{m}$$

$$S_{\omega,4} = \frac{t_4^2}{y_4} = \frac{12.508^2}{0.362} = 432.2\,\text{m}, \qquad S_{\omega,5} = \frac{t_5^2}{u_5} = \frac{12.508^2}{0.2106} = 742.9\,\text{m}$$

With the sum of the torsional stiffnesses

$$S_\omega = \sum_{i=1}^{f+m} S_{\omega,i} = 2375\,\text{m} \qquad \{3.55\}$$

the torsional apportioners can now be determined. Choosing, say, Bracing Unit 2, its torsional apportioner is

$$q_{\omega,2} = \frac{S_{\omega,2}}{S_\omega} = \frac{137}{2375} = 0.0577 \qquad \{3.56\}$$

The torsional moment share on Bracing Unit 2 is

$$m_{z,2} = q_{\omega,2} m_z = 0.0577 \cdot 55.48 = 3.20 \text{ kNm/m} \qquad \{3.57\}$$

The maximum rotation of the building can now be determined using Bracing Unit 2. As Bracing Unit 2 is a shear wall, the equation reduces to its first term (and $I_{f\omega}$ reduces to I_ω):

$$\varphi_{\max} = \varphi_2(H) = \frac{m_{z,2} H^4}{8EI_{f\omega,2}} = \frac{3.2 \cdot 30^4}{8 \cdot 25 \cdot 10^6 \cdot 0.45 \cdot 5.371^2} = 0.000998 \text{ rad} \qquad \{3.59\}$$

6.5.2.4 The maximum deflection of the building

The maximum deflection occurs at the right-hand corner of the building at

$$x_{\max} = L - \bar{x}_o = 18 - 6.629 = 11.371 \text{ m}$$

where the uniform translation and the additional translation due to the rotation of the building add up:

$$v_{\max} = v(H) = v_o + \varphi x_{\max} = 0.0278 + 0.000998 \cdot 11.371 = 0.039 \text{ m} \qquad \{3.66\}$$

The recommended maximum deflection of the building is

$$v_{ASCE} = \frac{H}{500} = 0.060 \text{ m} \qquad \{3.67\}$$

Figure 3.1 (also the front cover) shows the deflected shape of the building.

6.5.3 Fundamental frequency

The mass distribution factor is obtained using Table 4.1 as a function of the number of storeys: $r_f = 0.911$.

With $\gamma = 4.0$, the mass density per unit length for the building is

$$m = \rho A = \frac{\gamma A}{g} = \frac{4 \cdot 18 \cdot 12}{9.81} = 88.07 \text{ kg/m} \qquad \{4.1\}$$

6.5.3.1 Individual bracing units

Before the basic (lateral and torsional) frequencies are determined, and then their coupling is taken into account, the stiffness characteristics and the related frequencies of the individual bracing units are needed.

Bracing Unit 1 (U-core)
The fundamental frequencies in the principal directions are

$$f_{1x} = \frac{0.56 r_f}{H^2}\sqrt{\frac{EI}{m}} = \frac{0.56 \cdot 0.911}{30^2}\sqrt{\frac{25 \cdot 10^6 \cdot 14.42}{88.07}} = 1.147\,\text{Hz} \qquad \{2.75\}$$

$$f_{1y} = \frac{0.56 r_f}{H^2}\sqrt{\frac{EI}{m}} = \frac{0.56 \cdot 0.911}{30^2}\sqrt{\frac{25 \cdot 10^6 \cdot 2.257}{88.07}} = 0.454\,\text{Hz} \qquad \{2.75\}$$

Bracing Units 2 and 5 (shear wall W)
Only the in-plane fundamental frequency will be taken into account:

$$f_2 = \frac{0.56 r_f}{H^2}\sqrt{\frac{EI}{m}} = \frac{0.56 \cdot 0.911}{30^2}\sqrt{\frac{25 \cdot 10^6 \cdot 0.45}{88.07}} = 0.2026\,\text{Hz} \qquad \{2.75\}$$

Bracing Unit 3 (frame F7)
The stiffnesses calculated in the previous section for the deflection analysis can be used again. With K_3 = 53333 kN, the square of the frequency associated with the "original" shear stiffness is

$$f_{s',3}^2 = \frac{1}{(4H)^2}\frac{r_f^2 K_3}{m} = \frac{0.911^2 \cdot 53333}{(4 \cdot 30)^2 \cdot 88.07} = 0.0349\,\text{Hz}^2 \qquad \{2.42\}$$

The global second moment of area of the cross-sections of the columns is unchanged at $I_{g,3} = 11.52$ m^4 and the square of the frequency associated with the global full-height bending vibration of the frame is

$$f_{g,3}^2 = \frac{0.313 r_f^2 E_c I_{g,3}}{H^4 m} = \frac{0.313 \cdot 0.911^2 \cdot 25 \cdot 10^6 \cdot 11.52}{30^4 \cdot 88.07} = 1.049\,\text{Hz}^2 \qquad \{2.43\}$$

The square of the effectiveness factor is

$$s_{f,3}^2 = \frac{f_{g,3}^2}{f_{g,3}^2 + f_{s',3}^2} = \frac{1.049}{1.049 + 0.0349} = 0.9678 \qquad \{2.46\}$$

Using the effectiveness factor leads to the effective shear stiffness:

$$K_{e,3} = K_3 s_{f,3}^2 = 53333 \cdot 0.9678 = 51616\,\text{kN} \qquad \{2.45\}$$

The square of the frequency associated with the effective shear stiffness is

$$f_{s,3}^2 = \frac{1}{(4H)^2}\frac{r_f^2 K_{e,3}}{m} = \frac{0.911^2 \cdot 51616}{(4 \cdot 30)^2 \cdot 88.07} = 0.0338\,\text{Hz}^2 \qquad \{2.44\}$$

With $I_3 = 0.0048\ \text{m}^4$, the fundamental frequency which is associated with the local bending stiffness is

$$f_{b,3}^2 = \frac{0.313 r_f^2 EI}{H^4 m} = \frac{0.313 \cdot 0.911^2 \cdot 25 \cdot 10^6 \cdot 0.0048}{30^4 \cdot 88.07} = 0.000437\,\text{Hz}^2 \qquad \{2.47\}$$

As a function of the non-dimensional parameter

$$k = H\sqrt{\frac{K_e}{EI}} = 30\sqrt{\frac{51616}{25 \cdot 10^6 \cdot 0.0048}} = 19.68 \qquad \{2.51\}$$

the frequency parameter is obtained using Table 4.2 as

$$\eta = 5.153 + \frac{5.278 - 5.153}{20 - 19.5}(19.68 - 19.5) = 5.198 \qquad \{\text{Table } 4.2\}$$

Finally, the fundamental frequency of Bracing Unit 3 (frame F7) is

$$f_3 = \sqrt{f_b^2 + f_s^2 + \left(\frac{\eta^2}{0.313} - \frac{k^2}{5} - 1\right) s_f f_b^2} \qquad \{2.52\}$$

$$= \sqrt{0.000437 + 0.03378 + \left(\frac{5.198^2}{0.313} - \frac{19.68^2}{5} - 1\right) 0.9838 \cdot 0.000437} = 0.1939\,\text{Hz}$$

Bracing Unit 4 (frame F1)

The stiffnesses calculated in the previous section for the deflection analysis can be used again. With $K_4 = 28445$ kN, the square of the frequency associated with the "original" shear stiffness is

$$f_{s',4}^2 = \frac{1}{(4H)^2}\frac{r_f^2 K_4}{m} = \frac{0.911^2 \cdot 28445}{(4 \cdot 30)^2 \cdot 88.07} = 0.0186\,\text{Hz}^2 \qquad \{2.42\}$$

The global second moment of area of the cross-sections of the columns is unchanged at $I_{g,4} = 2.88\ \text{m}^4$ and the square of the frequency associated with the global full-height bending vibration of the frame is

$$f_{g,4}^2 = \frac{0.313 r_f^2 E_c I_{g,4}}{H^4 m} = \frac{0.313 \cdot 0.911^2 \cdot 25 \cdot 10^6 \cdot 2.88}{30^4 \cdot 88.07} = 0.2622 \text{ Hz}^2 \qquad \{2.43\}$$

The square of the effectiveness factor is

$$s_{f,4}^2 = \frac{f_{g,4}^2}{f_{g,4}^2 + f_{s',4}^2} = \frac{0.2622}{0.2622 + 0.0186} = 0.9337 \qquad \{2.46\}$$

Using the effectiveness factor leads to the effective shear stiffness:

$$K_{e,4} = K_4 s_{f,4}^2 = 28445 \cdot 0.9337 = 26559 \text{ kN} \qquad \{2.45\}$$

The square of the frequency associated with the effective shear stiffness is

$$f_{s,4}^2 = \frac{1}{(4H)^2} \frac{r_f^2 K_{e,4}}{m} = \frac{0.911^2 \cdot 26559}{(4 \cdot 30)^2 \cdot 88.07} = 0.0174 \text{ Hz}^2 \qquad \{2.44\}$$

The square of the frequency which is associated with the local bending stiffness is

$$f_{b,4}^2 = \frac{0.313 r_f^2 EI}{H^4 m} = \frac{0.313 \cdot 0.911^2 \cdot 25 \cdot 10^6 \cdot 0.003413}{30^4 \cdot 88.07} = 0.000311 \text{ Hz}^2 \qquad \{2.47\}$$

As a function of the non-dimensional parameter

$$k = H\sqrt{\frac{K_e}{EI}} = 30\sqrt{\frac{26559}{25 \cdot 10^6 \cdot 0.003413}} = 16.74 \qquad \{2.51\}$$

the frequency parameter is obtained using Table 4.2 as

$$\eta = 4.408 + \frac{4.532 - 4.408}{17 - 16.5}(16.74 - 16.5) = 4.468 \qquad \{\text{Table 4.2}\}$$

Finally, the fundamental frequency of Bracing Unit 4 (frame F1) is

$$f_4 = \sqrt{f_b^2 + f_s^2 + \left(\frac{\eta^2}{0.313} - \frac{k^2}{5} - 1\right) s_f f_b^2} \qquad \{2.52\}$$

$$= \sqrt{0.000311 + 0.0174 + \left(\frac{4.468^2}{0.313} - \frac{16.74^2}{5} - 1\right) 0.9663 \cdot 0.000311} = 0.1405 \text{ Hz}$$

6.5.3.2 Lateral vibration in direction x

The participating bracing units in direction x are Units 1, 4 and 5.

As there is only one frame among the participating units—F1—the original shear stiffness, the effective shear stiffness, the effectiveness factor and the square of the frequency associated with the effective shear stiffness are those of frame F1:

$$K = \sum_{i=1}^{f} K_i = K_4 = 28445 \text{ kN} \qquad \{4.6\}$$

$$K_e = \sum_{1}^{f} K_i s_i^2 = 28445 \cdot 0.9337 = 26559 \text{ kN} \qquad \{4.11\}$$

$$s_f = \sqrt{\frac{K_e}{K}} = \sqrt{\frac{26559}{28445}} = 0.9663 \qquad \{4.12\}$$

and

$$f_s^2 = \frac{1}{(4H)^2} \frac{r_f^2 K_e}{m} = \frac{0.911^2 \cdot 26559}{(4 \cdot 30)^2 \cdot 88.07} = 0.0174 \text{ Hz}^2 \qquad \{4.13\}$$

With EI already available (from the stability analysis), the square of the frequency of the system in bending is obtained from

$$f_b^2 = \frac{0.313 r_f^2 EI}{H^4 m} = \frac{0.313 \cdot 0.911^2 \cdot 371.8 \cdot 10^6}{30^4 \cdot 88.07} = 1.354 \text{ Hz}^2 \qquad \{4.15\}$$

With the non-dimensional parameter

$$k = H\sqrt{\frac{K_e}{EI}} = 30\sqrt{\frac{26559}{371.8 \cdot 10^6}} = 0.2536 \qquad \{4.17\}$$

the frequency parameter is obtained from Table 4.2 as

$$\eta = 0.5638 + \frac{0.569 - 0.5638}{0.3 - 0.2}(0.2536 - 0.2) = 0.5666 \qquad \{\text{Table 4.2}\}$$

The lateral frequency of the building in direction x is

$$f_x = \sqrt{f_b^2 + f_s^2 + \left(\frac{\eta^2}{0.313} - \frac{k^2}{5} - 1\right) s_f f_b^2} = \qquad \{4.18\}$$

$$= \sqrt{1.354 + 0.0174 + \left(\frac{0.5666^2}{0.313} - \frac{0.2536^2}{5} - 1 \right) 0.9663 \cdot 1.354} = 1.178\,\text{Hz}$$

6.5.3.3 Lateral vibration in direction y

The participating bracing units in direction y are Units 1, 2 and 3.

The situation is the same as with lateral vibration in direction x in that there is only one frame—F7—among the participating units. It follows that the original shear stiffness, the effective shear stiffness, the effectiveness factor and the square of the frequency associated with the effective shear stiffness do not need calculation as they are those of frame F7:

$$K = \sum_{i=1}^{f} K_i = K_3 = 53333\,\text{kN} \qquad \{4.6\}$$

$$K_e = \sum_{1}^{f} K_i s_i^2 = 53333 \cdot 0.9678 = 51616\,\text{kN} \qquad \{4.11\}$$

$$s_f = \sqrt{\frac{K_e}{K}} = \sqrt{\frac{51616}{53333}} = 0.9838 \qquad \{4.12\}$$

and

$$f_s^2 = \frac{1}{(4H)^2} \frac{r_f^2 K_e}{m} = \frac{0.911^2 \cdot 51616}{(4 \cdot 30)^2 \cdot 88.07} = 0.0338\,\text{Hz}^2 \qquad \{4.13\}$$

The square of the frequency of the system in bending is obtained from

$$f_b^2 = \frac{0.313 r_f^2 EI}{H^4 m} = \frac{0.313 \cdot 0.911^2 \cdot 67.68 \cdot 10^6}{30^4 \cdot 88.07} = 0.246\,\text{Hz}^2 \qquad \{4.15\}$$

With the non-dimensional parameter

$$k = H\sqrt{\frac{K_e}{EI}} = 30\sqrt{\frac{51616}{67.68 \cdot 10^6}} = 0.828 \qquad \{4.17\}$$

the frequency parameter is obtained from Table 4.2 as

$$\eta = 0.6223 + \frac{0.6376 - 0.6223}{0.9 - 0.8}(0.828 - 0.8) = 0.6266 \qquad \{\text{Table 4.2}\}$$

The lateral frequency of the building in direction y is

$$f_y = \sqrt{f_b^2 + f_s^2 + \left(\frac{\eta^2}{0.313} - \frac{k^2}{5} - 1\right) s_f f_b^2} \qquad \{4.18\}$$

$$= \sqrt{0.246 + 0.0338 + \left(\frac{0.6266^2}{0.313} - \frac{0.828^2}{5} - 1\right) 0.9838 \cdot 0.246} = 0.555 \text{ Hz}$$

6.5.3.4 Pure torsional vibration

All five bracing units participate in torsional vibration.

The shear centre of the bracing system (Figure 6.10) is located using the fundamental frequency of the bracing units:

$$\bar{x}_o = \frac{\sum_1^{f+m} f_{y,i}^2 \bar{x}_i}{\sum_1^{f+m} f_{y,i}^2} = \frac{3 \cdot 0.454^2 + 12 \cdot 0.2026^2 + 18 \cdot 0.1939^2}{0.454^2 + 0.2026^2 + 0.1939^2} = 6.28 \text{ m} \qquad \{4.19\}$$

$$\bar{y}_o = \frac{\sum_1^{f+m} N_{x,i} \bar{y}_i}{\sum_1^{f+m} N_{x,i}} = \frac{-1.125 \cdot 1.147^2 + 12(0.2026^2 + 0.1405^2)}{1.147^2 + 0.2026^2 + 0.1405^2} = -0.545 \text{ m} \qquad \{4.19\}$$

A new x-y coordinate system is now established whose O origin is in the shear centre (Figure 6.10). The locations of the bracing units in this coordinate system are

$$t_{1,x} = \bar{x}_o - 3 = 3.28 \text{ m}, \qquad t_{1,y} = e_o - |\bar{y}_o| = 0.58 \text{ m}, \qquad t_2 = 12 - \bar{x}_o = 5.72 \text{ m}$$

$$t_3 = 18 - \bar{x}_o = 11.72 \text{ m}, \qquad t_4 = t_5 = B + |\bar{y}_o| = 12.545 \text{ m}$$

With the coordinates of the centroid in the coordinate system x-y

$$x_c = \frac{L}{2} - \bar{x}_o = \frac{18}{2} - 6.28 = 2.72 \text{ m}, \quad y_c = \frac{B}{2} - \bar{y}_o = \frac{12}{2} + 0.545 = 6.545 \text{ m} \qquad \{4.25\}$$

the distance between the shear centre and the centroid is

$$t = \sqrt{x_c^2 + y_c^2} = \sqrt{2.72^2 + 6.545^2} = 7.088 \text{ m} \qquad \{4.24\}$$

With the above data, the radius of gyration is

$$i_p = \sqrt{\frac{L^2 + B^2}{12} + t^2} = \sqrt{\frac{18^2 + 12^2}{12} + 7.088^2} = \sqrt{89.24} = 9.447 \text{ m} \qquad \{4.24\}$$

The eccentricities of the bracing system will also be needed later on:

$$\tau_x = \frac{x_c}{i_p} = \frac{2.72}{9.447} = 0.288 \text{ m}, \qquad \tau_y = \frac{y_c}{i_p} = \frac{6.545}{9.447} = 0.693 \text{ m} \qquad \{4.35\}$$

The "original" Saint-Venant stiffness of the system comes from the Saint-Venant stiffness of the U-core and the original shear stiffness of the two frames:

$$(GJ) = \sum_1^m GJ_k + \sum_1^f \left((K_i)_x y_i^2 + (K_i)_y x_i^2\right) \qquad \{4.30\}$$

$$= 10.4 \cdot 10^6 (0.032 + 2 \cdot 0.008) + 53333 \cdot 11.72^2 + 28445 \cdot 12.545^2 = 12.3 \cdot 10^6 \text{ kNm}^2$$

The real (effective) Saint-Venant stiffness is always smaller than the "original" one as the effect of the frames is restricted by their effectiveness:

$$(GJ)_e = \sum_1^m GJ_k + \sum_1^f \left((K_{e,i})_x y_i^2 + (K_{e,i})_y x_i^2\right) \qquad \{4.21\}$$

$$= 10.4 \cdot 10^6 (0.032 + 2 \cdot 0.008) + 51616 \cdot 11.72^2 + 26559 \cdot 12.545^2 = 11.77 \cdot 10^6 \text{ kNm}^2$$

The effectiveness of the Saint-Venant stiffness for the whole system can now be determined:

$$s_\varphi = \sqrt{\frac{(GJ)_e}{(GJ)}} = \sqrt{\frac{11.77}{12.30}} = 0.978 \qquad \{4.29\}$$

Neglecting the contribution of the columns of the two frames, the warping stiffness of the system comes from two sources: the own warping stiffness of the core and the bending stiffness of the walls:

$$EI_\omega = E_w \sum_1^m \left(I_{\omega,k} + (I_{w,k})_x y_k^2 + (I_{w,k})_y x_k^2\right) \qquad \{4.22\}$$

$$= 25 \cdot 10^6 [14.18 + 2.257 \cdot 3.28^2 + 14.42 \cdot 0.58^2 + 0.45(5.72^2 + 12.545^2)] = 3221 \cdot 10^6 \text{kNm}^4$$

With the above stiffnesses, the part torsional frequencies can now be determined.

The square of the pure torsional frequency associated with the Saint-Venant stiffness is

$$f_t^2 = \frac{r_f^2 (GJ)_e}{16 i_p^2 H^2 m} = \frac{0.911^2 \cdot 11.77 \cdot 10^6}{16 \cdot 89.24 \cdot 30^2 \cdot 88.07} = 0.0863 \text{ Hz}^2 \qquad \{4.28\}$$

The square of the pure torsional frequency associated with the warping stiffness is

$$f_\omega^2 = \frac{0.313 r_f^2 EI_\omega}{i_p^2 H^4 m} = \frac{0.313 \cdot 0.911^2 \cdot 3221 \cdot 10^6}{89.24 \cdot 30^4 \cdot 88.07} = 0.1314 \text{ Hz}^2 \qquad \{4.27\}$$

As a function of torsion parameter

$$k_\varphi = H\sqrt{\frac{(GJ)_e}{EI_\omega}} = 30\sqrt{\frac{11.77 \cdot 10^6}{3221 \cdot 10^6}} = 1.813 \qquad \{4.31\}$$

the frequency parameter is obtained from Table 4.2:

$$\eta_\varphi = 0.817 + \frac{0.8397 - 0.817}{1.9 - 1.8}(1.813 - 1.8) = 0.820 \qquad \{\text{Table 4.2}\}$$

With the above part frequencies and stiffness characteristics, the fundamental frequency for pure torsional vibration is

$$f_\varphi = \sqrt{f_\omega^2 + f_t^2 + \left(\frac{\eta_\varphi^2}{0.313} - \frac{k_\varphi^2}{5} - 1\right) s_\varphi f_\omega^2} \qquad \{4.26\}$$

$$= \sqrt{0.1314 + 0.0863 + \left(\frac{0.82^2}{0.313} - \frac{1.813^2}{5} - 1\right) 0.978 \cdot 0.1314} = 0.530 \text{ Hz}$$

6.5.3.5 Coupling of the basic frequencies: the fundamental frequency

As the centroid of the layout does not coincide with the shear centre, the coupling of the basic frequencies has to be considered. Using the squares of the basic frequencies f_x, f_y and f_φ and the eccentricity parameters τ_x and τ_y, the smallest root of the cubic equation

$$\left(f^2\right)^3 + a_2\left(f^2\right)^2 + a_1 f^2 - a_0 = 0 \qquad \{4.33\}$$

is the fundamental frequency of the building. With

$$a_0 = \frac{f_x^2 f_y^2 f_\varphi^2}{1-\tau_x^2-\tau_y^2} = \frac{1.178^2 \cdot 0.555^2 \cdot 0.53^2}{1-0.288^2-0.693^2} = 0.2749 \qquad \{4.34\}$$

$$a_1 = \frac{f_x^2 f_y^2 + f_\varphi^2 f_x^2 + f_\varphi^2 f_y^2}{1-\tau_x^2-\tau_y^2} = 2.069$$

$$a_2 = \frac{f_x^2 \tau_x^2 + f_y^2 \tau_y^2 - f_x^2 - f_y^2 - f_\varphi^2}{1-\tau_x^2-\tau_y^2} = -3.923$$

the fundamental frequency is

$$f = 0.467 \text{ Hz}$$

7

Accuracy and reliability

To highlight the importance of the accuracy and reliability of approximate methods, the procedure recommended to a practicing structural designer who faces the task of carrying out the structural analysis of a multi-storey building is summarized first:

- establish the main geometrical and stiffness characteristics by using simple approximate procedures (like those presented in this book)
- carry out a comprehensive, full-blown structural analysis using a good computer program [often based on the Finite Element (FE) method]
- check the main results of the computer-based analysis and perform a global analysis, again using approximate procedures

As the approximate procedures play an important role in the design procedure, it is essential that their accuracy and range of applicability are established. No respectable approximate method should be published, let alone used in practice, without a comprehensive numerical analysis of its suitability for practical application.

Such an accuracy analysis is presented in this chapter. The results obtained using the approximate formulae presented in the preceding chapters will be compared to those obtained from corresponding FE solutions. The Axis VM X5 (2019) finite element package will be used for the comparison, whose results are considered "exact". To facilitate like-for-like comparison, when the FE models are created, the assumptions listed in "Chapter 1: Introduction" will be observed. Accordingly, in order for the accuracy analyses to correspond to the theoretical assumption that "the floor slabs have great in-plane and small out-of-plane stiffness", the floor slabs of the buildings in the FE analysis are modelled using sets of horizontal bars interconnecting the vertical bracing units. These bars have very great cross-sectional areas and pinned ends. The bracing units (frames and shear walls) are modelled by bar elements.

Thirty-two individual bracing units are used for the accuracy analysis. The details of these bracing units are given in Section 7.1. The accuracy analysis is carried out in two main groups. The accuracy of the formulae developed for the analysis of a *single bracing unit* (frame) is checked first in Section 7.2. In this case, the bracing unit acts in a two-dimensional manner and a planar analysis is needed. Using the thirty-two bracing units, thirty-four three-dimensional bracing *systems* are then investigated in Section 7.3. The height of the structures in both groups is increased from five storeys to eighty storeys in eight steps—5, 10, 15, 20, 25, 30, 40, 60 and 80 storeys—making it possible to demonstrate how the accuracy of a given formula changes as a function of the height of the structure.

Altogether 1631 calculations are made where the approximate results and those obtained using the Axis VM X5 finite element package are compared. According to the findings, summarized in this chapter, the accuracy of the approximate methods presented in the preceding chapters and advocated for structural engineering practice is acceptable for the purposes outlined above.

The chapter concludes with a summary of the performance of the analytical procedures used in the 19 worked examples presented in the previous chapters.

7.1 BASIC CHARACTERISTICS OF THE BRACING UNITS

The key units of the accuracy analysis are ten reinforced concrete sway-frames (Figure 7.1) and fourteen braced steel frames (Figures 7.2 and 7.3).

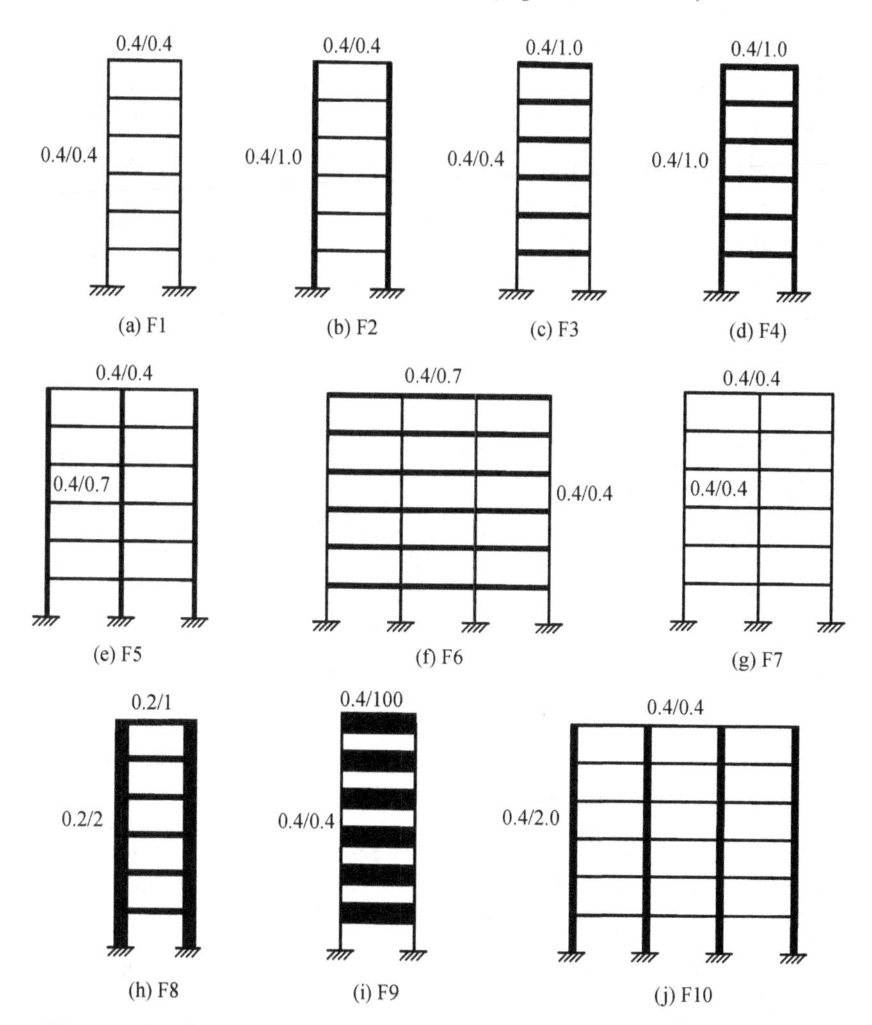

Figure 7.1 Rigid concrete sway-frames F1-F10 with cross-section size (width/depth) in metres.

The moduli of elasticity for the concrete and steel structures are $E = 25$ kN/mm^2 and $E = 200$ kN/mm^2, respectively. The bays of the one-, two- and three-bay frames are 6 metres and the storey height is $h = 3$ metres.

The rectangular cross-sections of the reinforced concrete frames F1-F10 are given in Figure 7.1/a to Figure 7.1/j in metres.

The cross-sections of the beams and columns are chosen in such a way that the structures cover a wide range of stiffnesses and even represent extreme special cases. Even the highly theoretical case of a frame with beams with a depth of 100 m (Figure 7.1/i) is included to model "pure" shear-type global deformation. The deflected shapes of the test structures represent predominant bending, mixed shear and bending, and predominant shear deformation.

Standard UK sections are used for the steel braced frames (Figures 7.2 and 7.3), whose cross-sectional characteristics are given in Table 7.1.

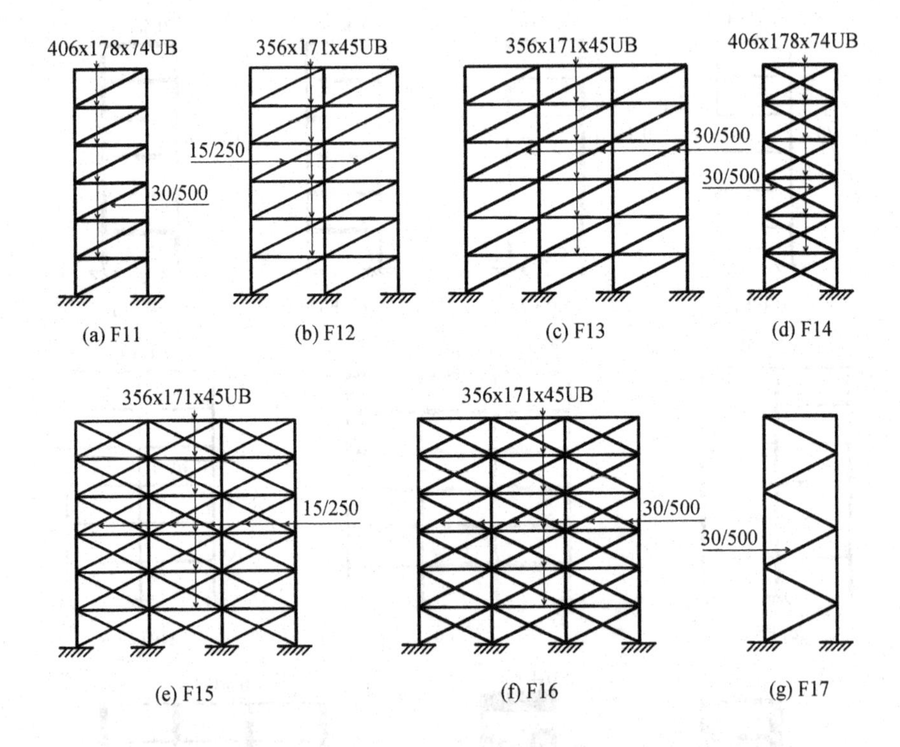

Figure 7.2 Steel braced sway-frames F11-F17 with full-length diagonals.

The sections of the columns are in each case 305x305xUC137 and they are not marked in the figures. The sections of the beams and diagonals are given in Figures 7.2 and 7.3. The rectangular cross-sections of the diagonals are in millimetres.

The fourteen steel braced frames are divided into two sub-groups. Frames F11-F17 (Figure 7.2) have diagonals that join the columns and beams of the frames in nodal points (in a 6x3 metres rectangular network). The diagonals of these frames will be referred to as "full-length" diagonals. The diagonals of frames F21-F27 (Figure 7.3) follow a different pattern: one (or both) of their ends joins the beams of the frame at a point that is not a nodal point in the 6x3 metres rectangular network. These diagonals will be referred to as "short" diagonals. Shear stiffness K for the braced frames can be determined using Tables 2.3 and 2.4.

Table 7.1 Cross-sectional characteristics for the steel braced frames.

Cross-section	All columns	Beams		Diagonals	
	305x305UC137	356x171x45UB	406x178x74UB	15/250	30/500
A [m^2]	$1.74 \cdot 10^{-2}$	$5.73 \cdot 10^{-3}$	$9.45 \cdot 10^{-3}$	$3.75 \cdot 10^{-3}$	$1.5 \cdot 10^{-2}$
I [m^4]	$3.281 \cdot 10^{-4}$	$1.207 \cdot 10^{-4}$	$2.731 \cdot 10^{-4}$	$1.953 \cdot 10^{-5}$	$3.125 \cdot 10^{-4}$

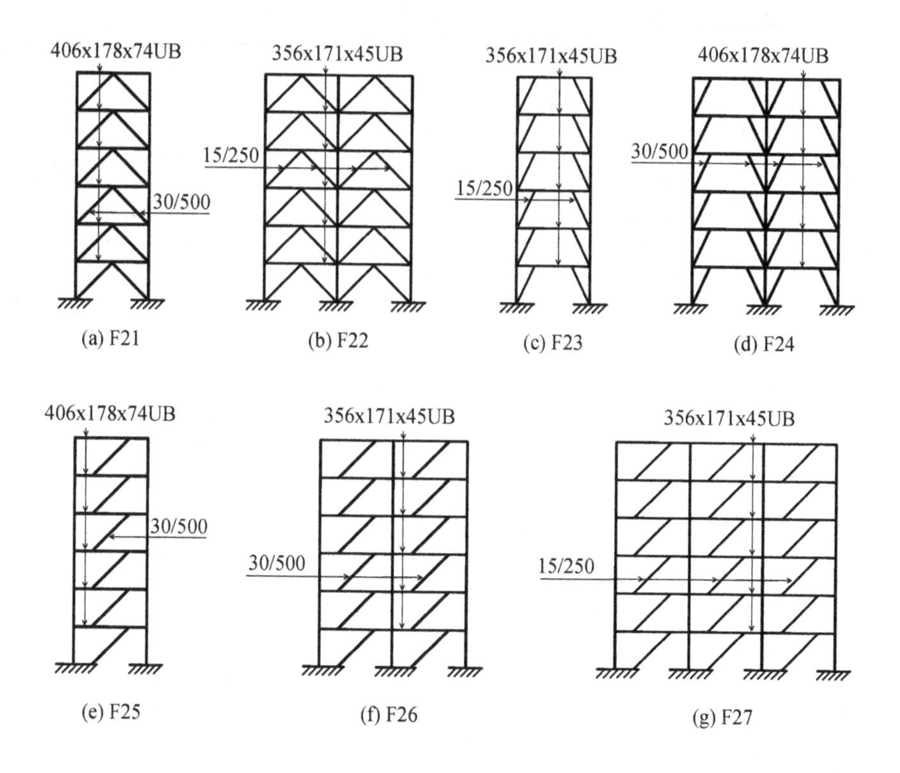

Figure 7.3 Steel braced sway-frames F21-F27 with short diagonals.

The accuracy analysis is carried out in several sections. The accuracy of the formulae developed for the analysis of *individual* bracing units (frames) is presented first. In this case, the bracing unit acts in a two-dimensional manner and a planar analysis is needed. The ten rigid sway-frames and fourteen braced sway-frames are used as "standard" units. As a rule, fixed supports will be considered. However, some of the frames are also investigated with pinned supports and also with longer ground floor columns.

When the accuracy of the formulae derived for bracing *systems* is investigated, the twenty-four frames are supplemented with seven shear walls (Figure 7.4) and a U-core (Figure 7.5). The accuracy of the analytical formulae related to these individual shear walls and the core is not investigated as these classic formulae are considered exact. The shear walls and the core are only needed to create realistic three-dimensional bracing systems with a mixture of bracing units. These three-dimensional bracing systems are then grouped into two sections: rotation-free systems are investigated in Section 7.3.1 and full three-dimensional analyses are carried out in Section 7.3.2. The accuracy of the formulae derived for the determination of the top deflection, the fundamental frequency and the critical load is established.

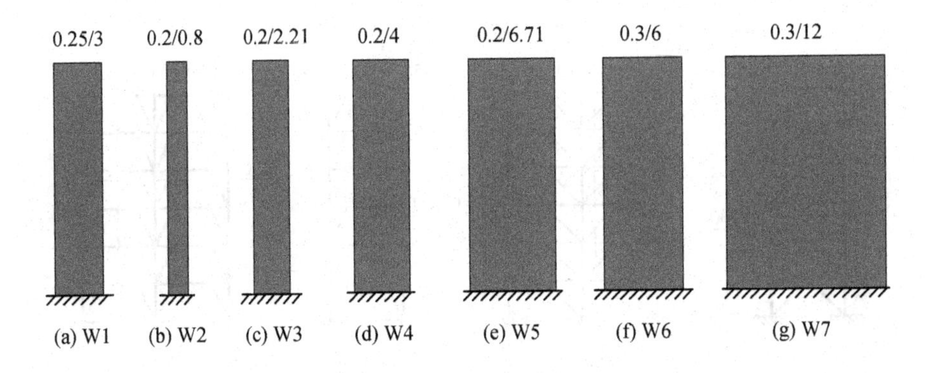

Figure 7.4 Shear walls W1-W7 for the accuracy analysis.

The dimensions of shear walls W1-W7 are shown above the shear walls in Figure 7.4 and the stiffness characteristics of the core are given in Figure 7.5. The torsional characteristics of the U-core have been calculated using Table 2.6.

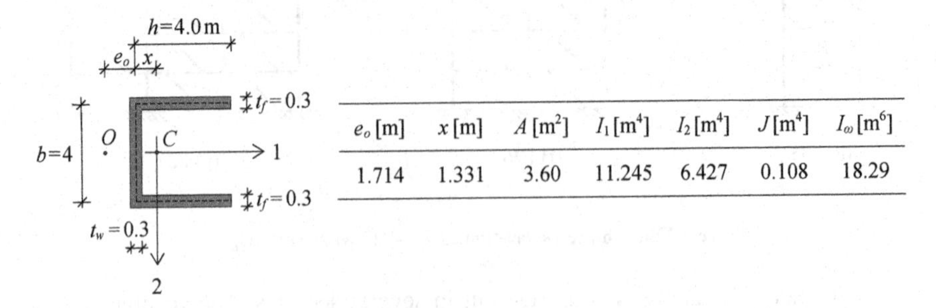

e_o [m]	x [m]	A [m^2]	I_1 [m^4]	I_2 [m^4]	J [m^4]	I_ω [m^6]
1.714	1.331	3.60	11.245	6.427	0.108	18.29

Figure 7.5 U-core for the accuracy analysis.

7.2 STRUCTURAL ANALYSIS OF INDIVIDUAL BRACING UNITS

When an individual bracing unit is considered, the structure behaves in a two-dimensional manner and a simple planar analysis is needed. The relevant formulae are presented in Chapter 2.

7.2.1 Maximum deflection

The accuracy of the formulae derived for the maximum deflection of rigid sway-frames on fixed supports is investigated first.

The accuracy of Equation (2.24) is checked with the determination of the maximum deflection of the ten sway-frames F1-F10 (Figure 7.1). Varying the height of the frames from five to eighty storeys in eight steps (at 5, 10, 15, 20, 25, 30, 40, 60 and 80 storeys) leads to ninety test structures. The summary of the accuracy analysis is given in Table 7.2 where the term "difference" means the difference between the continuum solution by Equation (2.24) and the "exact" (Finite Element based) solution, related to the "exact" solution. Positive difference means the overestimation of the deflection.

Table 7.2 Accuracy of Equation (2.24) for maximum deflection of rigid sway-frames F1-F10.

Rigid sway-frames F1-F10	Range of difference [%]	Average absolute difference [%]	Maximum difference [%]
Continuum solution [Equation (2.24)]	–5 to 7	1.2	7

In addition to the data given in Table 7.2, it is also important to see how the difference varies as the height of the structure changes. Figure 7.6 shows the difference as a function of height for the ten frames.

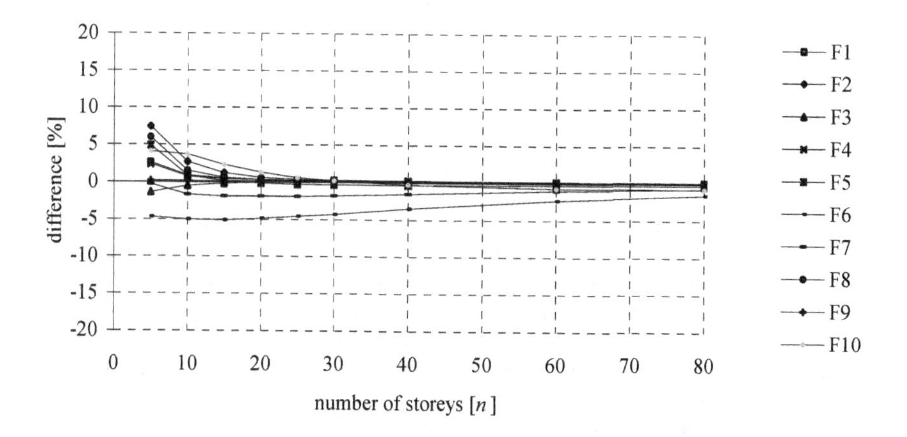

Figure 7.6 Accuracy of Equation (2.24) for the maximum deflection of rigid sway-frames.

The results summarised in Table 7.2 and shown in Figure 7.6 demonstrate the accuracy of the continuum solution. The difference range is between –5% (unconservative) and 7% (conservative). In the 90 cases, the average absolute difference between the results of the continuum method and those of the finite element solution is 1.2%.

Equation (2.24) can also be used to calculate the top deflection of braced frames. Two groups will be considered here. The first group contains 63 frames. These frames—F11-F17—have full-length diagonals (Figure 7.2). The results (Table 7.3) are quite spectacular: the difference range (between the continuum and FE solutions) is –4% to 3% and the average absolute difference is less than 1%. Figure 7.7 shows how accuracy changes over the height of the structure.

Table 7.3 Accuracy of Equation (2.24) for the maximum deflection of braced frames F11-F17.

Braced frames F11-F17	Range of difference [%]	Average absolute difference [%]	Maximum difference [%]
Continuum solution [Equation (2.24)]	–4 to 3	0.98	4

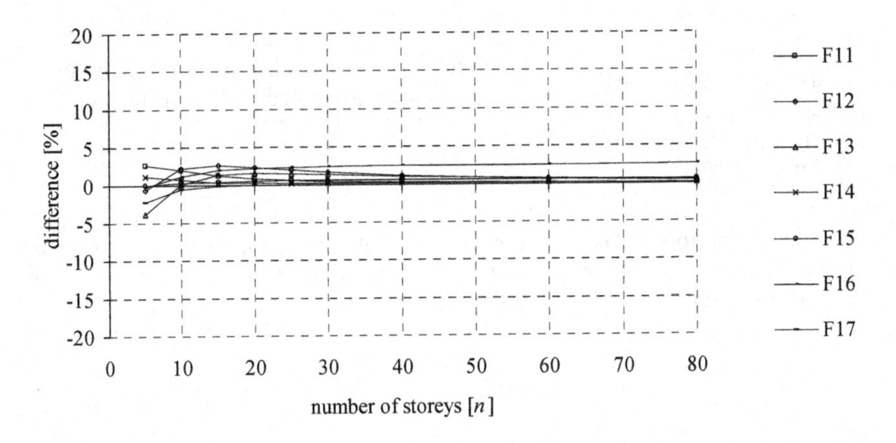

Figure 7.7 Accuracy of Equation (2.24) for the maximum deflection of braced frames with full-length diagonals.

The situation with the second group of 63 braced frames is different. In addition to the vertical columns and horizontal beams, frames F21 to F27 also have shorter than bay-size diagonals (Figure 7.3). These diagonals—as opposed to the bay-size diagonals of frames F11-F17—have some contribution to the local and global bending stiffness of the frame. As this contribution cannot be established in an exact manner, some approximation has to be made. When the local bending stiffness of such a frame is calculated, the diagonals are taken into account as if they were columns, i.e. their second moments of area are added to those of the columns. When the global bending stiffness of such a frame is calculated, the contribution of the diagonals is ignored.

Table 7.4 Accuracy of Equation (2.24) for the maximum deflection of braced frames F21-F27.

Braced frames F21-F27	Range of difference [%]	Average absolute difference [%]	Maximum difference [%]
Continuum solution [Equation (2.24)]	1 to 25	10.0	25

The results are summarized in Table 7.4 and Figure 7.8. According to the results, due to the above approximations, the accuracy of Equation (2.24) is not as good as in the previous cases.

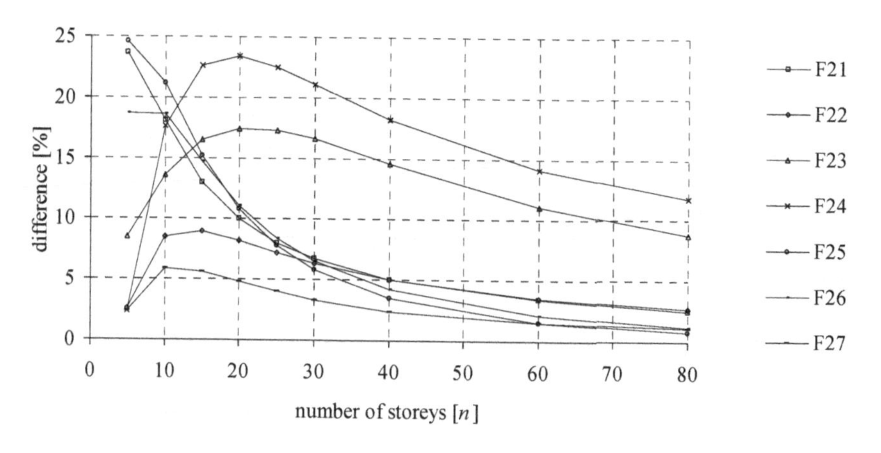

Figure 7.8 Accuracy of Equation (2.24) for the maximum deflection for braced frames with short diagonals.

As the graphs in Figure 7.8 demonstrate, these approximations lead to conservative estimates. The method is best used for situations when an upper limit to the maximum deflection of braced frames with short diagonals is sufficient.

7.2.2 Fundamental frequency

The accuracy of Equation (2.52) is checked with the determination of the fundamental frequency of the same twenty-four frames used in the previous section where the accuracy of the formula for the maximum deflection was investigated. Again, the height of the frames varies from five to eighty storeys in eight steps leading to 216 test structures altogether.

Sway-frames F1-F10 are used first. The summary of the accuracy analysis is given in Table 7.5 where "difference" again means the difference between the "exact" (FE) solution and the continuum solution by Equation (2.52), related to the "exact" solution. Positive difference means the underestimation of the fundamental frequency.

Table 7.5 Accuracy of Equation (2.52) for the fundamental frequency of rigid sway-frames F1-F10.

Rigid sway-frames F1-F10	Range of difference [%]	Average absolute difference [%]	Maximum difference [%]
Fundamental frequency [Equation (2.52)]	–3 to 7	1.6	7

In addition to the data given in Table 7.5, it is also important to see how the difference (between the continuum and FE solutions) varies as the height of the structure changes. Figure 7.9 shows the difference as a function of height for the ten frames. According to the results, the use of Equation (2.52) leads to exceptionally good estimates of the fundamental frequency of the structures.

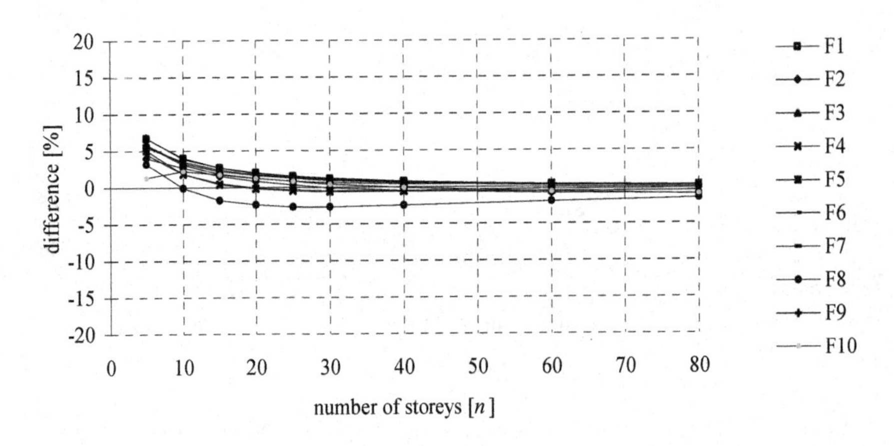

Figure 7.9 Accuracy of Equation (2.52) for the fundamental frequency of sway-frames F1-F10.

Equation (2.52) will now be used to calculate the fundamental frequency of braced frames in two groups. The first group contains frames F11-F17 (Figure 7.2). These frames have full-length diagonals. The results are collected in Table 7.6.

Table 7.6 Accuracy of Equation (2.52) for the fundamental frequency of braced frames F11-F17.

Braced frames F11-F17	Range of difference [%]	Average absolute difference [%]	Maximum difference [%]
Fundamental frequency [Equation (2.52)]	–1 to 17	1.7	17

Figure 7.10 shows how the difference changes as a function of height. The results are similar to those of the previous case: as the height of the structures increases, the difference between the approximate and "exact" values rapidly decreases. For frames over twenty storeys high, the difference becomes negligible.

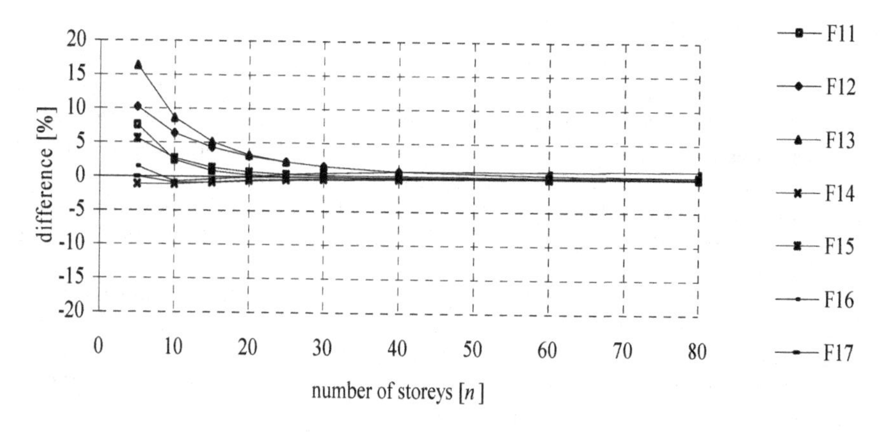

Figure 7.10 Accuracy of Equation (2.52) for the fundamental frequency of frames F11-F17.

Finally, Equation (2.52) will be used to calculate the fundamental frequency of braced frames F21-F27 (Figure 7.3). These frames have short diagonals. The results are collected in Table 7.7 and in Figure 7.11. Regarding the performance of Equation (2.52) applied to braced frames with short diagonals, it is interesting to note that Equation (2.52) underestimates the fundamental frequency in all 63 cases.

Table 7.7 Accuracy of Equation (2.52) for the fundamental frequency of braced frames F21-F27.

Braced frames F21-F27	Range of difference [%]	Average absolute difference [%]	Maximum difference [%]
Fundamental frequency [Equation (2.52)]	0 to 14	5.7	14

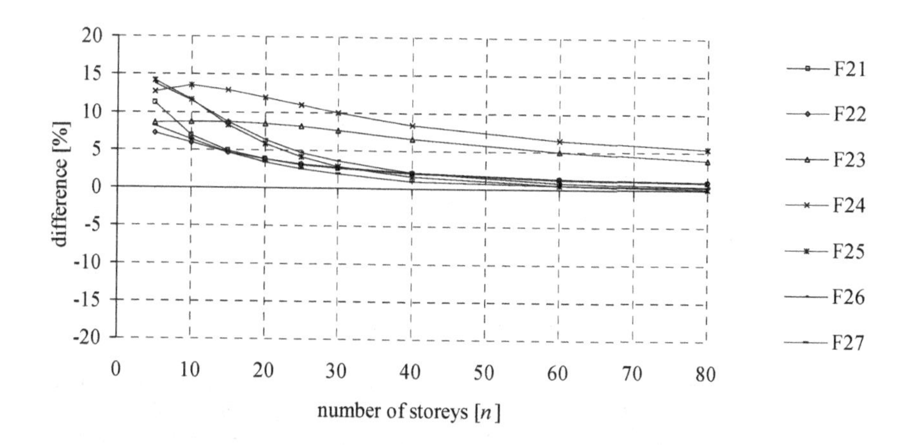

Figure 7.11 Accuracy of Equation (2.52) for the fundamental frequency of frames F21-F27.

7.2.3 Critical load

Two methods were presented in Sections 2.3.1 and 2.4.1 for the determination of the critical load of frames subjected to uniformly distributed load on the beams. The application of the thick sandwich model led to a comprehensive method of general validity—Equation (2.60)—while using the thin sandwich column model resulted in a simpler albeit less accurate solution—Equation (2.63)—with a limited range of validity. The accuracy of both Equation (2.60) and Equation (2.63) is investigated in this section.

The same twenty-four frames (F1-F10, F11-F17 and F21-27) are used that were used in the previous sections. Again, the height of the frames varies from five to eighty storeys in eight steps leading to 216 test structures. Positive difference (between the continuum and FE solutions) means that the approximate formula underestimates the value of the critical load.

The accuracy of Equation (2.60) is investigated first.

Table 7.8 Accuracy of Equation (2.60) for the critical load of sway-frames F1-F10.

Rigid sway-frames F1-F10	Range of difference [%]	Average absolute difference [%]	Maximum difference [%]
Critical load [Equation (2.60)]	–5 to 14	3.3	14

The first group contains sway-frames F1-F10. The results are collected in Table 7.8 and in Figure 7.12.

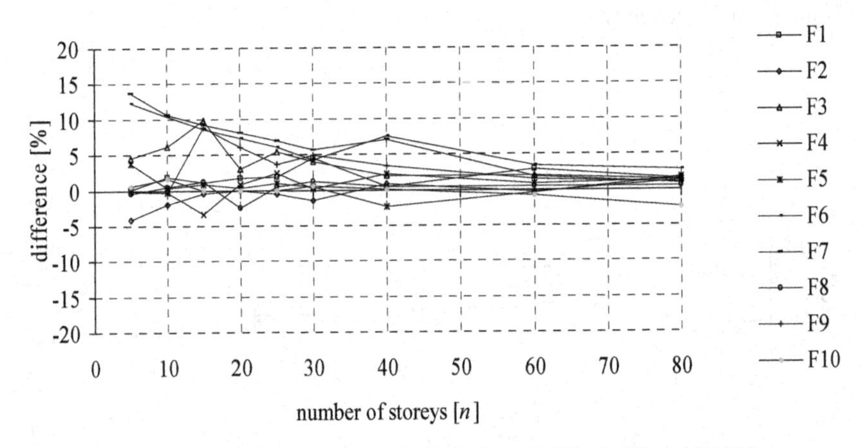

Figure 7.12 Accuracy of Equation (2.60) for the critical load of frames F1-F10.

Equation (2.60) can also be used to determine the critical load of braced frames. Frames F11-F17 are braced frames with full-length diagonals. Table 7.9 and Figure 7.13 contain the result of the comparison of the continuum solution and the FE results.

Table 7.9 Accuracy of Equation (2.60) for the critical load of braced frames F11-F17.

Braced frames F11-F17	Range of difference [%]	Average absolute difference [%]	Maximum difference [%]
Critical load [Equation (2.60)]	−13 to 6	2.8	13

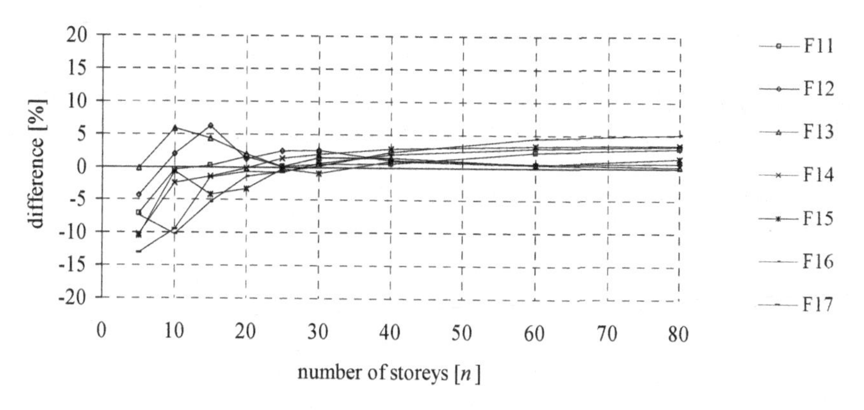

Figure 7.13 Accuracy of Equation (2.60) for the critical load of braced frames F11-F17.

It should be noted here that the continuum method—that leads to Equation (2.60) and Equation (2.63)—assumes global buckling failure and does not take into consideration the possibility that the structure may lose stability through the local buckling of one of its members. This is perfectly all right with the great majority of multi-storey frames as, as a rule, they lose stability through global buckling. In such cases, the smallest eigenvalue [given by Axis VM X5 (2019)] of the stability problem belongs to the global critical load and therefore characterises the smallest critical load of the structure. In some cases with this group of braced frames—F11-F17—however, the structure loses stability because of local buckling and the critical load—the first eigenvalue of the eigenvalue problem—belongs to local buckling. The thirteen cases that belong to this subgroup are F13(5,10,15), F14(5), F15(5,10,15,20), F16(5,10,15) and F17(5,10), where the numbers in parentheses represent the number of storeys. For the sake of like-for-like comparison, the "exact" critical load by Axis VM X5 (2019) is also taken as the one that belongs to the global buckling of the structures (and the eigenvalues that belong to local buckling are ignored). It is important to realise that in practical structural engineering design it is always the smallest critical load—the first eigenvalue—that is of interest. (With frames F1-F10 above, in all the 90 cases failure occurred through global buckling.)

Frames F21-F27 are braced frames with short diagonals. Table 7.10 and Figure 7.14 show the performance of Equation (2.60). The situation here is similar to the case with frames F11-F17 in that twelve frames—F21(5), F22(5,10,15), F25(5,10), F26(5,10) and F27(5,10,15,20)—lose stability with local buckling. Again, for the sake of like-for-like comparison, the critical loads that correspond to local buckling are ignored and the critical loads that belong to global buckling are taken into consideration and are used as the "exact" critical loads.

Table 7.10 Accuracy of Equation (2.60) for the critical load of braced frames F21-F27.

Braced frames F21-F27	Range of difference [%]	Average absolute difference [%]	Maximum difference [%]
Critical load [Equation (2.60)]	-1 to 20	7.3	20

As with the previous cases with frames F21-F27, the continuum solution tends to give conservative results: as a rule, the critical loads of these frames are underestimated. The reason for this underestimation is linked to the fact that the contribution of the short diagonals is ignored when the global banding stiffness of the model is calculated.

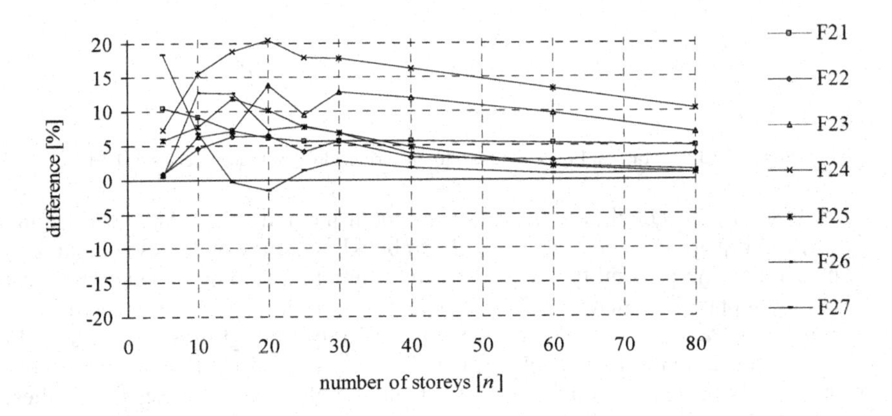

Figure 7.14 Accuracy of Equation (2.60) for the critical load of frames F21-F27.

Finally, the accuracy of Equation (2.63) is investigated. Equation (2.63) is a simpler alternative to Equation (2.60) which can only be used in certain cases. It was derived with the assumption that the effect of the local bending stiffness of the structure on the value of the critical load is small (and negligible). This is clearly the case with braced frames F11-F17. The results of the accuracy analysis are summarized in Table 7.11 and Figure 7.15.

Table 7.11 Accuracy of Equation (2.63) for the critical load of braced frames F11-F17.

Braced frames F11-F17	Range of difference [%]	Average absolute difference [%]	Maximum difference [%]
Critical load [Equation (2.63)]	-14 to 18	2.7	18

The behaviour of braced frames is dominated by their global bending stiffness and shear stiffness. As the data in Table 7.11 and the graphs in Figure 7.15 show, Equation (2.63) can safely be used for the determination of the critical load of braced frames with full-length diagonals.

What was said in connection with local versus global buckling above is also relevant here, i.e., when the comparison between the continuum solution [Equation (2.63)] and the "exact solution" (Axis VM X5, 2019) is made, only global buckling is considered and all the critical loads that belong to failure by local buckling are ignored with the exact solution. The thirteen frames that are affected are F13(5,10,15), F14(5), F15(5,10,15,20), F16(5,10,15) and F17(5,10).

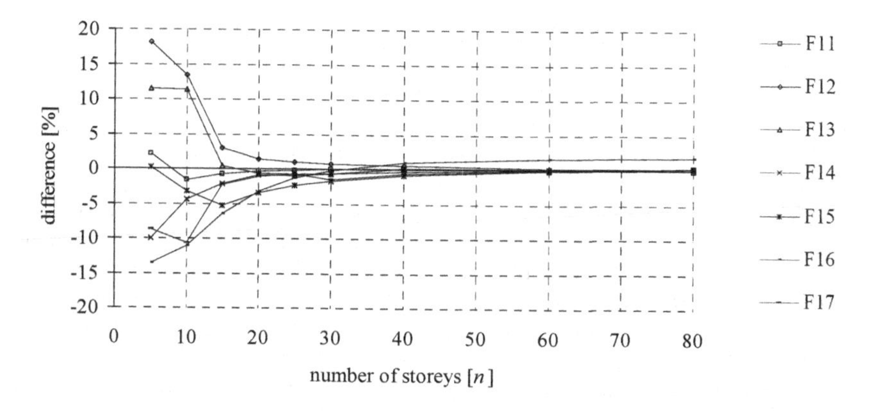

Figure 7.15 Accuracy of Equation (2.63) for the critical load of braced frames F11-F17.

The situation is slightly more complicated with braced frames F21-F27 with short diagonals. They do fulfil the assumption that the effect of their local bending stiffness is small. However, as with the deflection and frequency analyses, the effect of the diagonals is ignored when their global bending stiffness is calculated. This presumably "pushes" the procedure into conservative territory. Indeed, according to Table 7.12 and Figure 7.16, Equation (2.63) gives conservative estimates in practically every case.

Table 7.12 Accuracy of Equation (2.63) for the critical load of braced frames F21-F27.

Braced frames F21-F27	Range of difference [%]	Average absolute difference [%]	Maximum difference [%]
Critical load [Equation (2.63)]	-1 to 44	12	44

As in the previous cases, when a frame loses stability through local buckling, the critical load given by the "exact" (FE) method is ignored and the one that belongs to global buckling is taken into account. The twelve cases that are affected are F21(5), F22(5,10,15), F25(5,10), F26(5,10) and F27(5,10,15,20), where the numbers in parentheses represent the number of storeys.

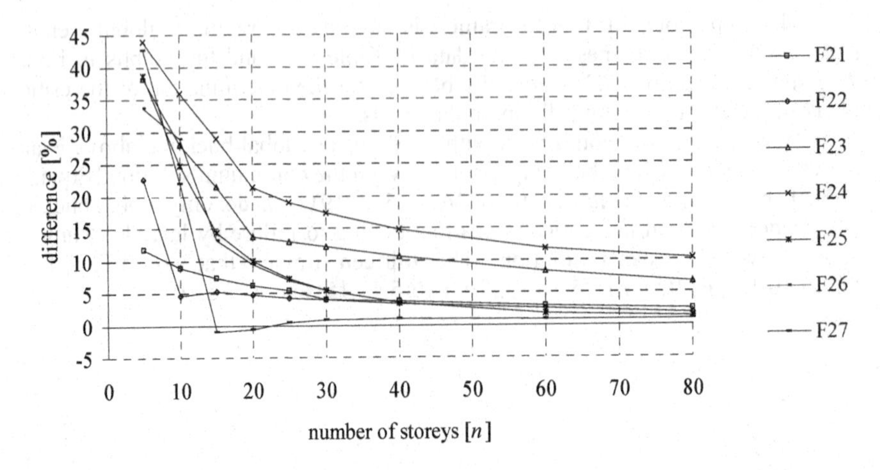

Figure 7.16 Accuracy of Equation (2.63) for the critical load of frames F21-F27.

The assumption that the effect of the local bending stiffness is small is clearly not fulfilled for many rigid sway-frames. However, in order that some guidance can be created regarding the applicability of Equation (2.63), Equation (2.63) will now be applied to frames F1-F10 as well. It is emphasised here that, apart from some special cases, Equation (2.63) is *not* recommended for use with rigid sway-frames. With all the accuracy data related to F1-F10, F11-F17 and F21-F27, a simple criterion will then be given as to *when* Equation (2.63) is recommended to be used. See Equation (2.66) in Section 2.4.1 and later in this section.

Table 7.13 and Figure 7.17 summarize the results for frames F1-F10.

Table 7.13 Performance of Equation (2.63) for the critical load of braced frames F1-F10.

Braced frames F1-F10	Range of difference [%]	Average absolute difference [%]	Maximum difference [%]
Critical load [Equation (2.63)]	-0.4 to 89	21	89

Table 7.13 and Figure 7.17 clearly show that, as a rule, Equation (2.63) is not ideal for use with rigid sway-frames. This is in line with the fact that in the case of some of the frames F1-F10 the local bending stiffness has sizeable contribution to the critical load (e.g. F2, F4, F5, F8 and F10). To have the data regarding the performance of Equation (2.63) is still useful, as, together with the other data in Tables 7.11 and 7.12 and Figures 7.15 and 7.16, they offer enough information for the creation of simple conditions as to the applicability of Equation (2.63).

It seems obvious that if all the frames that have "relatively great" local bending stiffness are left out of the accuracy analysis, the accuracy of Equation (2.63) should improve. The question is how the term "relatively great" can be turned into a condition (or—as it will turn out to be—into a set of two simple conditions).

The two parameters defined by Equation (2.62) can help to create such conditions. The numerical analysis of the 216 pairs of critical load [F1-F10, F11-F17 and F21-F27 at 5, 10, 15, 20, 25, 30, 40, 60 and 80 storeys, determined both by Equation (2.63) and using Axis VM X5 (2019)] and the corresponding 216 difference figures shown in Figures 7.15, 7.16 and 7.17 show that a simple set of two conditions can be established as to the applicability of Equation (2.63).

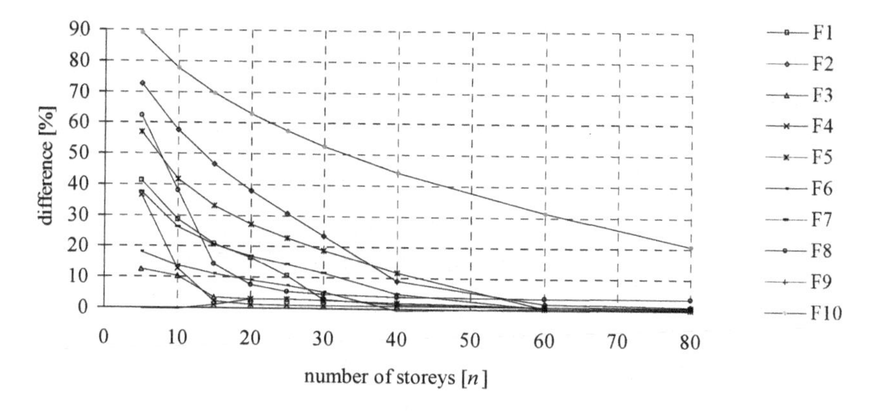

Figure 7.17 Performance of Equation (2.63) for the critical load of frames F1-F10.

According to the data, Equation (2.63) can be used with acceptable accuracy for the determination of the critical load of multi-storey frames (and coupled shear walls) if either of the conditions

$$\frac{I}{I_f} \le 0.0003 \qquad \text{or} \qquad \frac{KH^2}{EI_f r_s} \ge 3 \qquad (2.66)$$

is met where

- I is the local second moment of area of the columns of the frame [Equation (2.58)]
- I_g is the global second moment of area of the columns of the frame [Equation (2.57)]
- I_f is the sum of the local and global second moments of area of the columns of the frame ($I_f = I + I_g$)
- K is the shear stiffness
- H is the height of the frame
- E is the modulus of elasticity
- r_s is the load distribution factor [Equation (2.59), Figure 2.15, Table 5.1]

Going through the 216 frames (F1-F10, F11-F17 and F21-F27) again and applying conditions (2.66), the conditions eliminate 61 out of the 216 frames. In the case of the remaining 155 frames, Equation (2.63) performs well. The difference range is -13% to +21% and the absolute average difference is 4%. The results are summarized in Table 7.14 and in Figure 7.18.

Table 7.14 Accuracy of Equation (2.63) for the critical load of frames that fulfil condition (2.66).

155 frames that fulfil condition (2.66)	Range of difference [%]	Average absolute difference [%]	Maximum difference [%]
Critical load [Equation (2.63)]	-13 to 21	4	21

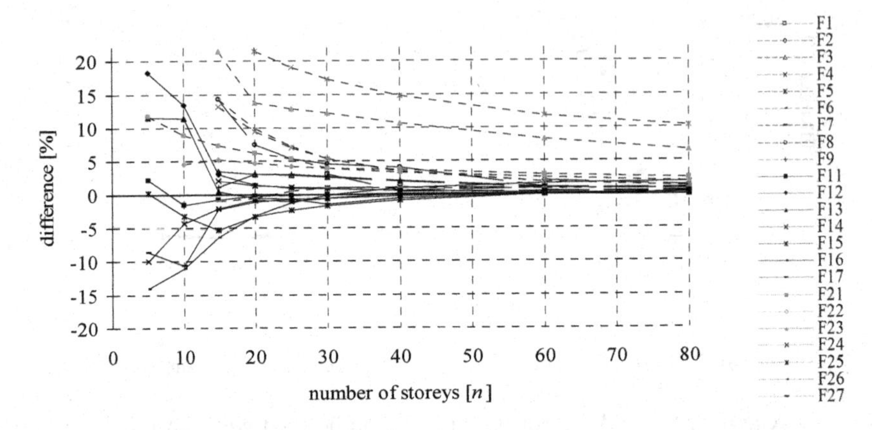

Figure 7.18 Accuracy of Equation (2.63) for the critical load of frames that fulfil condition (2.66).

To conclude this section, the accuracy of Equation (2.63) is investigated when it is applied to frames on pinned supports. Frames on pinned supports do not fulfil one of the regularity assumptions as their shear stiffness is not constant over the height of the structure. Nevertheless, Equation (2.63) can still be useful to produce a rough estimate of the critical load.

Frames F1*, F3*, F4*, F6*, F7* and F9* are used for the accuracy analysis. Apart from their supports, these frames are identical to their equivalents F1, F3, F4, F6, F7 and F9 shown in Figure 7.1. The only difference is in their supports: the columns of frames F1*, F3*, F4*, F6*, F7* and F9* have pinned support on ground floor level. The results of the 54 calculations are summarized in Table 7.15 and in Figure 7.19. (Frames F2*, F5*, F8* and F10*—equivalents of F2, F5, F8 and F10—are not included in the accuracy analysis as their columns are too wide and, because of their size, they would not make it possible to construct realistic pinned supports on ground floor level.)

Table 7.15 Accuracy of Equation (2.63) for the critical load of frames on pinned supports.

Frames F1*, F3*, F4*, F6*, F7*, F9* on pinned supports	Range of difference [%]	Average absolute difference [%]	Maximum difference [%]
Critical load [Equation (2.63)]	-0.4 to 22	5.5	22

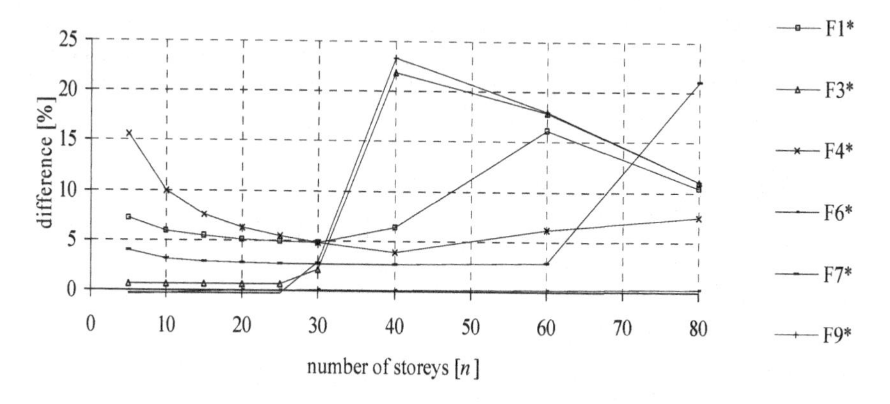

Figure 7.19 Accuracy of Equation (2.63) for the critical load of frames on pinned supports.

Equation (2.63) can also be used for the calculation of the critical load of frames on pinned support which have longer first-storey columns (h^*). The accuracy of Equation (2.63) was also investigated using frames F1*, F3*, F4*, F6*, F7* and F9* but with longer (4-metre long) first-storey columns. The results were very similar to those summarized in Table 7.15 and Figure 7.19.

7.3 STRUCTURAL ANALYSIS OF SYSTEMS OF BRACING UNITS

Frames F1-F10 (Figure 7.1), F11-F17 (Figure 7.2) and F21-F27 (Figure 7.3) will be supplemented with shear walls W1-W7 (Figure 7.4) and a U-core (Figure 7.5) to create 34 systems of bracing units. The accuracy of the relevant equations will be checked with the determination of the maximum deflection, the fundamental frequency and the critical load of bracing systems. These systems are created in such a way that they represent characteristic practical situations and, in some cases, to model extreme cases in order to establish realistic error ranges for the equations in question.

The accuracy analysis is carried out in three areas.

The subject of the first area is the maximum deflection. In structural engineering practice, the determination of the maximum deflection of a building is very important as the designer must ensure that, in line with the relevant Code, it does not exceed the maximum allowable value. The 34 bracing systems are divided into two groups: Section 7.3.1 contains 22 torsion-free systems and in Section 7.3.2 the remaining 12 asymmetric systems are investigated.

When the accuracy of the equations for the fundamental frequency and the critical load is investigated, only the 12 asymmetric systems are used, in Sections 7.3.3 and 7.3.4, respectively.

7.3.1 Maximum deflection of symmetric, torsion-free bracing systems

When the bracing system subjected to horizontal load is symmetric and the resultant of the horizontal load passes through the shear centre of the bracing system, no rotation occurs and the maximum deflection of the shear centre axis is also the maximum deflection of the building. Depending on whether the bracing system consists of frames only, or a mixture of frames and shear walls/cores, the accuracy of the relevant equations will be examined in two groups.

The bracing system contains frames only

Using some of the frames introduced in Section 7.1, eight symmetric bracing systems were created:

S1: F2-F5-F5-F2
S2: F2-F5-F10-F10-F5-F2
S3: F3-F6-F6-F3
S4: F3-F6-F12-F12-F6-F3
S5: F1-F7-F7-F1
S6: F1-F6-F7-F7-F6-F1
S7: F1-F6-F7-F10-F10-F7-F6-F1
S8: F1-F6-F7-F11-F15-F15-F11-F7-F6-F1

When the bracing system only contains frames (and no shear walls/cores) only Method “A” (The simple method) is applicable. The corresponding equation is Equation (3.23).

Using the Axis VM X5 (2019) Finite Element program, the maximum deflection of each system was determined and compared to the result of the analytical solution. The accuracy of Equation (3.23) is demonstrated in Figure 7.20 and in Table 7.16 which show the difference between the analytical and the FE solutions.

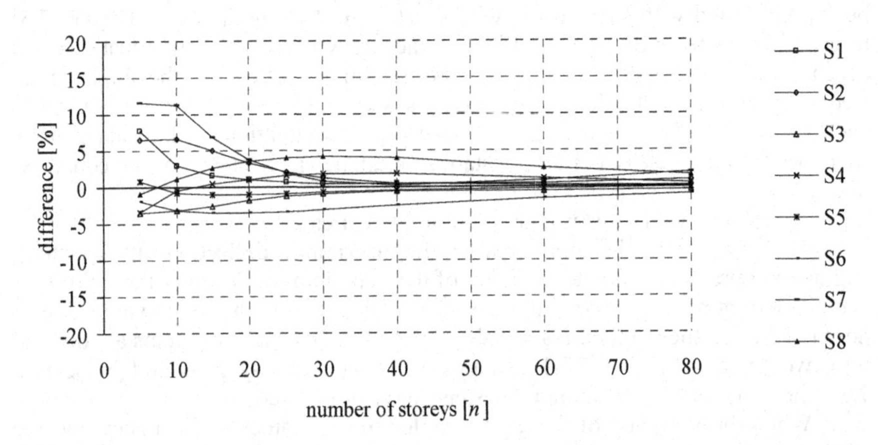

Figure 7.20 Accuracy of Equation (3.23) for the maximum deflection of symmetric torsion-free frame-systems S1-S8.

As in the previous cases, positive difference means that the continuum method [Equation (3.23)] overestimates the maximum deflection.

The performance of Equation (3.23) is good: the range of difference (between the continuum and Finite Element solutions) is -4% to 12%, the average absolute difference is 2.3% and the maximum difference is 12%.

Table 7.16 Accuracy of Equation (3.23) for the maximum deflection of symmetric torsion-free frame-systems S1-S8.

Torsion-free systems S1-S8	Range of difference [%]	Average absolute difference [%]	Maximum difference [%]
Maximum deflection [Equation (3.23)]	-4 to 12	2.3	12

The bracing system consists of frames and shear walls

In most practical cases, the bracing system of a multi-storey building contains a mixture of shear walls/cores and frames. In such cases, both methods introduced in Section 3.2 are available for the analysis. As their names suggest, Method "A" (The simple method) is recommended when a quick solution is needed and Method "B" (The more accurate method) when the level of accuracy is a priority. The accuracy of both methods is investigated in this section.

Using some of the frames and shear walls from Section 7.1, the following fourteen systems are included in the investigation:

S9: F1-W1-W1-F1
S10: F2-W1-W1-F2
S11: F3-W1-W1-F3
S12: F6-W7-W7-F6
S13: F11-W4-W4-F11
S14: F2-F5-W1-W1-F5-F2
S15: F2-F5-F9-W6-W6-F9-F5-F2
S16: F3-F6-W4-W4-F6-F3
S17: F3-F6-F12-W6-W6-F12-F6-F3
S18: F1-F7-W4-W4-F7-F1
S19: F1-F6-F7-W6-W6-F7-F6-F1
S20: F1-F6-F7-F10-W6-W6-F10-F7-F6-F1
S21: F1-F6-F7-F11-F15-W4-W4-F15-F11-F7-F6-F1
S22: F6-F10-W7-W7-F10-F6

Table 7.17 Accuracy of Method "A" for the maximum deflection of wall-frame systems S9-S22.

Torsion-free systems S9-S22	Range of difference [%]	Average absolute difference [%]	Maximum difference [%]
Maximum deflection [Equation (3.23) or (3.25)]	0 to 18	7.1	18

Each system includes one shear wall, but as it is demonstrated in Section 3.2.3 the number of shear walls does not affect the accuracy of the methods, as any number of shear walls can be incorporated into a single shear wall for the analysis.

Equation (3.23) or (3.25) is the relevant equation when Method "A" is used. The results are summarized in Table 7.17 and in Figure 7.21. Even when this simple method is used, accuracy is acceptable, although not spectacular. It is interesting to note that in all the 126 cases the approximate method offers conservative estimates. Method "A" tends to offer conservative estimates.

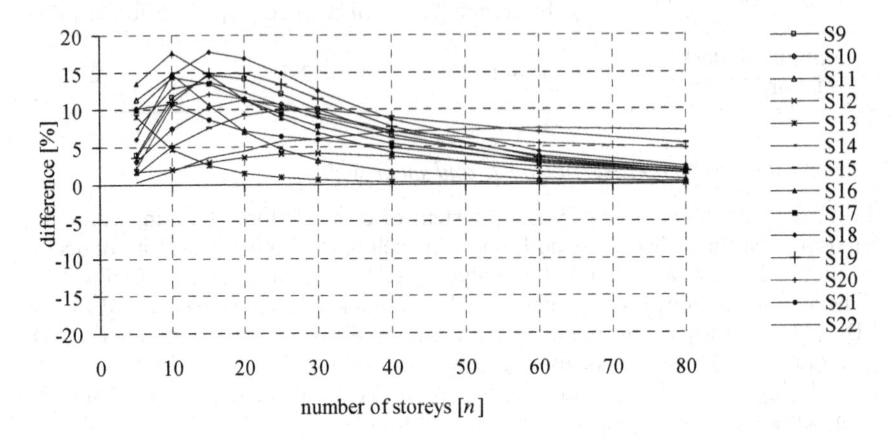

Figure 7.21 Accuracy of Method "A" for the maximum deflection of wall-frame systems S9-S22.

If more accurate results are needed, Method "B" should be used. The trade-off is that the procedure behind Method "B" takes more time to carry out, as it involves more calculations.

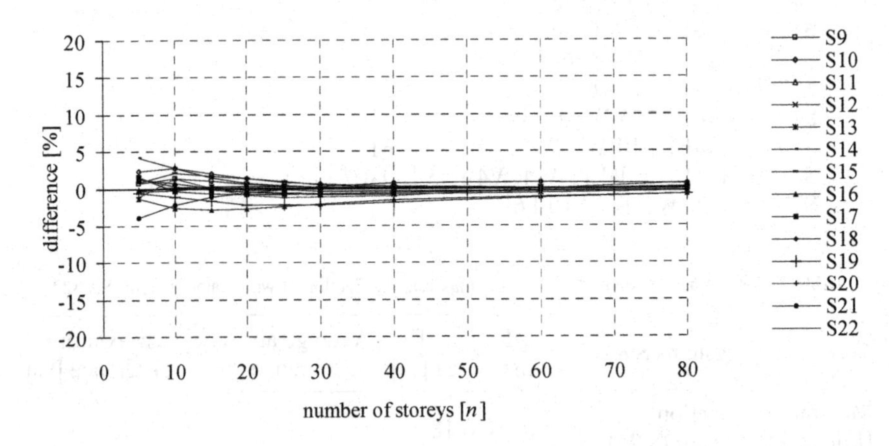

Figure 7.22 Accuracy of Method "B" for the maximum deflection of wall-frame systems S9-S22.

When the calculations are repeated using Method "B" and applying Equation (3.45), accuracy improves considerably. The range of difference is -4% to +4% and the average absolute difference is quite spectacular: less than 1%. The graphs in Figure 7.22 show the performance of the method over the height. The results are summarized in Table 7.18.

Table 7.18 Accuracy of Method "B" for the maximum deflection of wall-frame systems S9-S22.

Torsion-free systems S9-S22	Range of difference [%]	Average absolute difference [%]	Maximum difference [%]
Maximum deflection [Equation (3.45)]	-4 to 4	0.8	4

7.3.2 Maximum deflection of asymmetric bracing systems

Using some of the frames, shear walls and the U-core introduced in Section 7.1, twelve asymmetric bracing systems are created (Figure 7.23):

S23:	F7+F7+F7+W7+W7	Figure 7.23/a
S24:	F3+F3+F5+F7+W4+W4+W4+W4	Figure 7.23/b
S25:	F3+F3+F7+F7+W6+W6	Figure 7.23/c
S26:	F1+F5+F1+U+W6+W6	Figure 7.23/d
S27:	F1+F5+F7+U+W6+W6	Figure 7.23/e
S28:	F1+F5+F7+U+W4+W4+W6	Figure 7.23/f
S29:	F1+F3+F5+F6+W6+W6+W6	Figure 7.23/g
S30:	F1+F1+F2+F3+F6+W6+W6+W6	Figure 7.23/h
S31:	F1+F3+F6+F7+W6+W6+W6	Figure 7.23/i
S32:	F1+F2+F3+F3+F6+F6+F7+W6+W6+W6	Figure 7.23/j
S33:	F1+F2+F3+F4+F5+W5+W6+W6+W6+W7+W7	Figure 7.23/k
S34:	F1+F1+F3+F5+F6+F7+W4+W6+W6+W7	Figure 7.23/l

The twelve systems at storeys 5, 10, 15, 20, 25, 30, 40, 60 and 80 lead to 108 test structures. The situation with these systems is different from the previous case (Systems S1-S22) in that in the case of Systems S23-S34 the centre of the layout and the shear centre of the bracing system do not coincide. It follows that the procedure of the determination of the maximum deflection becomes more complex. As explained in Sections 3.3 and 3.4, two separate calculations have to be carried out: first, the maximum deflection of the shear centre axis is calculated, then the maximum rotation has to be determined, which causes additional deflection. To obtain the maximum deflection of the building, the two deflection components have to be added up. This procedure has a consequence regarding accuracy: as both deflection components result from an approximate formula, the final result tends to be less accurate.

In addition to the accuracy analysis regarding the maximum deflection in this section, Systems S23-S34 will also be used in Sections 7.3.3 and 7.3.4 for checking the accuracy of the relevant formulae of the fundamental frequency and the global critical load of the bracing system. It should be noted that with systems S23-S26 the shear centre of the bracing system lies on axis x, systems S27-S34 are asymmetric in both (x and y) directions. This fact is important, as in the case of the doubly eccentric systems S27-S34 the bracing system develops triple coupling (as opposed to double coupling with systems S23-S26).

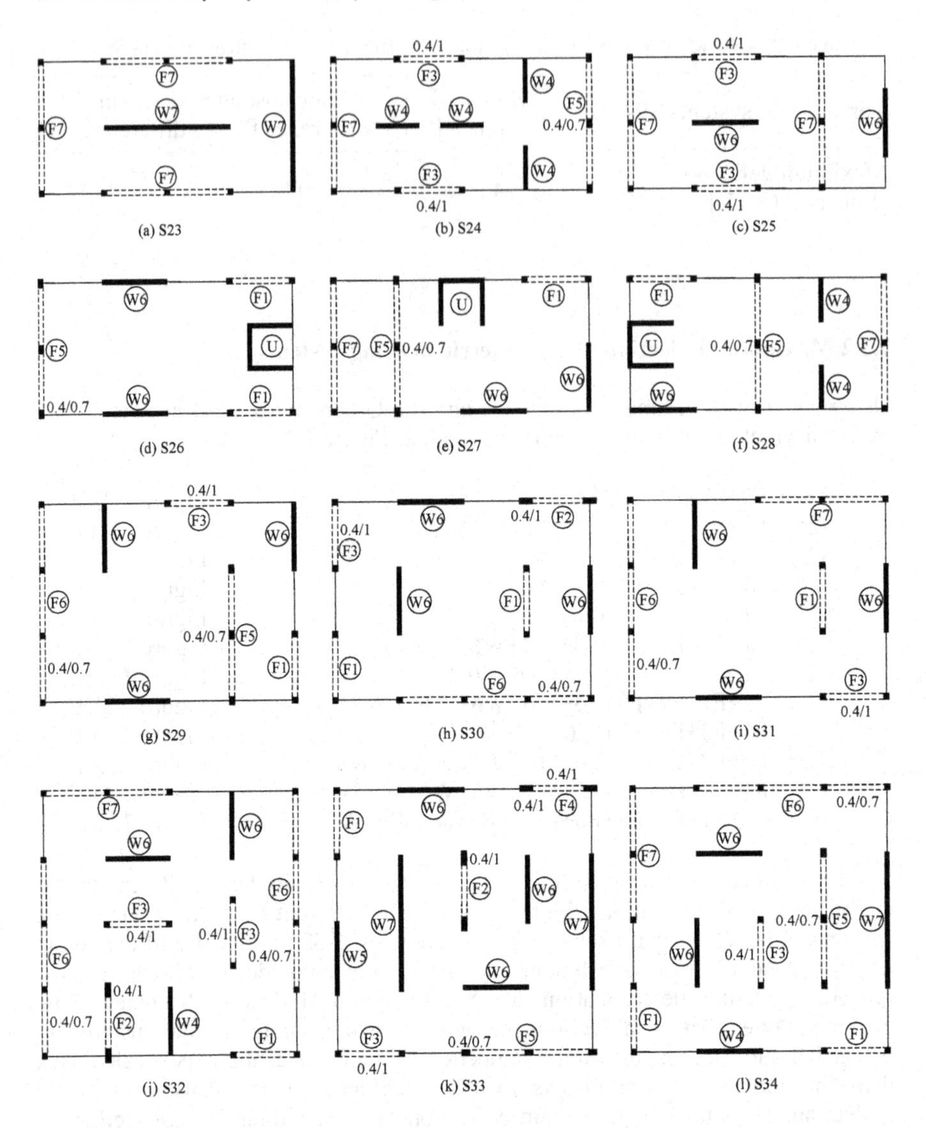

Figure 7.23 Asymmetric bracing systems S23-S34 for the three-dimensional accuracy analysis.

When the maximum deflection of an asymmetric system is determined, Equation (3.66) is used. To be precise, Equation (3.66) amalgamates two equations: Equation (3.45) is responsible for the maximum deflection of the shear centre axis and Equation (3.59) gives the maximum rotation of the building that causes additional deflection.

The accuracy of Equation (3.66) is demonstrated in Table 7.19 and in Figure 7.24. As in the previous investigations, positive difference indicates that the approximate solution overestimates the maximum deflection.

Table 7.19 Accuracy of Equation (3.66) for the maximum deflection of asymmetric systems S23-S34.

Asymmetric systems S23-S34	Range of difference [%]	Average absolute difference [%]	Maximum difference [%]
Maximum deflection [Equation (3.66)]	-8 to 16	5.2	16

At 5.2%, the average absolute difference of the 108 calculations offers useful indication what to expect when the approximate method is used for the determination of the maximum deflection. In the 108 cases, the maximum difference is 16% with a difference range of -8% to +16%.

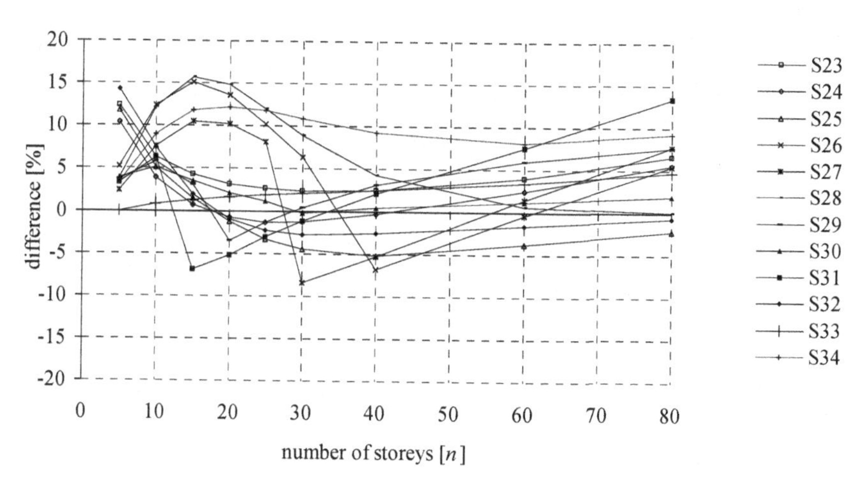

Figure 7.24 Accuracy of Equation (3.66) for the maximum deflection of asymmetric systems S23-S34.

7.3.3 Fundamental frequency

The same 12 systems (S23-S34) at nine heights are used to check the accuracy of the continuum solution developed in Chapter 4 for the determination of the fundamental frequency. Equation (4.33) is the "final" equation which eventually furnishes the value of the fundamental frequency. However, Equation (4.33) uses the results of two or three previous calculations, depending on the coupling of the basic frequencies, which use Equations (4.18) and (4.26).

The results are summarized in Table 7.20 where "difference" means the difference between the "exact" (FE) solution and the continuum solution, related to the "exact" solution. Positive difference means the underestimation of the fundamental frequency. The performance of the continuum solution is quite spectacular: the difference range is -9% to +6% and the average absolute difference is just over 3%. Figure 7.25 shows how accuracy changes as a function of the height of the building.

Table 7.20 Accuracy of Equation (4.33) for the fundamental frequency of asymmetric systems S23-S34.

Asymmetric systems S23-S34	Range of difference [%]	Average absolute difference [%]	Maximum difference [%]
Fundamental frequency [Equation (4.33)]	-9 to 6	3.3	9

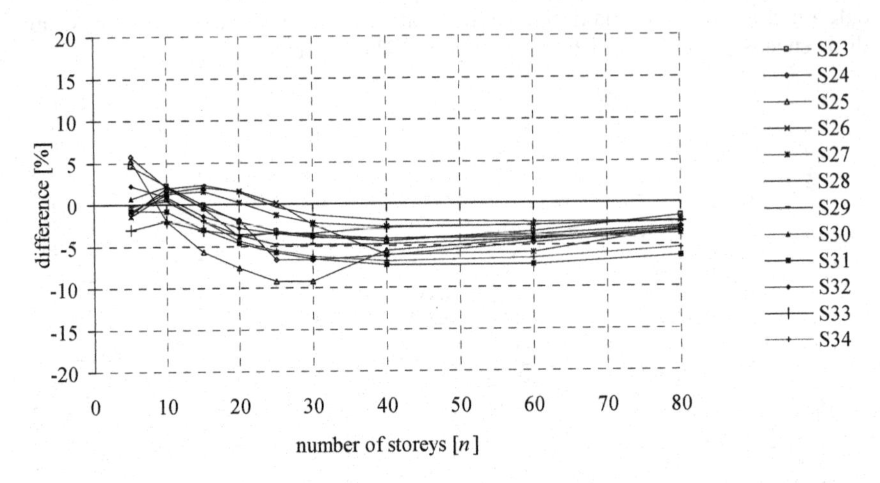

Figure 7.25 Accuracy of Equation (4.33) for the fundamental frequency of asymmetric systems S23-S34.

7.3.4 Critical load

Finally, the same 12 systems (S23-S34) are used to check the accuracy of the continuum solution derived in Chapter 5 for the determination of the global critical load. The method is based on several equations: the basic critical loads (sway buckling in the two principle directions and pure torsional buckling) have to be determined first using Equations (5.18) and (5.34). Equation (5.41) is the "final" equation which, by taking into account the coupling of the basic modes, eventually furnishes the value of the critical load.

The results are summarized in Table 7.21 where "difference" means the difference between the "exact" (FE) solution and the continuum solution, related to the "exact" solution. Positive difference means the underestimation of the critical load. Figure 7.26 shows how accuracy changes as a function of the height of the building. For the sake of simplicity, when accuracy data are given in Table 7.21 and in Figure 7.26, Equation (5.41) is named as representative of the continuum solution.

According to Table 7.21 and Figure 7.26, the performance of the continuum solution is acceptable: the difference range is -17% to +11% and the average absolute difference is 5.5%. The performance is less spectacular than with the deflection and frequency analyses; however, the method is still highly recommended for practical use. It follows from the fact that in practical structural engineering investigations, the situation and expectations with the stability analysis are very different from those with the deflection and frequency analyses.

Table 7.21 Accuracy of Equation (5.41) for the critical load of asymmetric systems S23-S34.

Asymmetric systems S23-S34	Range of difference [%]	Average absolute difference [%]	Maximum difference [%]
Critical load [Equation (5.41)]	-17 to 11	5.5	17

With deflections and frequencies even a couple of percentage points may be important as far as accuracy is concerned. With stability, however, the structural engineer has to ensure a much greater safety margin. When the global critical load ratio is in the region of 4 to 10—see Chapter 6 for more details—inaccuracy of less than 20% practically does not influence a judgement concerning the suitability of a building regarding stability.

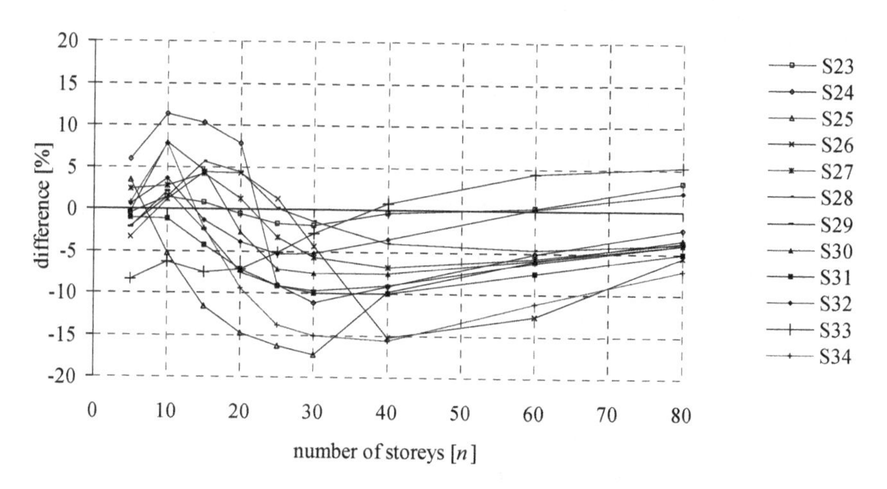

Figure 7.26 Accuracy of Equation (5.41) for the critical load of asymmetric systems S23-S34.

7.4 ACCURACY WITH THE NINETEEN WORKED EXAMPLES

This chapter's comprehensive accuracy analysis concludes with some information regarding the accuracy of the approximate solutions used in the case of the nineteen worked examples presented in the previous chapters/sections. These are the worked examples whose MathCad and Excel downloadable worksheets are listed in the Appendix. Table 7.22 shows how the approximate solutions perform compared to the "exact" solutions.

In each case the maximum deflection, the fundamental frequency and the critical load are determined both by the relevant analytical method and using the Finite Element based computer program Axis VM X5 (2019). Once again, the information in Table 7.22 indicates that the accuracy of the approximate solutions is well within the requirements of practical application.

Table 7.22 Difference between the approximate and "exact" (FE) solutions.

Which Section?	Structure	Difference [%]		
		Maximum deflection	Fundamental frequency	Critical load
2.1.5	10-storey frame F7	2	3	11
2.2.3	25-storey frame F6	5	1	6
2.3.2	25-storey frame F5	0	1	1
2.4.6	15-storey frame F12	3	4	3
2.5.2	30-storey c. shear walls CSW3	-5	6	4
3.7.1	28-storey symmetric building U-F6-F5-F5-F7-U	0	0	-1
3.7.2	28-storey asymmetric building W7-F7-F7-F7-F7-W4-W4-U	7	1	1
4.4.1	25-storey symmetric building F6-F6-W6-W6-W7-W7	0	-5	-9
4.4.2	20-storey asymmetric building F5-F5-W5-W5-F7-W7	7	-1	2
5.6.1	20-storey monosym. building F1-F1-F5-U-W6-W6	15	1	5
5.6.2	15-storey asymmetric building F6-F5-F1-F3-W6-W6-W6	3	-1	-1
6.3.2	8-storey building F6-F6-W6-W6: Case 1	-1	0	1
6.3.3	8-storey building F6-F6-W6-W6: Case 2	3	4	13
6.3.4	8-storey building F6-F6-W6-W6: Case 3	-5	2	5
6.4.1	5-storey building: Layout A	2	-1	0
6.4.2	5-storey building: Layout B	-1	-2	-4
6.4.3	5-storey building: Layout C	0	7	0
6.4.4	5-storey building: Layout D	3	2	5
6.5	10-storey asymmetric building U-W-F7-F1-W	12	1	5

Appendix:

List of worksheets

The practical use of the analytical methods developed in Chapters 2, 3, 4, 5 and 6 is first demonstrated in nineteen worked examples at the end of the relevant sections/chapters. These worked examples illustrate how the corresponding hand-calculations are carried out. To further simplify the computational work, special pieces of software can be used to create worksheets. Three sets of worksheets accompany the book. Mathcad Plus 6.0 (MathSoft, 1995) {filename.mcd}, Mathcad 14.0 (PTC, 2007) {filename.xmcd} and Microsoft Office Excel (2003 Service Pack 3) {filename.xls} were used for the preparation of the worksheets that are available for download at www.crcpress.com/9780367350253.

The worksheets are best used together with the worked examples in the book where figures and detailed descriptions offer explanation of the practical situation. Alternatively, the worksheets can also be considered stand-alone calculation tools and can also be used as templates to handle similar structural engineering problems. The numbers in the title of the worksheets below refer to the relevant section where the worked examples are to be found.

215DefF7 (Section 2.1.5: Deflection of frame F7)
The worksheet calculates the maximum deflection of a two-bay, ten-storey sway-frame under uniformly distributed horizontal load. In modifying the input data (modulus of elasticity, number/size of bays, number of storeys, storey height, size of the cross-sections of the columns/beams, intensity of the horizontal load), the maximum deflection of any multi-storey sway-frame on fixed supports can be determined at once. The MathCad worksheet also produces the deflection shape, the shear flow and the normal force in the columns of the structure.

223FreF6 (Section 2.2.3: Fundamental frequency of frame F6)
The worksheet calculates the fundamental frequency of a twenty-five storey three-bay sway-frame subjected to uniformly distributed mass on floor levels. In modifying the input data (modulus of elasticity, number/size of bays, number of storeys, storey height, size of the cross-sections of the columns/beams, magnitude of mass), the fundamental frequency of any multi-storey frame on fixed supports can be determined at once.

232StaF5 (Section 2.3.2: Critical load of frame F5)
The worksheet calculates the global critical load of a twenty-five storey, two-bay

sway-frame subjected to uniformly distributed vertical load on floor levels. In modifying the input data (modulus of elasticity, number/size of bays, number of storeys, storey height, size of the cross-sections of the columns/beams), the global critical load of any multi-storey frame on fixed supports can be determined at once.

246StF12 (Section 2.4.6: Critical load of steel frame F12 with single cross-bracing)
The worksheet calculates the global critical load of a fifteen-storey two-bay frame with single cross-bracing subjected to uniformly distributed vertical load on floor levels. In modifying the input data (modulus of elasticity, number/size of bays, number of storeys, storey height, size of the cross-sections of the columns/beams/diagonals, type of cross-bracing), the global critical load of any multi-storey frame with cross-bracing can be determined at once.

252_CSW3 (Section 2.5.2: Maximum deflection, fundamental frequency and critical load of coupled shear walls CSW3)
The worksheet calculates the maximum deflection, the fundamental frequency and the global critical load of the thirty-storey, three-bay coupled shear walls. In modifying the input data (modulus of elasticity, number/size of bays, number of storeys, storey height, size of the cross-sections of the wall-sections/beams and magnitude of load), the maximum deflection, the fundamental frequency and the global critical load of any multi-storey, multi-bay coupled shear walls can be determined at once. When modifying the input data, attention should be paid to the calculation of the shear stiffness (K_b and K_c), as the structure in the worked example has beams and wall sections of different size.

371Build (Section 3.7.1: Maximum deflection, fundamental frequency and critical load of twenty-eight storey symmetric building 2F5+2F7+2U)
The worksheet calculates the maximum deflection of a twenty-eight storey symmetric building braced by four three-bay frames and two U-cores, under uniformly distributed horizontal load. In modifying the input data (modulus of elasticity, number/size of bays, number of storeys, storey height, size of the cross-sections of the columns/beams, size of shear wall, size of U-core, size of building and magnitude of load), and adding/removing bracing units in a symmetric manner, the maximum deflection of any symmetric system of frames, coupled shear walls and shear walls/cores can be determined at once. In addition to the maximum deflection, the fundamental frequency and the critical load of the building are also determined.

372Build (Section 3.7.2: Maximum deflection, fundamental frequency and critical load of twenty-eight storey building with one axis of symmetry, braced by 4F7+2W4+W7+U)
The worksheet calculates the deflection of and the rotation around the shear centre axis, then the maximum deflection of a twenty-eight storey asymmetric building braced by four two-bay frames, three shear walls and one U-core, under uniformly distributed horizontal load. In modifying the input data (modulus of elasticity, number/size of bays, number of storeys, storey height, size of the cross-sections of the columns/beams, size of shear wall, size of U-core, size of layout, location of bracing units and magnitude of load), and adding/removing bracing units in a

monosymmetric manner, the maximum deflection of any similar system of frames, coupled shear walls, shear walls and cores can be determined at once. In addition to the maximum deflection, the fundamental frequency and the critical load of the building are also calculated.

441Build (Section 4.4.1: Fundamental frequency, maximum deflection and critical load of doubly symmetric building braced by 2F6+2W6+2W7)
The worksheet calculates the fundamental frequency of a twenty-five storey doubly symmetric building braced by two three-bay frames and four shear walls. In modifying the input data (modulus of elasticity, number/size of bays, number of storeys, storey height, size of the cross-sections of the columns/beams, size of shear walls, size of layout, magnitude of mass), and adding/removing bracing units in a doubly symmetric arrangement, the fundamental frequency of any doubly symmetric building can be determined at once. In addition to the fundamental frequency, the maximum deflection and the critical load of the building are also calculated.

442Build (Section 4.4.2: Fundamental frequency, maximum deflection and critical load of twenty-storey building with one axis of symmetry, braced by 2F5+F7+2W5+W7)
The worksheet calculates the fundamental frequency of the building, braced by three two-bay frames and three shear walls, developing lateral-torsional vibration. In modifying the input data (modulus of elasticity, number/size of bays, number of storeys, storey height, size of the cross-sections of the columns/beams, size of shear walls, size of layout, location of bracing units and magnitude of mass), and adding/removing bracing units in a monosymmetric manner, the fundamental frequency of any similar building can be determined at once. In addition to the fundamental frequency, the maximum deflection and the critical load of the building are also calculated.

561Build (Section 5.6.1: Critical load, maximum deflection and fundamental frequency of twenty-storey monosymmetric building braced by 2F1+F5+2W6+U)
The worksheet calculates the global critical load and the global critical load ratio of the building braced by two one-bay frames, one two-bay frame, two shear walls and a U-core. In modifying the input data (modulus of elasticity, number/size of bays, number of storeys, storey height, size of the cross-sections of the columns/beams, size of shear walls, size and type of core, and size of layout), and adding/removing bracing units in a monosymmetric arrangement, the critical load of any similar monosymmetric building can be determined at once. In addition to the critical load, the maximum deflection and the fundamental frequency of the building are also calculated.

562Build (Section 5.6.2: Critical load, maximum deflection and fundamental frequency of fifteen-storey asymmetric building braced by F1+F3+F5+F6+3W6)
The worksheet calculates the global critical load and the global critical load ratio of the asymmetric building braced by a three-bay frame, a two-bay frame, two one-bay frames and three shear walls, developing three-dimensional sway-torsional buckling. In modifying the input data (modulus of elasticity, number/size of bays,

number of storeys, storey height, size of the cross-sections of the columns/beams, size of shear walls and size of layout), and adding/removing bracing units, the global critical load and the global critical load ratio of any building can be determined at once. In addition to the critical load, the maximum deflection and the fundamental frequency of the building are also calculated.

632Case1, 633Case2 and 634Case3 (Sections 6.3.2, 6.3.3 and 6.3.4: Eight-storey building braced by 2F6+2W6)
The three worksheets carry out a comprehensive global structural analysis of the same eight-storey building, braced by two frames and two shear walls. In the three cases the bracing system consists of the same bracing units (the two three-bay frames and the two shear walls) but their arrangement is different. The global critical load, the fundamental frequency and the maximum rotation and deflection of the building are calculated. The global critical load ratio is used as a performance indicator to characterize the overall behaviour of the building. The worksheets can be used as templates for the global structural analysis of similar buildings.

641Kol_A, 642Kol_B, 643Kol_C and 644Kol_D (Sections 6.4.1, 6.4.2, 6.4.3 and 6.4.4: Kollár's five-storey building)
The worksheets carry out four comprehensive global structural analyses. The five-storey building is the same, the bracing system—a single U-core—is nearly the same: in two cases it is open and in the other two cases it is partially closed. The other difference is the location of the core. The maximum rotation and deflection, the fundamental frequency, global critical load and the global critical load ratio are determined. The global critical load ratio is used as a performance indicator to characterize the overall behaviour and structural suitability of the building. Other similar buildings braced by a single core can be investigated by replacing the U-core and/or changing the geometrical and stiffness characteristics.

65_Build (Section 6.5: Global structural analysis of ten-storey asymmetric building braced by F1+F7+2W+U)
The worksheet presents a comprehensive, global, three-dimensional structural analysis of the building. It calculates the location of the shear centre, the global critical load, the global critical load ratio, the fundamental frequency and the maximum rotation and deflection of the building. In modifying the input data (modulus of elasticity, number/size of bays, number of storeys, storey height, size of the cross-sections of the columns/beams, size of U-core and shear walls, size of layout, location of bracing units, intensity of vertical load on floor levels, intensity of horizontal load, magnitude of mass), and adding/removing bracing units, the comprehensive analysis can be repeated for other multi-storey buildings in minutes.

References

Achyutha, H., Injaganeri, S.S., Satyanarayanan, S. and Krishnamoorthy, C.S., 1994, Inelastic behaviour of brick infilled reinforced concrete frames. *Journal of Structural Engineering*, **21**, No. 2, pp. 107–115.

Allen, H.G., 1969, *Analysis and design of structural sandwich panels,* (Oxford: Pergamon Press).

Axis VM X5, 2019, Structural Analysis and Design Software, InterCAD, www.axisvm.com.

Barkan, D.D., 1962, *Dynamics of bases and foundations*, (London: McGraw-Hill).

Beck, H., 1956, Ein neues Berechnungsverfahren für gegliederte Scheiben, dargestellt am Beispiel des Vierendelträgers. *Der Bauingenieur,* **31**, pp. 436–443.

Brohn, D.M., 1996, Avoiding CAD: The Computer Aided Disaster. Symposium: Safer computing. *The Institution of Structural Engineers*, 30 January 1996.

Chitty, L., 1947, On the cantilever composed of a number of parallel beams interconnected by cross bars. *Philosophical Magazine*, London. Ser. 7, Vol. XXXVIII, pp. 685–699.

Chitty, L. and Wan, W.Y., 1948, Tall building structures under wind load. *Proceedings of the 7th International Congress for Applied Mechanics*. London, 22 January 1948, pp. 254–268.

Chwalla, E., 1959, Die neuen Hilfstafeln zur Berechnung von Spannungsproblemen der Theorie zweiter Ordnung und von Knickproblemen. *Bauingenieur*, **34**, (4, 6 and 8), p128, p240 and p299.

Coull, A., 1975, Free vibrations of regular symmetrical shear wall buildings. *Building Science*, **10**, pp. 127–133.

Coull, A., 1990, Analysis for structural design. In *Tall Buildings: 2000 and Beyond.* Council on Tall Buildings and Urban Habitat, pp. 1031–1047.

Coull, A. and Wahab, A.F.A., 1993, Lateral load distribution in asymmetrical tall building structures. *Journal of Structural Engineering, ASCE*, **119**, pp. 1032–1047.

Council on Tall Buildings, 1978, *Planning and Design of Tall Buildings*, a Monograph in 5 volumes, (New York: ASCE).

Csonka, P., 1950, *Eljárás elmozduló sarkú derékszögű keretek számítására (Procedure for rectangular sway-frames*). (Budapest: Tudományos Könyvkiadó Vállalat).

Danay, A., Glück, J. and Gellert, M., 1975, A generalized continuum method for dynamic analysis of asymmetric tall buildings. *Earthquake Engineering and Structural Dynamics*, **4**, pp. 179–203.

Despeyroux, J., 1972, Analyse statique et dynamique des contraventments par consoles. *Annales de l'Institut Technique du Bâtiment et des Travaux Publics*, No. 290.

Dowrick, D.J., 1976, Overall stability of structures. *The Structural Engineer*, **54**, pp. 399–409.
Ellis, R.B., 1980, An assessment of the accuracy of predicting the fundamental natural frequencies of buildings. *Proceedings of The Institution of Civil Engineers*, **69**, Part 2, September, pp. 763–776.
Ellis, R.B., 1986, The significance of dynamic soil-structure interaction in tall buildings. *Proceedings of The Institution of Civil Engineers*, **81**, Part 2, pp. 221–242.
EN 1992 (Eurocode 2), 2004, *Design of concrete structures*. (European Committee of Standardization).
EN 1993 (Eurocode 3), 2004, *Design of steel structures*. (European Committee of Standardization).
Fintel, M. (editor), 1974, *Handbook of concrete engineering*, (London: Van Nostrand Reinhold).
Gluck, J. and Gellert, M., 1971, On the stability of elastically supported cantilever with continuous lateral restraint. *International Journal of Mechanical Sciences*, **13**, pp. 887–891.
Glück, J., Reinhorn, A. and Rutenberg, A., 1979, Dynamic torsional coupling in tall building structures. *Proceedings of The Institution of Civil Engineers*, **67**, Part 2, pp. 411–424.
Goldberg, J.E., 1973, Approximate methods for stability and frequency analysis of tall buildings. *Regional Conference on Planning and Design of Tall Buildings*, Madrid, pp. 123–146.
Goschy, B., 1970. Räumliche Stabilität von Groβtafelbauten (Spatial stability of system buildings). *Die Bautechnik*, **47**, pp. 416–425.
Halldorsson, O.P. and Wang, C.K., 1968, Stability analysis of frameworks by matrix methods. *Journal of the Structural Division, ASCE,* **94**, ST7, p. 1745.
Hegedűs, I. and Kollár, L.P., 1984, Buckling of sandwich columns with thick faces subjecting to axial loads of arbitrary distribution. *Acta Technica Scientiarum Hungaricae*, **97**, pp. 123–132.
Hegedűs, I. and Kollár, L.P., 1999, Application of the sandwich theory in the stability analysis of structures. In *Structural stability in engineering practice*, edited by Kollár, L., (London: E & FN Spon), pp. 187–241.
Hoenderkamp, J.C.D., 1995, Approximate deflection analysis of non-symmetric high-rise structures. In *Habitat and the high-rise – Tradition and innovation. Proceedings of the Fifth World Congress*, edited by Lynn S. Beedle, (Bethlehem: Council on Tall Buildings and Urban Habitat, Lehigh University), pp. 1185–1209.
Hoenderkamp, J.C.D. and Stafford Smith, B., 1984, Simplified analysis of symmetric tall building structures subject to lateral loads. *Proceedings of the 3rd International Conference on Tall Buildings*, Hong Kong and Gaungzhou, pp. 28–36.
Howson, P., 2006, Global analysis: Back to the future. *The Structural Engineer*, **84**, (3), pp. 18–21.
Howson, W.P. and Rafezy, B., 2002, Torsional analysis of asymmetric proportional building structures using substitute plane frames. *Proceedings of the 3rd International Conference on Advances in Steel Structures,* Volume II, (Hong Kong: Elsevier), pp. 1177–1184.

Irwin, A.W., 1984, *Design of shear wall buildings*, Report 102, (London: Construction Industry Research and Information Association).

Jeary, A.P. and Ellis, B.R., 1981, The accuracy of mathematical models of structural dynamics. *International Seminar on Dynamic Modelling*, (Watford, UK: Building Research Establishment/The Institution of Civil Engineers), PD112/81.

Kollár, L., 1977, Épületek merevítése elcsavarodó kihajlás ellen (Bracing of buildings against torsional buckling), *Magyar Építőipar*, pp. 150–154.

Kollár, L. (editor), 1999, *Structural stability in engineering practice*, (London: E & FN Spon).

Kollár, L.P., 1986, Buckling analysis of coupled shear walls by the multi-layer sandwich model. *Acta Technica Scientiarum Hungaricae*, **99**, pp. 317–332.

Kollár, L.P., 1992, Calculation of plane frames braced by shear walls for seismic load. *Acta Technica Scientiarum Hungaricae*, **104**, (1–3), pp. 187–209.

Kollbrunner, C.F. and Basler, K., 1969, *Torsion in structures*, (Berlin, New York: Springer-Verlag).

MacLeod, I.A., 1990, *Analytical modelling of structural systems*, (London: Ellis Horwood).

MacLeod, I.A., 2005, *Modern structural analysis. Modelling process and guidance*, (London: Thomas Telford).

MacLeod, I.A. and Marshall, J., 1983, Elastic stability of building structures. *Proceedings of 'The Michael R. Horne Conference: Instability and plastic collapse of steel structures'*, edited by Morris, L.J. (London: Granada), pp. 75–85.

MacLeod, I.A. and Zalka, K.A., 1996, The global critical load ratio approach to stability of building structures. *The Structural Engineer*, **74**, (15), pp. 249–254.

Madan, A., Reinhorn, A.M., Mander, J.B. and Valles, R.E., 1997, Modeling of masonry infill panels for structural analysis. *Journal of Structural Engineering, ASCE,* **123**, (10), pp. 1295–1302.

Martin, L. and Purkiss, J., 2008, *Structural design of steelwork to EN 1993 and EN 1994*, (Oxford: Butterworth-Heinemann).

Mainstone, R.J. and Weeks, G.A., 1972, *The influence of a bounding frame on the racking stiffness and strengths of brick walls*, (Watford: Building Research Station, Current Paper 3/72).

MathSoft, 1995, *Mathcad Plus 6.0. User's Guide,* (Cambridge, USA: MathSoft).

Nadjai, A. and Johnson, D., 1998, Torsion in tall buildings by a discrete force method. *Structural Design of Tall Buildings*, **7**, (3), pp. 217–231.

Ng, S.C. and Kuang, J.S., 2000, Triply coupled vibration of asymmetric wall-frame structures. *Journal of Structural Engineering, ASCE*, **126**, No. 8, pp. 982–987.

Pearce, D.J. and Matthews, D.D., 1971, *Shear walls. An appraisal of their design in box-frame structures*, (London: Property Services Agency, Department of the Environment).

Plantema, F.J., 1961, *Sandwich construction. The bending and buckling of sandwich beams, plates and shells*, (London: McGraw-Hill).

Polyakov, S.V., 1956, *Kamennaya Kladka v Karkasnykh zdaniyakh. Issledovanie prochnosti i zhestkosti kamennogo zapolneniya,* (Moscow). (*Masonry in framed buildings. An investigation into the strength and stiffness of masonry infilling*). English translation by G. L. Cairns, National Lending Library for Science and

Technology, Boston, Yorkshire, England, 1963.

Potzta, G. and Kollár, L.P., 2003, Analysis of building structures by replacement sandwich beams. *Solids and Structures*, **40**, pp. 535–553.

PROSEC, 1994, Section Properties, version 4.05 or higher, (London: PROKON Software Consultants Ltd).

PTC, 2007, *Mathcad 14.0*, (Needham, USA: Parametric Technology Corporation)

Riddington, J.R. and Stafford Smith, B., 1977, Analysis of infilled frames subject to racking with design recommendations. *The Structural Engineer*, **55**, pp. 263–268.

Rosman, R., 1960, Beitrag zur statischen Berechnung waagerecht belasteter Querwände bei Hochbauten (On the structural analysis of tall cross-wall buildings under horizontal load). *Der Bauingenieur*, **4**, pp. 133–141.

Rosman, R., 1973, Dynamics and stability of shear wall building structures. *Proceedings of The Institution of Civil Engineers*, Part 2, **55**, pp. 411–423.

Rosman, R., 1974, Stability and dynamics of shear-wall frame structures. *Building Science*, **9**, pp. 55–63.

Rosman, R., 1981, Buckling and vibrations of spatial building structures. *Engineering Structures*, **3**, (4), pp. 194–202.

Rutenberg, A., 1975, Approximate natural frequencies of coupled shear walls. *Earthquake Engineering and Structural Dynamics*, **4**, pp. 95–100.

Schueller, W., 1977, *High-rise building structures*, (New York: John Wiley & Sons).

Schueller, W., 1990, *The vertical building structure*, (New York: Van Nostrand Reinhold).

Smart, R.A., 1997, Computers in the design office: boon or bane. *The Structural Engineer*, **75**, (3), p. 52.

Southwell, R.V., 1922, On the free transverse vibration of a uniform circular disc clamped at its centre; and on the effects of rotation. *Proceedings of the Royal Society of London. Ser. A*, **101**, pp. 133–153.

Stafford Smith, B., 1966, The composite behaviour of infilled frames. In *Proceedings of a Symposium on Tall Buildings with particular reference to shear wall structures*, edited by Coull, A. and Stafford Smith, B., University of Southampton, Department of Civil Engineering. (Oxford: Pergamon Press), pp. 481–492.

Stafford Smith, B. and Carter, C., 1969, A method of analysis for infilled frames. *Proceedings of the Institution of Civil Engineers*, **44**, pp. 31–48.

Stafford Smith, B. and Coull, A., 1991, *Tall building structures. Analysis and design*, (New York: John Wiley & Sons), pp. 213–282 and 372–387.

Stafford Smith, B., Kuster, M. and Hoenderkamp, J.C.D., 1981, Generalized approach to the deflection analysis of braced frame, rigid frame and coupled shear wall structures. *Canadian Journal of Civil Engineers*, **8**, (2), pp. 230–240.

Stevens, L.K., 1983, The practical significance of the elastic critical load in the design of frames. *Proceedings of 'The Michael R. Horne Conference: Instability and plastic collapse of steel structures'*, edited by Morris, L.J., (London: Granada), pp. 36–46.

Tarnai, T., 1999, Summation theorems concerning critical loads of bifurcation. In *Structural stability in engineering practice*, edited by Kollár, L., (London: E & FN Spon), pp. 23–58.

Timoshenko, S., 1928, *Vibration problems in engineering*, (London: D. Van Nostrand Company, Inc).

Vértes, G., 1985, *Structural dynamics*, (New York: Elsevier).

Vlasov, V.Z., 1961, *Tonkostennye uprugie sterzhni* (*Thin-walled elastic beams*), (Jerusalem: Israeli Program for Scientific Translations).

Zalka, K.A., 1979, Buckling of a cantilever subjected to distributed normal loads, taking the shearing deformation into account. *Acta Technica Hung.* **89**, 3-4, 497-508.

Zalka, K.A., 1994, *Dynamic analysis of core supported buildings*, N127/94, (Watford, UK: Building Research Establishment).

Zalka, K.A., 2000, *Global structural analysis of buildings*, (London: E & FN Spon).

Zalka, K.A., 2001, A simplified method for the calculation of the natural frequencies of wall-frame buildings. *Engineering Structures*, **23**, No. 12, pp. 1544–1555.

Zalka, K.A., 2002, Buckling analysis of buildings braced by frameworks, shear walls and cores. *The Structural Design of Tall Buildings*, **11**, No. 3, 197–219.

Zalka, K.A., 2009, A simple method for the deflection analysis of tall wall-frame building structures under horizontal load. *The Structural Design of Tall and Special Buildings*, **18**, No. 3, 291–311.

Zalka, K.A., 2013, Maximum deflection of symmetric wall-frame buildings. *Periodica Polytechnica, Civil Engineering,* **57/2**, 173–184. doi: 10.3311/PPci7172.

Zalka, K.A., 2014, Maximum deflection of asymmetric wall-frame buildings under horizontal load. *Periodica Polytechnica, Civil Engineering,* **58/4**, 387–396. doi: 10.3311/PPci7084.

Zalka, K.A. and Armer, G.S.T., 1992, *Stability of large structures*, (Oxford: Butterworth-Heinemann. Download: http://zalkak.hu/index_files/Books.html).

Zbirohowski-Koscia, K., 1967, *Thin walled beams. From theory to practice*, (London: Crosby Lockwood and Son).

Subject index

Author index

Printed in the United States
by Baker & Taylor Publisher Services